Science Notebook
Student Edition

GLENCOE

CHEMISTRY
MATTER AND CHANGE

About the Consultant

Douglas Fisher, PhD, is a Professor in the Department of Teacher Education at San Diego State University. He is the recipient of an International Reading Association Celebrate Literacy Award as well as a Christa McAuliffe award for Excellence in Teacher Education. He has published numerous articles on reading and literacy, differentiated instruction, and curriculum design as well as books, such as *Improving Adolescent Literacy: Strategies at Work* and *Responsive Curriculum Design in Secondary Schools: Meeting the Diverse Needs of Students*. He has taught a variety of courses in SDSU's teacher credentialing program as well as graduate-level courses on English language development and literacy. He also has taught classes in English, writing, and literacy development to secondary school students.

Cover SPL/Photo Researchers

The McGraw·Hill Companies

Glencoe

Send all inquiries to:
Glencoe/McGraw-Hill
8787 Orion Place
Columbus, OH 43240-4027

ISBN: 978-0-07-896415-2
MHID: 0-07-896415-6

Printed in the United States of America.

10 11 12 13 14 15 16 17 18 19   LHS   22 21 20 19 18 17

Table of Contents

Table of Contents

Using Your Science Notebook

This note-taking guide is designed to help you succeed in learning science content. Chapters include:

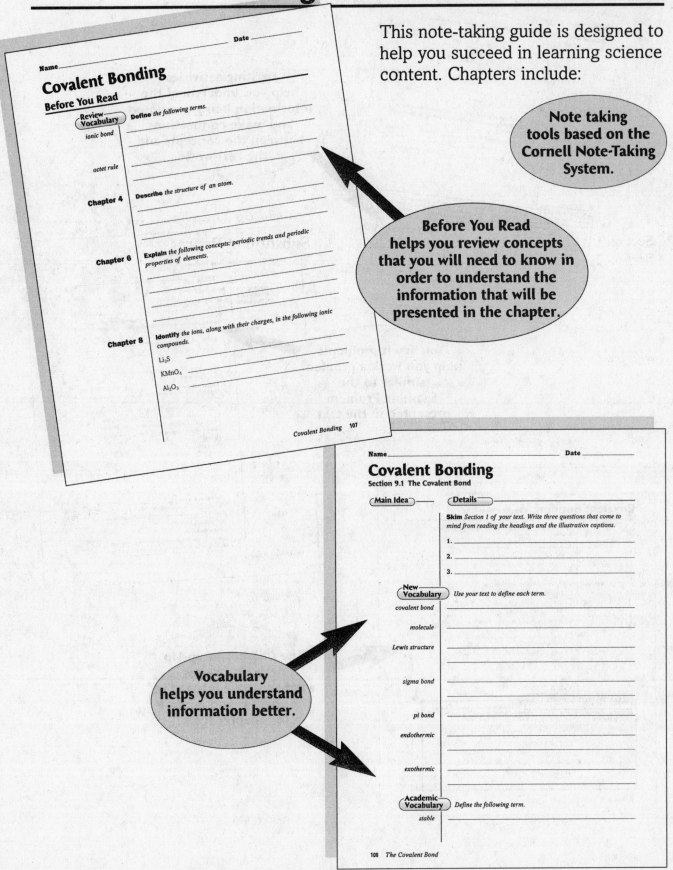

Note taking tools based on the Cornell Note-Taking System.

Before You Read helps you review concepts that you will need to know in order to understand the information that will be presented in the chapter.

Vocabulary helps you understand information better.

Name_____ Date_____

Covalent Bonding

Before You Read

Review Vocabulary — Define the following terms.

ionic bond

octet rule

Chapter 4 — Describe the structure of an atom.

Chapter 6 — Explain the following concepts: periodic trends and periodic properties of elements.

Chapter 8 — Identify the ions, along with their charges, in the following ionic compounds.

Li_2S

$KMnO_4$

Al_2O_3

Covalent Bonding **107**

Name_____ Date_____

Covalent Bonding

Section 9.1 The Covalent Bond

Main Idea — **Details**

Skim *Section 1 of your text. Write three questions that come to mind from reading the headings and the illustration captions.*

1. _____
2. _____
3. _____

New Vocabulary — Use your text to define each term.

covalent bond

molecule

Lewis structure

sigma bond

pi bond

endothermic

exothermic

Academic Vocabulary — Define the following term.

stable

108 *The Covalent Bond*

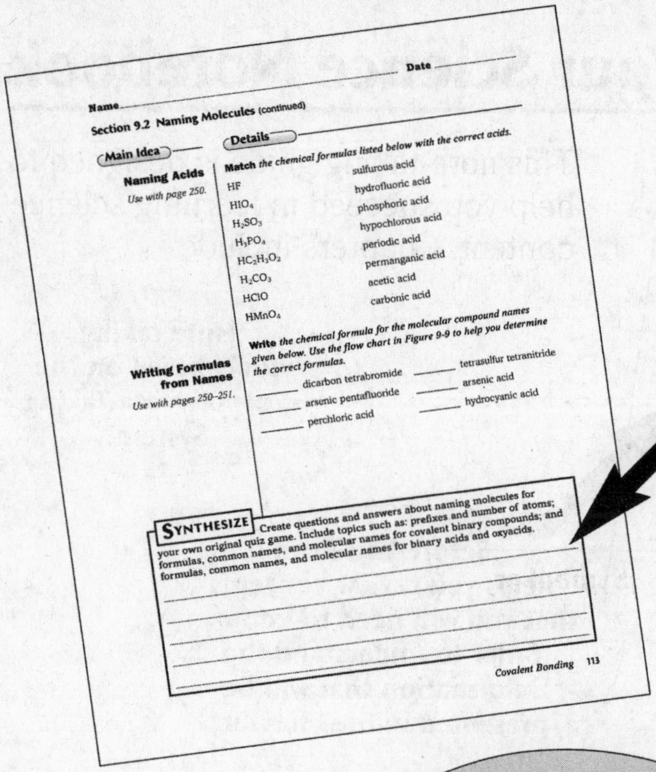
Name ______________________ Date ______

Section 9.2 Naming Molecules (continued)

Main Idea | Details

Naming Acids
Use with page 250.

Match *the chemical formulas listed below with the correct acids.*

HF	sulfurous acid
HIO_4	hydrofluoric acid
H_2SO_3	phosphoric acid
H_3PO_4	hypochlorous acid
$HC_2H_3O_2$	periodic acid
H_2CO_3	permanganic acid
HClO	acetic acid
$HMnO_4$	carbonic acid

Writing Formulas from Names
Use with pages 250–251.

Write *the chemical formula for the molecular compound names given below. Use the flow chart in Figure 9-9 to help you determine the correct formulas.*

_____ dicarbon tetrabromide _____ tetrasulfur tetranitride
_____ arsenic pentafluoride _____ arsenic acid
_____ perchloric acid _____ hydrocyanic acid

SYNTHESIZE Create questions and answers about naming molecules for your own original quiz game. Include topics such as: prefixes and number of atoms; formulas, common names, and molecular names for covalent binary compounds; and formulas, common names, and molecular names for binary acids and oxyacids.

Covalent Bonding 113

Writing activities help you understand the information being presented and make connections between the concepts and the real-world.

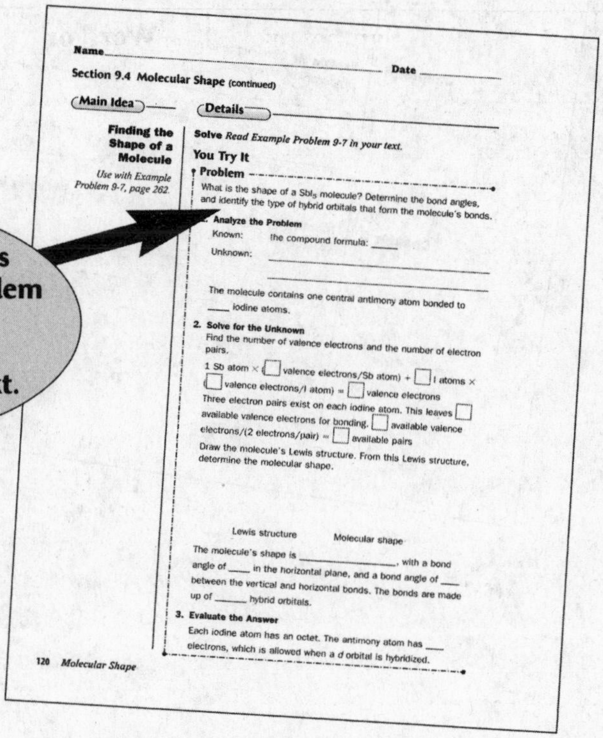
Name ______________________ Date ______

Section 9.4 Molecular Shape (continued)

Main Idea | Details

Finding the Shape of a Molecule
Use with Example Problem 9-7, page 262.

Solve *Read Example Problem 9-7 in your text.*

You Try It

Problem

What is the shape of a SbI_5 molecule? Determine the bond angles, and identify the type of hybrid orbitals that form the molecule's bonds.

1. **Analyze the Problem**
Known: the compound formula: ______
Unknown: ______

The molecule contains one central antimony atom bonded to ____ iodine atoms.

2. **Solve for the Unknown**
Find the number of valence electrons and the number of electron pairs.
1 Sb atom × (☐ valence electrons/Sb atom) + ☐ I atoms × (☐ valence electrons/I atom) = ☐ valence electrons
Three electron pairs exist on each iodine atom. This leaves ☐ available valence electrons for bonding. ☐ available valence electrons/(2 electrons/pair) = ☐ available pairs
Draw the molecule's Lewis structure. From this Lewis structure, determine the molecular shape.

Lewis structure   Molecular shape

The molecule's shape is ______, with a bond angle of ____ in the horizontal plane, and a bond angle of ____ between the vertical and horizontal bonds. The bonds are made up of ____ hybrid orbitals.

3. **Evaluate the Answer**
Each iodine atom has an octet. The antimony atom has ____ electrons, which is allowed when a *d* orbital is hybridized.

120 Molecular Shape

You Try It problems help you work a problem similar to the Example Problem presented in the text.

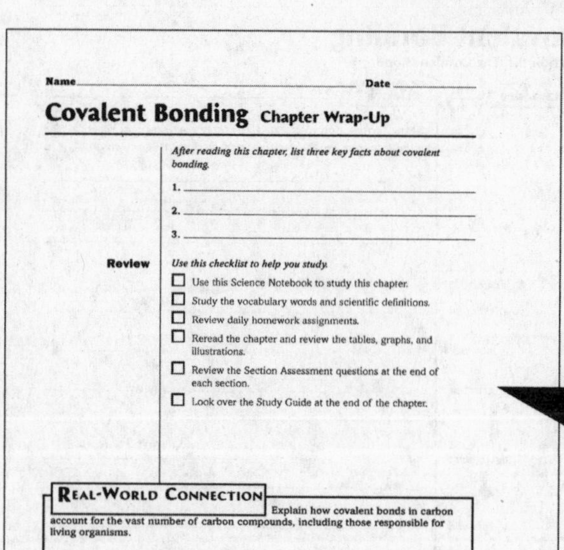
Name ______________________ Date ______

Covalent Bonding Chapter Wrap-Up

After reading this chapter, list three key facts about covalent bonding.

1. ______
2. ______
3. ______

Review *Use this checklist to help you study.*

- ☐ Use this Science Notebook to study this chapter.
- ☐ Study the vocabulary words and scientific definitions.
- ☐ Review daily homework assignments.
- ☐ Reread the chapter and review the tables, graphs, and illustrations.
- ☐ Review the Section Assessment questions at the end of each section.
- ☐ Look over the Study Guide at the end of the chapter.

REAL-WORLD CONNECTION Explain how covalent bonds in carbon account for the vast number of carbon compounds, including those responsible for living organisms.

124 Chapter Wrap-Up

The Chapter Wrap-Up helps you assess what you have learned in the chapter and prepare for chapter tests.

Note-Taking Tips

Your notes are a reminder of what you learned in class. Taking good notes can help you succeed in science. The following tips will help you take better classroom notes.

- Before class, ask what your teacher will be discussing in class. Review mentally what you already know about the concept.

- Be an active listener. Focus on what your teacher is saying. Listen for important concepts. Pay attention to words, examples, and/or diagrams you teacher emphasizes.

- Write your notes as clear and concise as possible. The following symbols and abbreviations may be helpful in your note-taking.

Word or Phrase	Symbol or Abbreviation	Word or Phrase	Symbol or Abbreviation
for example	e.g.	and	+
that is	i.e.	approximately	≈
with	w/	therefore	∴
without	w/o	versus	vs

- Use a symbol such as a star (★) or an asterisk (∗) to emphasize important concepts. Place a question mark (?) next to anything that you do not understand.

- Ask questions and participate in class discussion.

- Draw and label pictures or diagrams to help clarify a concept.

- When working out an example, write what you are doing to solve the problem next to each step. Be sure to use your own words.

- Review you notes as soon as possible after class. During this time, organize and summarize new concepts and clarify misunderstandings.

Note-Taking Don'ts

- **Don't** write every word. Concentrate on the main ideas and concepts.
- **Don't** use someone else's notes as they may not make sense.
- **Don't** doodle. It distracts you from listening actively.
- **Don't** lose focus or you will become lost in your note-taking.

Name _____ Date _____

Introduction to Chemistry

Before You Read

Science Journal

Before you read the chapter, write down four facts you know about chemistry.

1. _____

_____

2. _____

_____

3. _____

_____

4. _____

_____

Write three questions about scientific methods and research.

1. _____

_____

2. _____

_____

3. _____

_____

Introduction to Chemistry
Section 1 A Story of Two Substances

(Main Idea) —— **(Details)** ——————————————

Scan *Section 1 of your text. Use the checklist below as a guide.*

- Read all section titles.

- Read all boldfaced words.

- Read all tables and graphs.

- Look at all pictures and read the captions.

- Think about what you already know about ozone and chlorofluorocarbons (CFCs).

Write *four facts you discovered about ozone and CFCs.*

1. _____

2. _____

3. _____

4. _____

(New Vocabulary) *Use your text to define each term.*

chemistry _____

_____

substance _____

_____

Section 1 A Story of Two Substances (continued)

⟨ Main Idea ⟩ ——— ⟨ Details ⟩ ———————————————————

The Ozone Layer

Use with pages 5–7.

Detail *the ozone layer by completing the following paragraph.*

Overexposure to _____ causes sunburn, is harmful to

_____, lowers _____, and disrupts

_____. When _____ is exposed to ultraviolet

radiation in the upper regions of the _____, a chemical called

_____ is formed. About _____ of Earth's ozone is

spread out in a layer that surrounds and _____ our planet.

Ozone forms over the _____ and flows toward the _____.

Sequence *the steps necessary for the formation of ozone.*

1. _____

_____

2. _____

_____

3. _____

Illustrate *the balance between oxygen gas and ozone levels in the stratosphere, using Figure 3 in your text as a model. Give it a title and label the parts of your model.*

Section 1 A Story of Two Substances (continued)

⟨ **Main Idea** ⟩ ——— ⟨ **Details** ⟩ ——————————————————

Chlorofluoro-carbons

Use with pages 7–8.

Analyze *the graph in Figure 6. Write a brief description of the concentration of CFCs from 1979 through 2010.*

_____

_____

_____

Analyze *chlorofluorocarbons by completing the following table.*

CFCs Were First Developed Because:	Facts about CFCs	Uses of CFCs

REAL-WORLD CONNECTION

Infer from your reading the potential connection between CFCs and the ozone layer. Use Figures 5 and 6 to draw your conclusions.

_____

_____

Introduction to Chemistry
Section 2 Chemistry and Matter

(Main Idea) ———— **(Details)** ————————————————————

Skim *Section 2 of your text. Write four facts that come to mind from reading the headings, boldfaced words, and the illustration captions.*

1. _____

2. _____

3. _____

4. _____

New Vocabulary   *Use your text to define each term.*

mass _____

_____

weight _____

_____

model _____

_____

Section 2 Chemistry and Matter (continued)

⸨**Main Idea**⸩ ——— ⸨**Details**⸩ ————————————————————————

Matter and its Characteristics

Use with pages 9–10.

Compare and contrast *mass and weight using the Venn diagram below.*

- does not reflect gravitational pull on matter
- a measure of the effect of gravitational pull on matter
- a measurement that reflects the amount of matter in an object

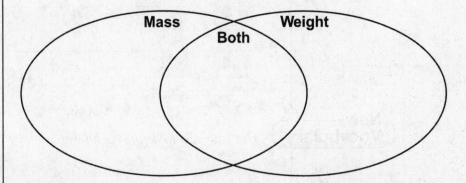

Chemistry: The Central Science

Use with page 11.

Identify *six substances mentioned in the book that are important in everyday life and are made of chemicals.*

1. _____   4. _____

2. _____   5. _____

3. _____   6. _____

Section 2 Chemistry and Matter (continued)

Main Idea ———— **Details** ——————————————————————

Organize *the following terms by arranging them from largest to smallest.*

macroscopic, submicroscopic, microscopic

_____

Explain *a chemical model by completing the following sentences.*

The _____, composition, and _____ of all matter can

be explained on a _____ level. All that we observe depends

on _____ and the _____ they undergo.

_____ seeks to explain the submicroscopic events that lead to

_____. One way to do this is by making a chemical

model, which is a _____ of a

_____.

REAL-WORLD CONNECTION

Analyze the importance of chemistry in our society using the branches of chemistry as examples.

_____

_____

_____

_____

_____

Introduction to Chemistry
Section 3 Scientific Methods

Main Idea ——   **Details** ——————————————

Skim *Section 3 of your text. Write three questions that come to mind from reading the headings, boldface terms, and illustration captions.*

1. _____

2. _____

3. _____

New Vocabulary   *Use your text to define each term.*

scientific method _____

_____

qualitative data _____

quantitative data _____

hypothesis _____

experiment _____

independent variable _____

dependent variable _____

control _____

conclusion _____

theory _____

scientific law _____

Section 3 Scientific Methods (continued)

Main Idea ———— **Details** ———————————————————————

A Systematic Approach

Use with pages 12–15.

Compare *the terms qualitative data and quantitative data.*

_____

_____

_____

_____

Compare *the terms independent variable and dependent variable.*

_____

_____

_____

Analyze *whether the characteristics listed below represent qualitative data, quantitative data, or both.*

Characteristic	Type of Data
the rate at which a candle burns	
a blanket with varying degrees of softness	
sand with a reddish-brown color	

Sequence *the steps of this scientific method.*

_____ Plan and set up one or more experiments to test one variable at a time.

_____ Gather information using both qualitative data and quantitative data.

_____ Observe, record, and analyze experimental data.

_____ Develop a hypothesis, or tentative explanation based on observations.

_____ Develop a theory or a scientific law.

_____ Compare findings to the hypothesis, and form a conclusion.

Section 3 Scientific Methods (continued)

Main Idea ———— **Details** ————————————————

Use with page 15.

Analyze *Figure 13 and the caption information on Molina and Rowland's model. Explain in words what the model visually predicts about the effect of ultraviolet radiation on CFCs.*

_____

_____

_____

_____

SYNTHESIZE

Design a simple experiment using a scientific method. Give your experiment a descriptive title. Limit the number of variables you test. Write the steps of the experiment based on the scientific method, including but not limited to hypothesis, analysis, and conclusions. Draw a simple sketch of your experiment, if appropriate, and label the independent, dependent, and control variables.

Title: _____

Steps: _____

_____

_____

_____

_____

_____

_____

_____

Independent variable(s): _____

Dependent variable(s): _____

Control variable(s): _____

Introduction to Chemistry

Section 4 Scientific Research

Main Idea ———— **Details** ————————————————

Skim *Section 4 of your text. Write three questions that come to mind from reading the headings, boldfaced terms, and illustration captions.*

1. _____

2. _____

3. _____

New Vocabulary   *Use your text to define each term.*

pure research _____

_____

applied research _____

_____

Word Origin   *Define the following term.*

recover _____

_____

Section 4 Scientific Research (continued)

Main Idea	Details

Types of Scientific Investigations

Use with pages 17–18.

Describe *scientific investigations by completing the following sentences.*

Pure research becomes _____ when scientists develop a

hypothesis based on the data and try to solve a specific problem.

_____ have been made when a scientist reaches a

conclusion far different than anticipated. Some wonderful scientific

discoveries have been made _____.

Students in the Laboratory

Use with pages 18–19.

Review *Table 2 in your text. Write an A if you agree with the statement. Write a D if you disagree with the statement.*

_____ Return unused chemicals to the stock bottle.

_____ It is not safe to wear contact lenses in the lab.

_____ Only a major accident, injury, incorrect procedure, or damage to equipment needs to be reported.

_____ Graduated cylinders, burettes, or pipettes should be heated with a laboratory burner.

Analyze *laboratory safety by responding to the following situations.*

1. Explain in your own words why safety goggles and a laboratory apron must be worn whenever you are in the lab.

   _____

   _____

2. State why bare feet or sandals are not permitted in the lab.

   _____

   _____

   _____

Section 4 Scientific Research (continued)

⟨ **Main Idea** ⟩ —— ⟨ **Details** ⟩ ————————————————————————

3. Describe how you would explain to another student why you should not return unused chemicals to the stock bottle.

_____

_____

_____

4. Explain why is it important to keep the balance area clean.

_____

_____

_____

SYNTHESIZE

Some students are conducting an experiment that involves combining sodium and water. Too much sodium is added, which causes a fire. A student reacts by throwing water on the fire, but this only causes the fire to spread. The teacher finally puts the fire out. Based on what you now know about chemistry and lab safety, explain how this could have been avoided.

_____

_____

_____

_____

Introduction to Chemistry Chapter Wrap-Up

Now that you have read the chapter, review what you have learned. Fill in the blanks below with the correct word or phrase.

Chemistry is the study of _____.

Matter is anything that has _____ and takes up _____. Mass is

_____ and differs from weight in that

it does not measure the effect of _____ on matter.

The steps of a scientific process include:

_____

_____

Two types of scientific investigation are:

_____

Review   *Use this checklist to help you study.*

☐ Study your Science Notebook for this chapter.

☐ Study the vocabulary words and scientific definitions.

☐ Review daily homework assignments.

☐ Reread the chapter and review the tables, graphs, and illustrations.

☐ Answer the Section Review questions at the end of each section.

☐ Look over the Study Guide at the end of the chapter.

REAL-WORLD CONNECTION

Explain three ways you use chemistry in daily life.

1. _____

2. _____

3. _____

Analyzing Data

Before You Read

Review Vocabulary *Define the following terms.*

qualitative data
_____

_____

quantitative data
_____

_____

variable
_____

analysis
_____

_____

Chapter 1 *You and a friend are making sweetened iced tea. You both have different opinions about how much sugar to add and at what temperature is best to add the sugar. Design an experiment to find out how much sugar will dissolve at three different temperatures. In your experiment, identify the following:*

Qualitative data
_____

Quantitative data
_____

_____

Independent variable
_____

Dependent variable
_____

_____

Analyzing Data
Section 1 Units and Measurements

⟨Main Idea⟩ ——— **⟨Details⟩** ——————————————

Skim *Section 1 of your text. Write a question you have about each of the two types of units discussed in this section.*

1. _____

2. _____

⟨New Vocabulary⟩   *Use your text to define each term.*

base unit   _____

_____

derived unit   _____

_____

density   _____

_____

Match *the SI base units below with their functions.*

second ———————————→ distance

meter ———⟋ temperature

kilogram   time

kelvin   mass

liter   volume

Section 1 Units and Measurements (continued)

Main Idea	Details

Units

Use with page 32.

Identify *five items around your home that use SI units of measurement.*

1. _____

2. _____

3. _____

4. _____

5. _____

Base Units and SI Prefixes

Use with pages 33–35.

Sequence *these prefixes from smallest to largest.*

_____ pico   _____ giga

_____ micro   _____ nano

_____ deci   _____ milli

_____ kilo   _____ centi

_____ mega

Temperature

Use with pages 34–35.

Compare and contrast *the Kelvin scale and the Celsius scale.*

_____

_____

_____

Derived Units

Use with pages 35–37.

Explain *density by completing the following statement and equation.*

Density is a _____ that _____ the _____ of an object to

its _____.

density = ————

Section 1 Units and Measurements (continued)

⟨Main Idea⟩ ──────── ⟨Details⟩ ──────────────────────────

Using Density and Volume to Find Mass

Use with Example Problem 1, page 38.

Solve *Read Example Problem 1 in your text.*

You Try It

Problem

Determine the mass of an object that, when placed in a 25-mL graduated cylinder containing 14 mL of water, causes the level of the water to rise to 19 mL. The object has a density of 3.2 g/mL.

1. Analyze the Problem

Known: _____

Unknown: _____

You know the density and the volume of an object and must determine its mass; therefore, you will calculate the answer using the density equation.

2. Solve for the Unknown

Write the density equation.

= ————

Rearrange the density equation to solve for mass.

_____

Substitute the known values for _____ and _____ into the equation.

_____

Multiply the values and units. The mL units will cancel out.

mass = _____ × _____ = _____

3. Evaluate the Answer

The two sides of the equation should be _____

density = _____

_____

If you divide 16 g by 5.0 mL, you get _____

Analyzing Data
Section 2 Scientific Notation and Dimensional Analysis

Main Idea —— | —— **Details** ————————————————————————

Scan *Section 2 of your text. Use the checklist below as a guide.*

- Read all section titles.
- Read all boldfaced words.
- Read all tables and graphs.
- Look at all pictures and read the captions.
- Think about what you already know about this subject.

Write *three facts you discovered about scientific notation and dimensional analysis.*

1. _____

2. _____

3. _____

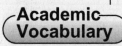

Use your text to define each term.

scientific notation   _____

_____

dimensional analysis   _____

_____

conversion factor   _____

Academic Vocabulary   *Define the following term.*

sum   _____

Section 2 Scientific Notation and Dimensional Analysis (continued)

Main Idea ——— Details ———————————————————————

Scientific Notation

Use with Example Problem 2, page 41.

Solve *Read Example Problem 2 in your text.*

You Try It

Problem

Change the following data into scientific notation:

a. The distance between Pluto and the Sun is 5,913,000 km.

b. The density of nitrogen gas, a major component of Pluto's atmosphere, is .0012506 g/cm^3.

1. **Analyze the Problem**

Known: _____

Unknown: _____

You are given two measurements. In both cases, the answers will be factors between 1 and 10 that are multiplied by a power of ten.

2. **Solve for the Unknown**

Move the decimal point to produce a factor between 1 and 10. Count the number of places the decimal point moved and the direction.

   a.   **b.** _____

The decimal point moved   *The decimal point moved*
_____ *places to the* ____.   _____ *places to the* _____.

Remove the extra zeros at the end or beginning of the factor.

Multiply the result by 10^n where n equals the _____.

_____. When the decimal point moves to the left, n is a

_____ number. When the decimal point moves to the right,

n is a _____ number. Remember to add units to the answers.

a. _____

b. _____

3. **Evaluate the Answer**

The answers have _____ factors. The first factor is a number

between _____ and _____. In answer a, because the distance to Pluto is

a large number, 10 has a _____ exponent. In answer b, because

the density of nitrogen gas is a very small number, the exponent is

_____.

Section 2 Scientific Notation and Dimensional Analysis (continued)

Main Idea	Details

Using Conversion Factors

Use with Example Problem 4, page 46.

Solve *Read Example Problem 4 in your text.*

You Try It

Problem

The Cassini probe heading toward Saturn will reach speeds of 5.2 kilometers per second. How many meters per minute would it travel at this speed?

1. **Analyze the Problem**

   Known: _____

   Unknown: _____

   You need conversion factors that relate kilometers to meters and seconds to minutes. A conversion factor is a _____ of _____ used to express _____ in _____.

2. **Solve for the Unknown**

   First convert kilometers to meters. Set up the conversion factor so that the kilometer units will cancel out.

   $$\frac{5.2 \text{ km}}{s} \times \frac{1000 \text{ m}}{1 \text{ km}} = \frac{\quad}{s} \text{ m}$$

   Next convert seconds to minutes. Set up the conversion factor so that the seconds will cancel out.

   $$\frac{5200 \text{ m}}{s} \times \frac{60 \text{ s}}{1 \text{ min}} = \frac{\quad}{\text{min}} \text{ m}$$

3. **Evaluate the Answer**

   To check your answer, you can do the steps in reverse order.

   $$\frac{5.2 \text{ km}}{s} \times \frac{60 \text{ s}}{1 \text{ min}} = \frac{312 \text{ km}}{\text{min}} \times \frac{1000 \text{ m}}{\text{km}} = \frac{\quad}{\text{min}} \text{ m}$$

Analyzing Data
Section 3 Uncertainty in Data

Main Idea —————   **Details** ———————————————————

Skim *Section 3 of your text. Focus on the headings, subheadings, boldfaced words, and main ideas. Summarize the main ideas of this section.*

_____

_____

_____

_____

New Vocabulary   *Use your text to define each term.*

accuracy   _____

_____

precision   _____

_____

error   _____

_____

percent error   _____

_____

significant figure   _____

_____

Section 3 Uncertainty in Data (continued)

Main Idea	Details

Error and Percent Error

Use with pages 48–49.

Explain *percent error by completing the statement and equation below.*

Percent error is the _____ of an _____ to an _____.

Percent error = _____ × ____

Solve *Read Example Problem 5 in your text.*

Calculating Percent Error

Use with Example Problem 5, page 49.

You Try It

Problem

Calculate the percent errors. Report your answers to two places after the decimal point. The table below summarizes Student B's data.

Trial	Density (g/cm^3)	Error (g/cm^3)
1	1.4	−0.19
2	1.68	0.09
3	1.45	−0.14

1. **Analyze the Problem**

   Known: _____

   _____

   Unknown: _____

   Use the accepted value for density and the errors to calculate percent error.

2. **Solve for the Unknown**

   Substitute each error into the percent error equation.

   $$\text{percent error} = \frac{}{\text{accepted value}} \times 100$$

   $$\text{percent error} = \frac{}{1.59 \text{ g/cm}^3} \times 100 = \boxed{}$$

   $$\text{percent error} = \frac{}{1.59 \text{ g/cm}^3} \times 100 = \boxed{}$$

   $$\text{percent error} = \frac{}{1.59 \text{ g/cm}^3} \times 100 = \boxed{}$$

3. **Evaluate the Answer**

   The percent error is greatest for trial ____ which had the largest error, and smallest for trial ____, which was closest to the accepted value.

Section 3 Uncertainty in Data (continued)

Main Idea ———— **Details** ————————————————————

Significant Figures

Use with page 50.

Rounding Numbers

Use with page 52.

Identify *the significant numbers below by drawing a circle around them. Use the five rules for recognizing significant digits on page 51 for reference.*

0.025   325,078   5600

Explain *the rules for rounding numbers by completing the following sentences. Then complete the example of each rule for rounding numbers.*

1. If the digit to the immediate right of the last significant figure is less than five, _____

   3.751 _____

2. If the digit to the immediate right of the last significant figure is greater than five, _____

   4.127 _____

3. If the digit to the immediate right of the last significant figure is equal to five and is followed by a nonzero digit, _____

   _____

   8.3253 _____

4. If the digit to the immediate right of the last significant figure is equal to five and is not followed by a nonzero digit, look at the last significant figure. _____

   _____

   1.4750 = _____ ; 1.4650 = _____

Analyzing Data
Section 4 Representing Data

Main Idea ——— **Details** ——————————————————

Scan *Section 4 of your text. Use the checklist below as a guide.*

• Read all section titles.

• Read all tables and graphs.

• Look at all pictures and read the captions.

• Think about what you already know about data analysis.

Write *facts you learned about representing data as you scanned the section.*

1. _____

2. _____

3. _____

New Vocabulary   *Use your text to define each term.*

graph _____

Section 4 Representing Data (continued)

Main Idea ———— **Details** ————————————————

Graphing

Use with pages 55–56.

Draw *and label (a) a circle graph and (b) a bar graph using the information in the table below.*

Student Budget	
Budget items	**Percent**
Car insurance	45
Movies	6
Books	5
Clothing	30
Miscellaneous	4
Gas	10

Student Budget bar graph   Student Budget circle graph

The _____ best displays the data in the Student Budget table

because _____

_____ .

Section 4 Representing Data (continued)

Main Idea ——— **Details** ——————————————————

Line Graphs

Use with pages 56–57.

Identify *each of the following slopes.*

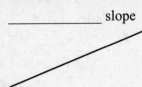 _____ slope

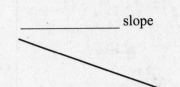 _____ slope

Analyze *whether the following sequences will likely plot as linear or nonlinear relationships.*

Sequence A:	Sequence B:
Result 1: 2	Result A: 31
Result 2: 4	Result B: 27
Result 3: 7	Result C: 49
Result 4: 10	Result D: 45
Answer: _____	Answer: _____

Interpreting Graphs

Use with pages 57–58.

Organize *information about interpreting graphs by completing the sentences below.*

Information on a graph typically consists of _____ types of

variables: _____ variables and _____ variables. The

relationship between the variables may reflect either a _____ or a

_____ slope.

When reading the graph, you use either interpolation for _____

_____ or _____ for estimated

values beyond the plotted points.

Analyzing Data Chapter Wrap-Up

Now that you have read the chapter, review what you have learned. Write out the key equations and relationships.

density =

percent error = $\times\ 100$

slope =

Conversion between temperature scales:

°C + _____ = _____

K − _____ = _____

Review   *Use this checklist to help you study.*

☐ Study your Science Notebook for this chapter.

☐ Study the definitions of vocabulary words.

☐ Review daily homework assignments.

☐ Reread the chapter and review the tables, graphs, and illustrations.

☐ Answer the Section Review questions at the end of each section.

☐ Look over the Study Guide at the end of the chapter.

SUMMARIZE

If you were a scientist, what precautionary guidelines would you use to ensure the accuracy of your data and to provide a clear representation of that data?

_____

_____

_____

_____

Matter—Properties and Changes

Before You Read

Review Vocabulary *Define the following terms.*

matter _____

significant figure _____

Chapter 2

Measure *the height and arm length for five friends or family members. In the space below, create an appropriate graph to represent the data you collected.*

Compare and contrast *circle, bar and line graphs.*

_____

_____

_____

_____

_____

_____

_____

_____

_____

Matter—Properties and Changes
Section 1 Properties of Matter

Main Idea ———— **Details** —————————————————————

Skim *Section 1 of your text. Write three questions that come to mind from reading the headings and the illustration captions.*

1. _____

2. _____

3. _____

New Vocabulary *Use your text to define each term.*

states of matter _____

vapor _____

_____

physical property _____

_____

extensive property _____

intensive property _____

chemical property _____

_____

Match *each of the following states of matter with its physical description*

solid   flows and fills the entire volume of its container

liquid   has definite shape and volume

gas   flows and has a constant volume

Academic Vocabulary *Define the following term.*

resource _____

Section 1 Properties of Matter (continued)

Main Idea ——— **Details** ———————————————————

States of Matter

Use with pages 71–72.

Compare *the way the three common states of matter fill a container.*

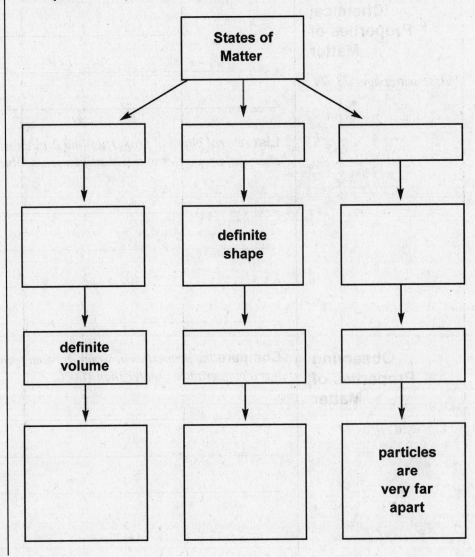

States of
Matter

definite
shape

definite
volume

particles
are
very far
apart

REAL-WORLD CONNECTION

Meteorologists (scientists who study weather) refer to water in the gaseous state in the atmosphere as water vapor. Explain why this term is used.

_____

_____

_____

Section 1 Properties of Matter (continued)

⟨ **Main Idea** ⟩ ———— ⟨ **Details** ⟩ ————————————————

Physical and Chemical Properties of Matter *Use with pages 73–74.*	**Contrast** *intensive and extensive physical properties.* _____ _____ _____
	List *several physical properties and explain why they are used more than chemical properties in the identification of objects.* _____ _____ _____ _____
Observing Properties of Matter *Use with pages 74–75.*	**Compare** *the properties of water at room temperature with water that has a temperature greater than 100°C.* _____ _____ _____

Name _____ Date _____

Matter—Properties and Changes
Section 2 Changes in Matter

Main Idea ———— **Details** ————————————————————————

Scan *Section 2 of your text. Use the checklist below as a guide.*

• Read all section titles.

• Read all boldfaced words.

• Read all tables and graphs.

• Look at all pictures and read the captions.

• Think about what you already know about this subject.

Write *three facts you discovered about changes in matter.*

1. _____

2. _____

3. _____

New Vocabulary   *Use your text to define each term.*

physical change _____

phase change _____

chemical change _____

*law of conservation
of mass* _____

_____

Section 2 Changes in Matter (continued)

Main Idea ———   **Details** ———————————————

Physical and Chemical Changes

Use with pages 76–77.

Determine *which type of change each statement represents. Use P for physical change and C for chemical change. Explain your answers.*

silver spoon tarnishes _____

    Explanation: _____

crushing an aluminum can _____

    Explanation: _____

freezing water _____

    Explanation: _____

burning wood _____

    Explanation: _____

copper turns a greenish color _____

    Explanation: _____

grind coffee beans _____

    Explanation: _____

Describe *how iron turns into a brownish-red powder. Name the reactants and product that are involved.*

_____

_____

_____

_____

_____

Section 2 Changes in Matter (continued)

Main Idea ——— **Details** ————————————————————

Conservation of Mass

Use with Example Problem 1, page 78.

Summarize *Fill in the blanks to help you take notes while you read Example Problem 1.*

Problem

The total _____ of the products must _____ the total mass of the

_____. This shows the law of _____.

1. **Analyze the Problem**

   Known: _____

   _____

   Unknown: _____

2. **Solve for the Unknown**

   Write an equation showing conservation of mass of reactants and products.

   mass of _____ = mass of _____ + mass of _____

   Write an equation to solve for the mass of oxygen.

   mass of _____ = mass of _____ − mass of _____

   Substitute known values and solve.

   Mass of oxygen = _____ g − _____ g

   Mass oxygen = _____ g

3. **Evaluate the Answer**

   Write an equation that shows mass of the two products equals the mass of the reactant.

   _____ g mercury + _____ g oxygen = _____ g mercury(II) oxide

Matter—Properties and Changes
Section 3 Mixtures of Matter

(Main Idea) ———   **(Details)** ————————————————

Scan *Section 3 of your text. Use the checklist below as a guide.*

• Read all section titles.

• Read all boldfaced words.

• Read all charts and graphs.

• Look at all pictures and read the captions.

List *three facts you have learned about mixtures.*

1. _____

2. _____

3. _____

New Vocabulary   *Use your text to find the correct term for each definition.*

mixture   _____

                       _____

heterogeneous mixture _____

homogeneous mixture   _____

solution   _____

filtration   _____

distillation   _____

                       _____

crystallization   _____

                       _____

sublimation   _____

chromatography   _____

                       _____

Section 3 Mixtures of Matter (continued)

Main Idea ———— **Details** ————————————————————

Mixtures

Use with pages 80–81.

Describe *how mixtures relate to substances.*

_____

_____

Compare *heterogeneous and homogeneous mixtures.*

_____

_____

_____

Describe *what an alloy is and why alloys are used.*

_____

_____

Separating Mixtures

Use with pages 82–83.

Identify *four techniques that take advantage of different physical properties in order to separate mixtures and describe how each is done.*

Technique 1: _____

How it is done: _____

_____

_____

Technique 2: _____

How it is done: _____

_____

Technique 3: _____

How it is done: _____

_____

_____

Section 3 Mixtures of Matter (continued)

⟨**Main Idea**⟩ —— ⟨**Details**⟩ _____

Technique 4: _____

How it is done: _____

_____

Sequence *the steps of separating a mixture of sand, salt, and iron filings. Identify which physical property you were using in each step.*

_____ Mix the sand and salt mixture with water.

Physical property used: _____

_____

_____ Boil the salt and water mixture, leaving the salt behind.

Physical property used:

_____

_____ Separate the iron filings from the sand and salt by using a magnet.

Physical property used: _____

_____

_____ Use filtration to separate the sand from the salt and water.

Physical property used: _____

_____

REAL-WORLD CONNECTION

Crude oil (petroleum) is a mixture of several materials, including gasoline, kerosene, diesel fuel, and heating oil. Describe whether you think distillation or filtration would be a better method to separate the products of crude oil. Hint: each of the products listed has a different boiling point.

_____

_____

_____

Matter–Properties and Changes

Section 4 Elements and Compounds

⟨Main Idea⟩ —————— **⟨Details⟩** ———————————————————

Scan *Review the periodic table of elements. Record some observations about how the table is organized and what information you can determine just by looking at the table.*

_____

_____

_____

New Vocabulary

Use your text to define each term.

_____

element

_____

_____

periodic table

_____

_____

compound

_____

_____

law of definite
proportions

_____

_____

percent by mass

_____

_____

law of multiple
proportions

_____

_____

Section 4 Elements and Compounds (continued)

Main Idea ——— **Details** ——————————————————

Elements and Compounds

Use with pages 84–87.

Discuss *elements and compounds by completing the following paragraph.*

There are _____ naturally occurring elements. Seventy-five percent of

the universe is _____. The Earth's crust and the human body are made

of different elements. But _____ is an element that is abundant in both.

Most objects are made of _____ with approximately ten million

known and over _____ being developed and discovered every _____.

Analyze *the concept map for matter in Figure 19. Write a brief description of the information the concept map is conveying.*

_____

_____

_____

Describe *how the periodic table organizes elements.*

_____

_____

_____

Explain *how Figure 18 illustrates the fact that the properties of a compound are different from the properties of its component elements.*

_____

_____

_____

_____

Section 4 Elements and Compounds (continued)

Main Idea ————— **Details** ————————————————————————

Law of Definite Proportions

Use with pages 87–88.

Describe *how to do percent by mass by completing the following paragraph.*

The _____ of a compound is _____ to the _____ of the masses

of the _____ that make up the compound. This demonstrates the law

of _____.

Analyze *the law of definite proportions by indicating whether the following examples are for identical or different compounds.*

Description	Analysis
Compound 1 consists of 24g of Na, and 36g of Cl. Compound 2 has 36g of Na and 54g of Cl.	
Compound 3 has 10.00g of lead and 1.55g of sulfur. Compound 4 has 10.00 g of lead, 1.55g of sulfur, and 1.55g of carbon.	

Law of Multiple Proportions

Use with pages 89–90.

Describe *the law of multiple proportions by completing the following statement.*

When different _____ are formed by combining the same

_____, different masses of one element combine with the same

_____ of the other element in a ratio of _____.

SYNTHESIZE

Carbon combines with oxygen to form two compounds, carbon monoxide and carbon dioxide. Based on the law of multiple proportions, describe how the proportions of oxygen in the two compounds relate to each other.

_____

_____

_____

Matter–Properties and Changes   Chapter Wrap-Up

After reading this chapter, list three things you have learned about the properties and changes in matter.

1. _____

2. _____

3. _____

Review   *Use this checklist to help you study.*

☐ Use this Science Notebook to study this chapter.

☐ Study the vocabulary words and scientific definitions.

☐ Review daily homework assignments.

☐ Reread the chapter and review the tables, graphs, and illustrations.

☐ Answer the Section Review questions at the end of each section.

☐ Look over the Study Guide at the end of the chapter.

REAL-WORLD CONNECTION

Explain how understanding the physical and chemical properties of matter can help find alternatives to the burning of fossil fuels, thus reducing the amount of harmful greenhouse gases released into the atmosphere.

_____

_____

_____

_____

_____

The Structure of the Atom

Before You Read

Review Vocabulary *Define the following terms.*

scientific law _____

_____

theory _____

_____

_____

element _____

_____

law of definite proportions _____

_____

law of multiple proportions _____

_____

_____

Describe *three things that you already know about the atom.*

1. _____

2. _____

3. _____

The Structure of the Atom
Section 1 Early Ideas About Matter

Main Idea ——— **Details** ———

Scan *Section 1 of your text. Use the checklist below as a guide.*

- Read all section titles.
- Read all boldfaced words.
- Read all tables and graphs.
- Look at all pictures and read the captions.
- Think about what you already know about this subject.

List *three things you expect to learn about while reading the section.*

1. _____

2. _____

3. _____

New Vocabulary   *Use your text to define each term.*

Dalton's atomic theory

_____

_____

_____

Section 1 Early Ideas About Matter (continued)

⟨**Main Idea**⟩ ————— ⟨**Details**⟩ ——————————————————

Greek Philosophers

Use with pages 102–103.

Summarize *the effect that Aristotle had on the atomic theory proposed by Democritus.*

_____

_____

_____

_____

John Dalton

Use with pages 104.

List *the main points of Dalton's atomic theory.*

1. _____

_____

2. _____

_____

_____

3. _____

_____

4. _____

_____

5. _____

_____

Discuss *Dalton's ideas by completing the following paragraph.*

After years of studying _____, Dalton was able to

accurately determine the _____ of the elements involved in the

reactions. His conclusions resulted in the _____, which helped to

explain that _____ in chemical reactions separate, _____, or

_____, but are not created, _____, or_____.

Section 1 Early Ideas About Matter (continued)

Main Idea ———— **Details** ——————————————————

Compare and contrast *the atomic theories of Democritus and Dalton. Mark an X under each name if a statement in the table applies to that person's theory.*

Statement	Democritus	Dalton
All matter is made of tiny pieces.		
Matter is made of empty space through which atoms move.		
Atoms cannot be divided.		
Atoms cannot be created.		
Atoms cannot be destroyed.		
Different atoms combine in whole-number ratios to form compounds.		
The properties of atoms vary based on shape, size, and movement.		
Different kinds of atoms come in different sizes and shapes.		

REAL-WORLD CONNECTION

The experiments of the alchemists revealed the properties of some metals and provided the foundation for the science of chemistry. Although not successful, alchemy proved beneficial to science. Explain how this example can be applied to modern research.

_____

_____

_____

Name _____ Date _____

The Structure of the Atom
Section 2 Defining the Atom

Main Idea ———— **Details** ————————————————

Scan *Section 2 of your text. Use the checklist below as a guide.*

- Read all section titles.
- Read all boldfaced words.
- Read all tables and graphs.
- Look at all pictures and read the captions.
- Think about what you already know about this subject.

Write *two facts you discovered about subatomic particles.*

1. _____

2. _____

New Vocabulary   *Use your text to define each term.*

atom   _____

cathode ray   _____

electron   _____

nucleus   _____

              _____

proton   _____

              _____

neutron   _____

              _____

The Atom

Use with pages 106–107.

Explain *an atom by completing the following statements.*

The atom is the _____

_____.

When a group of atoms _____ and act as a

_____, the result is known as a _____.

Section 2 Defining the Atom (continued)

Main Idea ———— **Details** ——————————————————

The Electron

Use with pages 107–110.

Summarize *the information you learned from cathode ray experiments. Use Figure 7 for reference.*

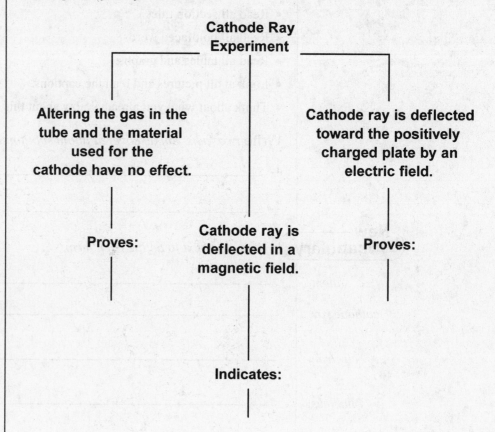

Identify *the major discoveries about subatomic particles made by the 19th century.*

1. _____

2. _____

3. _____

Section 2 Defining the Atom (continued)

Main Idea ———— **Details** ——————————————————————

The Nucleus
Use with pages 111–114.

Describe *Rutherford's model of the atom by completing the following statements.*

1. Most of an atom consists of _____ moving _____ through _____.

2. The electrons are _____ within the atom by their _____ to the positively charged _____.

3. The volume of _____ through which the electrons move is many times _____ than the volume of the _____.

Organize *the properties of subatomic particles by completing the table below. Use Table 3 for reference.*

	Electron	Proton	Neutron
Symbol			
Location			in nucleus
Relative electrical charge		1+	

Summarize *what you have learned about subatomic particles by completing the following paragraph.*

Atoms have a _____ shape. The _____ of an atom is made up of _____ that have a positive charge and _____ that have no _____. The nucleus makes up _____ of the mass of an atom. Most of the area of an _____ is made up of negatively charged _____ traveling around the _____ charged nucleus. The _____ are held in place by their _____ to the positive charge of the _____. The _____ of the protons and neutrons are almost _____ to each other while the _____ of the electrons is _____.

The Structure of the Atom

Section 3 How Atoms Differ

Main Idea ——— **Details** ——————————————————

Skim *Section 3 of your text. Focus on the headings, boldfaced words, and main ideas. Then summarize the main ideas of this section.*

1. _____

2. _____

3. _____

New Vocabulary

In the left margin, write the term defined below.

_____   the number of protons in an atom

_____   atoms with the same number of protons but different numbers of neutrons

_____   the sum of the number of protons and neutrons in the nucleus

_____   1/12 the mass of a carbon-12 atom; the standard unit of measurement for the mass of atoms

_____   the weighted average mass of the isotopes of an element

Academic Vocabulary

Define the following term.

specific   _____

Section 3 How Atoms Differ (continued)

Main Idea ————— **Details** —————————————————————

Atomic Number

Use with page 115.

Explain *how to use an atomic number to identify an element by completing the paragraph below.*

Each _____ of an element has a unique number of _____. Since

the overall charge of an atom is _____ the number of _____ equals

the number of _____. Atomic number = number of _____ = number

of _____. If you know how many one of the three an atom contains,

you also know the other _____. Once you know the _____,

the _____ can be used to find the name of the _____.

Atomic Number

Use with Example Problem 1, page 116.

Solve *Read Example Problem 1 in your text.*

You Try It

Problem

Given the following information about atoms, determine the name of each atom's element and its atomic number.

    a. Atom 1 has 11 protons   **b. Atom 2 has 20 electrons**

1. **Analyze the Problem**
   Apply the relationship among atomic number, number of protons, and number of electrons to determine the name and atomic number of each element.

2. **Solve for the Unknown**
   a. Atom 1

   Atomic number = number of protons = number of electrons

   Atomic number = _____ = number of electrons

   The element with an atomic number of 11 is _____

   b. Atom 2

   Atomic number = number of protons = number of electrons

   Atomic number = number of protons = _____

   The element with an atomic number of _____ is _____.

3. **Evaluate the Answer**

   The answers agree with _____ and element

   _____ given in the periodic table.

Section 3 How Atoms Differ (continued)

Main Idea ———— **Details** ————————————————————

Isotopes and Mass Number

Use with page 117.

Review *your understanding of isotopes and mass number by completing the following paragraph.*

Isotopes are elements with _____ but with

_____. The number of neutrons can be

determined by _____ the atomic number from the

_____ . The mass number is

_____ .

Use Atomic Number and Mass Number

Use with Example Problem 2, page 118.

Solve *Read Example Problem 2 in your text.*

You Try It

Problem

You are given two samples of carbon. The first sample, carbon-12, has a mass number of 12, the second sample, carbon-13, has a mass number of 13. Both samples have an atomic number of 6. Determine the number of protons, electrons, and neutrons in each sample.

1. **Analyze the Problem**

   Known:

   Carbon-12   **Carbon-13**

   Mass number is _____   Mass number is _____

   Atomic number is _____   Atomic number is _____

   Unknown:

   The number of protons, electrons, and neutrons in each sample.

2. **Solve for the Unknown**

   Number of protons = number of electrons = atomic number = _____

   Number of neutrons = mass number – atomic number

   The number of neutrons for carbon-12 = 12 – 6 = _____

   The number of neutrons for carbon-13 = 13 – 6 = _____

3. **Evaluate the Answer**

   The number of neutrons does equal the _____ minus

   the _____ , or the number of protons.

Section 3 How Atoms Differ (continued)

| Main Idea | Details |

Mass of Atoms

Use with pages 119–120.

Explain *why the mass number for chlorine is more than 35.*

_____

_____

_____

Calculate Atomic Mass

Use with Example Problem 3, page 121.

Isotope Abundance for Element X		
Isotope	**Mass (amu)**	**Percent abundance**
^6X	6.015	7.59%
^7X	7.016	92.41%

Summarize *Fill in the blanks to help you take notes while you read Example Problem 3.*

Problem

Given the _____ in the table in the left margin, _____ the

_____ of unknown element X. Then, _____ the unknown

_____, which is used _____ to treat some _____.

1. **Analyze the problem**

   Known: Unknown:

   For isotope ^6X: _____ of X = ? amu

   mass = _____   _____ of element X = ?

   abundance = _____

   For isotope ^7X:

   mass = _____

   abundance = _____

2. **Solve for the unknown**

   Mass contribution = (_____) (_____)

   For ^6X: Mass contribution = _____ = _____

   For ^7X: Mass contribution = _____ = _____

   Sum the mass contributions to find the atomic mass.

   _____ of X = _____ = _____

   Use the _____ to identify the element.

   The element with an atomic mass of 6.939 amu is _____.

3. **Evaluate the answer**

   The number of neutrons does equal the _____ minus the

   _____, or number of _____.

The Structure of the Atom
Section 4 Unstable Nuclei and Radioactive Decay

Main Idea ——— **Details** ———————————————

Skim *Section 4 of your text. Write two questions that come to mind from reading the headings, and the captions.*

1. _____

2. _____

New Vocabulary

Use your text to define each term.

radioactivity _____

radiation _____

nuclear reaction _____
_____

radioactive decay _____
_____

alpha radiation _____

alpha particle _____
_____

nuclear equation _____
_____

beta radiation _____

beta particle _____

gamma ray _____
_____

Section 4 Unstable Nuclei and Radioactive Decay (continued)

Main Idea	Details

Radioactivity and Types of Radiation

Use with pages 122–124.

Explain *radioactivity by completing the paragraph below.*

In chemical reactions, atoms may be _____, but their

_____ do not change. The rearrangement _____ only

the _____ of the atoms, not the _____.

_____ reactions are different. In nuclear reactions,

_____ gain stability by emitting _____. As a result of

_____ in the nuclei, the atoms' _____ change.

_____ will continue emitting _____ , in a process

called _____ , until stable nuclei, often of a

_____ , are formed.

Sequence *the steps of a nuclear reaction.*

_____ A stable, nonradioactive atom is formed.

_____ Radiation is emitted.

_____ The process of radioactive decay continues until the nucleus is stable.

_____ An atom has an unstable nucleus.

Distinguish *between alpha, beta, and gamma radiation by completing the table below.*

Radiation Type			
	Alpha	**Beta**	**Gamma**
Symbol	$\alpha\left(^4_2\text{He}\right)$		
Mass (amu)		1/1840	
Charge			0

Discuss *why some elements are radioactive while most elements are not.*

_____

_____

_____

The Structure of the Atom Chapter Wrap-Up

Now that you have read the chapter, review what you have learned. List three important things you learned about the structure of an atom.

_____

_____

_____

_____

Review *Use this checklist to help you study.*

☐ Study your Science Notebook for this chapter.

☐ Study the definitions of vocabulary words.

☐ Review daily homework assignments.

☐ Reread the chapter and review the tables, graphs, and illustrations.

☐ Answer the Section Review questions at the end of each section.

☐ Look over the Study Guide at the end of the chapter.

REAL-WORLD CONNECTION

Radioactive materials are used in power plants and for medical uses. Some people object to the widespread use of nuclear reactors and radioactive materials. Discuss how what you've learned in this chapter affects your view on the use of radioactive materials.

_____

_____

_____

_____

Electrons in Atoms

Before You Read

Chapter 4   **Review** *the structure of the atom by completing the following table.*

Part of the Atom	Description
proton	
	centrally located part of the atom that contains protons and neutrons
electron	
	subatomic particle with no charge found in the _____

Draw *a typical atom and label the structures.*

Identify *three facts about electrons.*

Example: Electrons are a part of the structure of an atom.

1. _____

2. _____

3. _____

Electrons in Atoms
Section 1 Light and Quantized Energy

Main Idea —————— **Details**

Scan *Section 1 of your text. Use the checklist below as a guide.*

• Read all section titles.

• Read all boldfaced words.

• Read all tables and graphs.

• Look at all pictures and read the captions.

Write three facts you discovered about light.

1. _____

2. _____

3. _____

New Vocabulary   *Use your text to define each term.*

electromagnetic radiation _____

wavelength _____

frequency _____

amplitude _____

electromagnetic spectrum _____

quantum _____

Planck's constant _____

photoelectric effect _____

_____

photon _____

atomic emission spectrum _____

Section 1 Light and Quantized Energy (continued)

⟨**Main Idea**⟩ ——— ⟨**Details**⟩ ——————————————————————————————

The Atom and Unanswered Questions

Use with pages 136.

List *the three reasons scientists found Rutherford's nuclear atomic model to be fundamentally incomplete.*

1. _____

_____

2. _____

_____

3. _____

_____

Wave Nature of Light

Use with pages 137–140.

Explain *the relationship shown by the figure below. Use the following terms: wavelength, frequency, amplitude, and speed.*

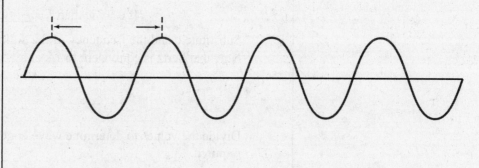

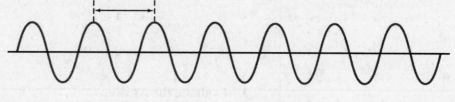

_____

_____

_____

_____

Section 1 Light and Quantized Energy (continued)

Main Idea —— **Details** ——————————————————

Calculating Wavelength of an EM Wave

Use with Example Problem 1, page 140.

Solve *Read Example Problem 1 in your text.*

You Try It

Problem

Radio waves are used to transmit information on various channels. What is the wavelength of a radio wave having the frequency of 5.40×10^{10} Hz?

1. **Analyze the Problem**
   Known: _____ $v =$ _____ and $c =$ _____

   Unknown: $\lambda =$ _____

   You know that because radio waves are part of the electromagnetic spectrum, their speed, frequency, and wavelength are related by the formula $c = \lambda v$.

2. **Solve for the Problem**

   Solve the equation relating the speed, frequency, and wavelength of an electromagnetic wave for wavelength (λ).

   If $c = \lambda v$, then $\lambda =$ _____

   Substitute c and the frequency of the radio wave, v, into the equation. Note that hertz is equivalent to 1/s or s^{-1}.

   $\lambda =$

   Divide the values to determine wavelength λ, and cancel units as required.

   $\lambda =$

3. **Evaluate the Answer**

   The answer is correctly expressed in a unit of _____. Both of the known values in the problem are expressed with _____ significant figures, so the answer must have _____ significant figures.

Section 1 Light and Quantized Energy (continued)

Main Idea ———— **Details** ——————————————————————

Particle Nature of Light

Use with pages 141–143.

Identify *two facts the wave model of light failed to explain.*

1. _____

_____

2. _____

_____

Describe *Planck's quantum concept by completing the following statement.*

The quantum concept concludes that matter can gain or lose _____ only

in small, specific amounts called _____. A quantum is the

minimum amount of energy that can be _____ or _____ by an

atom.

Atomic Emission Spectra

Use with pages 144–145.

Compare and contrast *Einstein's equation with Planck's equation by completing the following sentence.*

Planck's equation, _____, demonstrates mathematically that the

energy of a quantum is related to the _____ of the emitted radiation.

Einstein went further by explaining that, in addition to its wavelike

characteristics, a beam of light can be thought of as a stream of

_____ called _____.

Contrast *the continuous electromagnetic spectra and the atomic emission spectra.*

_____

_____

_____

_____

_____

Electrons in Atoms

Section 2 Quantum Theory and the Atom

Main Idea ———— **Details** ————————————————

Skim *Section 2 of your text. Write three questions that come to mind from reading the headings and the illustration captions.*

1. _____

2. _____

3. _____

New Vocabulary   *Use your text to define each term.*

ground state   _____

quantum number   _____

de Broglie equation   _____

                                      _____

Heisenberg uncertainty   _____
principle
                                      _____

quantum mechanical   _____
model of the atom

atomic orbital   _____

principal quantum number   _____

principal energy level   _____

energy sublevel   _____

Section 2 Quantum Theory and the Atom (continued)

Main Idea ———— **Details** ————————————————————

Bohr's Model of the Atom

Use with pages 146–148.

Classify *the characteristics of each series in hydrogen's line spectrum.*
Include the following information.
  1. Beginning orbit(s)/ending orbit
  2. Description of the spectral lines

Balmer	Paschen	Lyman
1.	1.	1.
2.	2.	2.

The Quantum Mechanical Model of the Atom

Use with page 149–150.

Sequence *de Broglie's process in developing his equation by completing the flow chart below.*

Whole _____ of _____ are allowed in a circular orbit of fixed _____.

Light has both _____ and _____ characteristics.

Can particles of matter, including electrons, behave like _____?

If an electron has _____ and is restricted to circular orbits of fixed radius, the _____ is allowed only certain possible wavelengths, _____, and _____.

Section 2 Quantum Theory and the Atom (continued)

Main Idea ———— **Details** ————————————————————

The Heisenberg Uncertainty Principle

Use with pages 151–152.

Discuss *how Heisenberg's principle influenced Schrödinger to develop his wave equation.*

_____

_____

_____

_____

_____

Hydrogen's Atomic Orbitals

Use with page 153.

Identify *four facts about atomic orbitals by completing the following statements.*

1. _____ indicate the relative sizes and energies of atomic orbitals.

2. The atom's major energy levels are called

   _____

3. Principal energy levels contain _____.

4. The number of _____ in a principal energy level _____ as *n* increases.

SUMMARIZE

Compare and contrast the Bohr and quantum mechanical models of the atom.

_____

_____

_____

_____

_____

Electrons in Atoms
Section 3 Electron Configuration

⟨**Main Idea**⟩ ——— ⟨**Details**⟩ ———————————————

Skim *Section 3 of your text. Focus on the headings, subheadings, boldfaced words, and figure captions. Summarize the main ideas of this section.*

_____

_____

_____

_____

⟨**New Vocabulary**⟩   *Use your text to define each term.*

electron configuration   _____

aufbau principle   _____

Pauli exclusion principle   _____

_____

Hund's rule   _____

_____

_____

valence electron   _____

electron-dot structure   _____

_____

_____

Section 3 Electron Configurations (continued)

Main Idea ———— **Details** ——————————————————

Ground-State Electron Configuration

Use with pages 156–157.

Organize *information about electron configurations by completing the following outline.*

Electron configuration is _____.

I. Ground–state electron configurations

A. Three rules define how electrons can be arranged in an atom's orbitals:

1. _____

2. _____

3. _____

Electron Arrangement

Use with pages 158–160.

B. The _____ methods for representing an atom's electron configuration

1. Orbital diagrams

a. An empty box represents an _____.

b. A box containing a single up arrow represents an orbital with

_____.

c. A box containing both up and down arrows represents a

_____.

d. Each box is labeled with the _____.

_____ and associated with the

orbital.

2. _____

a. This method designates the _____ and

_____ associated with each of the atom's

orbitals, and includes a _____.

_____

Valence Electrons

Use with pages 161.

C. Only valence electrons _____

_____.

1. Electron-dot structures consist of the _____,

which represents the _____

_____, surrounded by dots representing the

_____

Section 3 Electron Configurations (continued)

Main Idea	Details

Electron-Dot Structures

Use with Example Problem 3, Page 162.

Solve *Read Example Problem 3 in your text.*

You Try It

Problem

Ruthenium (Ru) is commonly used in the manufacture of platinum alloys. What is the ground-state electron configuration for an atom of ruthenium?

1. **Analyze the Problem**

   Known: _____

   Unknown: _____

   Determine the number of additional electrons a ruthenium atom has compared to the nearest preceding noble gas, and then write out ruthenium's electron configuration.

2. **Solve for the Problem**

   From the periodic table, ruthenium's atomic number is determined to be ☐. Thus a ruthenium atom contains ☐ electrons. The noble gas preceding ruthenium is krypton (Kr), which has an atomic number of 36. Represent ruthenium's first 36 electrons using the chemical symbol for krypton written inside brackets. _____

   The first 36 electrons have filled out the 1s, 2s, 2p, 3s, 3p, 4s, 3d and 4p sublevels. The remaining ☐ electrons of ruthenium's configuration need to be written out. Thus, the remaining ☐ electrons fill the _____ orbitals.

   Using the maximum number of electrons that can fill each orbital, write out the electron configuration. _____

3. **Evaluate the Answer**

   All ☐ electrons in a ruthenium atom have been accounted for. The correct preceding noble gas _____ has been used in the notation, and the order of orbital filling for the _____ is correct.

Electrons in Atoms Chapter Wrap-Up

Now that you have read the chapter, review what you have learned. List three important things you have learned about electrons in atoms.

_____

_____

_____

Review

Use this checklist to help you study.

☐ Study your Science Notebook for this chapter.

☐ Study the definitions for vocabulary words.

☐ Review daily homework assignments.

☐ Reread the chapter and review the tables, graphs, and illustrations.

☐ Answer the Section Review questions at the end of each section.

☐ Look over the Study Guide at the end of the chapter.

REAL-WORLD CONNECTION

Explain how advances in our understanding of the atom influence our daily lives.

_____

_____

_____

_____

_____

The Periodic Table and Periodic Law

Before You Read

Review Vocabulary   **Define** *the following terms.*

atom

_____

_____

electron configuration   _____

valence electrons   _____

electron-dot structure   _____

_____

_____

Chapter 4   **Distinguish** *between the subatomic particles in terms of relative charge.*

Subatomic Particle   Electrical Charge

_____   _____

_____   _____

Describe *how the subatomic particles are arranged.*

_____

_____

The Periodic Table and Periodic Law
Section 1 Development of the Modern Periodic Table

⟨Main Idea⟩ ———— **⟨Details⟩** ————————————————————

Skim *Section 1 of your text. Look at the headings, boldfaced words, figures and captions. Write two facts you discovered about the periodic table.*

1. _____

2. _____

⟨New Vocabulary⟩   *Use your text to define each term.*

periodic law _____

_____

group _____

_____

period _____

_____

representative element _____

transition element _____

metal _____

alkali metal _____

alkaline earth metal _____

transition metal _____

inner transition metal _____

nonmetal _____

halogen _____

noble gas _____

metalloid _____

Section 1 Development of the Modern Periodic Table (continued)

Main Idea ——— **Details** ———————————————————————————

Development of the Periodic Table

Use with pages 174–176.

Sequence *the events that helped develop the periodic table.*

1. In the 1790's, _____.

2. In 1864, _____
   and saw the properties of elements _____.

3. In 1869,_____
   _____. He left blank spaces
   _____.

4. In 1913, _____
   _____. He arranged
   elements by_____

The Modern Periodic Table

Use with pages 177–180.

Determine *where you can find each of the following groups of elements on the periodic table below:*

alkali metals   nonmetals   halogens

alkaline earth metals   representative elements   transition metals

inner transition metals   transition elements   noble gases

Hint: colored pencils might be helpful. Be sure to include a legend.

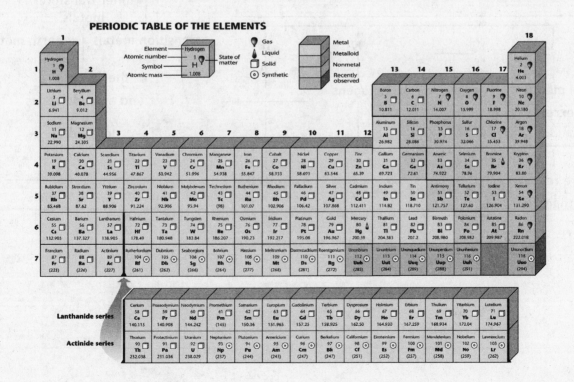

PERIODIC TABLE OF THE ELEMENTS

Section 1 Development of the Modern Periodic Table (continued)

⌒Main Idea⌒ ——— ⌒Details⌒ ——————————————————————

Organize *information about the periodic table by completing the concept map below.*

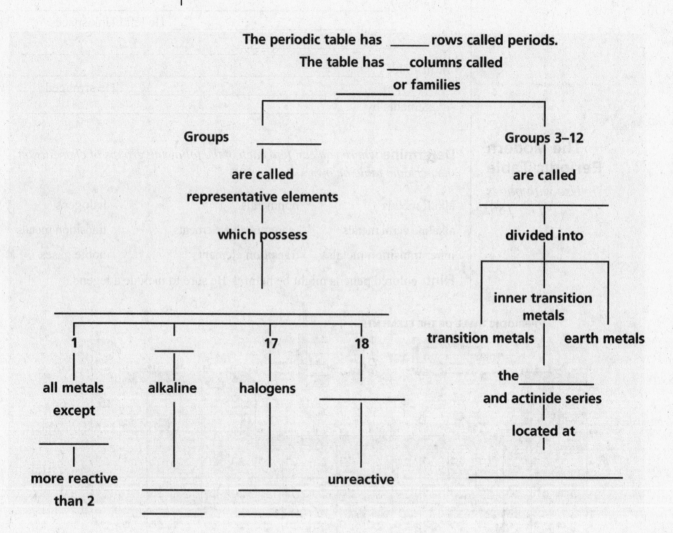

The periodic table has _____ rows called periods.
The table has ____ columns called
_____ or families

Groups _____   Groups 3–12

are called   are called

representative elements   divided into

which possess

                                                     inner transition
                                                    metals

1   17   18   transition metals   earth metals

all metals   alkaline   halogens   the _____
except   and actinide series

more reactive   unreactive   located at
than 2

_____

Section 1 Development of the Modern Periodic Table (continued)

⌐Main Idea⌐ ————	⌐Details⌐ —————————————————————

Identify *the information that is given on a typical box from the periodic table.*

1. _____

2. _____

3. _____

4. _____

5. _____

Match *the box color on the periodic table in Figure 5 with the class of element the box describes.*

blue   nonmetal

green   recently discovered

yellow   metalloid

gray   metal

REAL-WORLD CONNECTION

Describe how knowledge of the periodic table would be important in three different careers, based on what you've read.

_____

_____

_____

_____

_____

_____

The Periodic Table and Periodic Law
Section 2 Classification of the Elements

Main Idea ———— **Details** —————————————————

Scan *Section 2 of your text. Use the checklist below as a guide.*

- Read all section titles.

- Read all boldfaced words.

- Read all tables.

- Look at all pictures and read the captions.

- Think about what you already know about the shapes and arrangements of atoms in covalent compounds.

Write *three facts that you discovered about the relationship between electrons and an element's location on the periodic table.*

1. _____

2. _____

3. _____

New Vocabulary   *Define the following terms.*

structure   _____

Section 2 Classification of the Elements *(continued)*

⸨**Main Idea**⸩ ——— ⸨**Details**⸩ ————————————————————

Organizing the Elements by Electron Configuration

Use with pages 182–183.

Organize *information about electron configurations by completing the outline below.*

I. Electrons

   A. Valence electrons

      1. electrons in _____

      2. atoms in the _____ have _____

         _____.

   B. Valence electrons and period

      1. The _____ of an element's valence electrons

      indicates _____.

         a. Elements with valence electrons in energy level 2 are found in

         _____.

         b. Elements with _____

         are found in the fourth period.

   C. Valence electrons and group number

      1. Representative elements.

         a. All elements in group 1 have _____.

         b. All elements in group 2 have _____.

         c. Group 13 elements have _____,

         group 14 elements have _____, and so on.

      2. Helium, in group 18, is an _____.

Describe *the relationship between the number of valence electrons and the chemical properties of atoms.*

_____

_____

_____

Section 2 Classification of the Elements (continued)

| Main Idea | Details |

The s-, p-, d-, and f-Block Elements

Use with pages 183–185.

Distinguish *between s-, p-, d-, and f-block elements by completing the table below.*

	Periodic Table Groups	Orbitals	Type of Occupied Element
s-block			representative elements
p-block		p	
d-block	3 through 12		
f-block			

Electron Configuration and the Periodic Table

Use with Example Problem 1, page 186.

Summarize *Fill in the blanks to help you take notes while you read Example Problem 1.*

Problem

Without using the periodic table, determine the group, period, and block in which strontium is located on the periodic table.

1. Analyze the Problem
   Known: Unknown:

   _____   _____

   Use the electron configuration of strontium to determine its place.

2. Solve for the unknown

   Group: Strontium has a valence configuration of _____. All

   group _____ elements have the _____ configuration.

   Period: The _____ in $5s^2$ indicates that strontium is in _____.

   Block: The _____ indicates that strontium's valence

   electrons_____. Therefore, strontium is in the _____.

3. **Evaluate the answer**

   The relationships among _____ and

   _____ have been correctly applied.

The Periodic Table and Periodic Law
Section 3 Periodic Trends

Main Idea ———— **Details** ——————————————————

Scan Section 3 of your text. Use the checklist below as a guide.

- Read all section titles.
- Read all boldfaced words.
- Read all tables.
- Look at all pictures and read the captions.

Write *three facts you discovered about periodic trends.*

1. _____

2. _____

3. _____

New Vocabulary   *Use your text to define each term.*

ion   _____

                      _____

ionization energy   _____

                      _____

octet rule   _____

                      _____

electronegativity   _____

                      _____

Section 3 Periodic Trends (continued)

(Main Idea) ———— (Details) ————————————————————————

Atomic Radius

Use with pages 187–188.

Describe *how atomic size is defined.*

_____

_____

_____

Analyze *any trends that you observe in Figure 11 and how the trends relate to atomic mass.*

_____

_____

_____

Interpret Trends in Atomic Radii

Use with Example Problem 2, Page 189.

Summarize *Fill in the blanks to help you take notes while you read Example Problem 2.*

Problem

Which has the largest atomic radius: carbon (C), fluorine (F), beryllium (Be), or lithium (Li)? Explain your answer in terms of trends in atomic radii.

1. **Analyze the Problem**
   Known: periodic table information for four elements

   Unknown: which of the four has the _____

2. **Solve for the unknown**

   Use the _____ to determine if the elements are in the

   same group or period. All four elements are in _____.

   Order the elements from _____ across the period.

   _____

   Determine the largest based on trends of _____ .

   _____

3. **Evaluate the answer**

   The _____ in atomic radii have been correctly applied.

Section 3 Periodic Trends (continued)

Main Idea	Details

Ionic Radius

Use with pages 189–190.

Describe *atomic size and ionic change by completing the table below.*

Ionic Change	Ion Charge	Size of Atom
Atom _____ electrons	becomes positive	
atom gains electrons	becomes	increases

Identify *two reasons why the relative size of an atom becomes smaller due to the loss of electrons:*

1. _____

2. _____

_____

Explain *why atoms increase in size when the atom gains electrons.*

_____

_____

Ionization Energy

Use with pages 191–193.

Describe *ionization energy trends on the periodic table by completing the paragraphs below.*

Ionization energies generally _____ as you move left-to-right across a

_____ Increased nuclear charge leads to an _____ on

valance electrons. Ionization energy generally _____ when you move

down a _____. Less energy is required to remove

_____ because they are _____ from the nucleus.

The octet rule states that atoms tend to gain, lose, or share _____ in

order to acquire a full set of _____. First period

elements are the _____ to this rule.

Electronegativity

Use with page 194

Predict *what part of the periodic table has the greatest electronegativity. Use Figure 18 for reference.*

_____

The Periodic Table and Periodic Law   Chapter Wrap-Up

Review

Now that you have read the chapter, review what you have learned. List three facts about the periodic table and periodic law.

_____

_____

_____

_____

_____

_____

Use this checklist to help you study.

☐ Study your Science Notebook for this chapter.

☐ Study the definitions and vocabulary words.

☐ Review daily homework assignments.

☐ Reread the chapter and review the tables, graphs, and illustrations.

☐ Answer the Section Review questions at the end of each section.

☐ Look over the Study Guide at the end of the chapter.

REAL-WORLD CONNECTION

Explain how an understanding of the periodic table can help you gain confidence in studying chemistry.

_____

_____

_____

_____

Ionic Compounds and Metals

Before You Read

Review Vocabulary *Define the following terms.*

ion _____

ionization energy _____

_____

_____

noble gas _____

valance electron _____

_____

Chapter 5 **Create** *electron-dot diagrams for the following elements.*

aluminum: _____

calcium: _____

arsenic: _____

tellurium: _____

xenon: _____

Ionic Compounds and Metals

Section 1 Ion Formation

Main Idea ———— **Details** ————————————————

Skim *Section 1 of your text. Read the title and subheads. List three concepts that you think will be discussed in this section.*

1. _____

2. _____

3. _____

New Vocabulary   *Use your text to define each term.*

chemical bond   _____

cation   _____

_____

anion   _____

_____

Section 1 Ion Formation (continued)

Main Idea ——— **Details** ————————————————————

Valence Electrons and Chemical Bonds

Use with pages 206–209.

Organize *information about forming chemical bonds by completing the concept map below.*

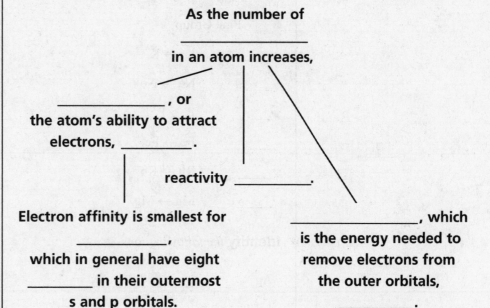

As the number of

_____

in an atom increases,

_____ , or
the atom's ability to attract
electrons, _____ .

reactivity _____ .

Electron affinity is smallest for
_____ ,
which in general have eight
_____ in their outermost
s and p orbitals.

_____ , which
is the energy needed to
remove electrons from
the outer orbitals,
_____ .

Write *the electron configuration of the most likely ion and the charge that is lost or gained by each of the following atoms. Indicate what the overall charge of the ion is, and whether it is a cation or an anion.*

Cs: $[Xe]6s^1$ _____

O: $[He]2s^2 2p^4$ _____

Ga: $[Ar]4s^2 3d^{10} 4p^1$ _____

Br: $[Ar]4s^2 3d^{10} 4p^5$ _____

Ag: $[Kr]5s^1 4d^{10}$ _____

Sc: $[Ar]4s^2 3d^1$ _____

Section 1 Ion Formation (continued)

⟨ **Main Idea** ⟩ —————   ⟨ **Details** ⟩ ————————————————————————

Sequence *the first group of elements in order of increasing ionization energy. Sequence the second group of elements in order of increasing electron affinity.*

First Group		Second Group

First Group

_____ $K \rightarrow K^+$

_____ $Ne \rightarrow Ne^+$

_____ $P \rightarrow P^{5+}$

_____ $Fe \rightarrow Fe^{2+}$

_____ $Rb \rightarrow Rb^+$

_____ $Mg \rightarrow Mg^{2+}$

Second Group

_____ $P \rightarrow P^{3-}$

_____ $O \rightarrow O^{2-}$

_____ $Xe \rightarrow Xe^-$

_____ $S \rightarrow S^{2-}$

_____ $I \rightarrow I^-$

_____ $F \rightarrow F^-$

Identify *the following ions.*

Ag^+ _____

Li^+ _____

Br^- _____

Ca^{2+} _____

S^{2-} _____

B^{3+} _____

As^{3-} _____

H^- _____

Cd^{2+} _____

Se^{2-} _____

Ionic Compounds and Metals

Section 2 Ionic Bonds and Ionic Compounds

Main Idea ———————— **Details** ————————————————————

Skim *Section 2 of your text.. Write three questions that come to mind from reading the headings and the illustration captions.*

1. _____

2. _____

3. _____

New Vocabulary *Use your text to define each term.*

ionic bond _____

_____

ionic compound _____

crystal lattice _____

electrolyte _____

_____

lattice energy _____

_____

Section 2 Ionic Bonds and Ionic Compounds (continued)

Main Idea	Details

Formation of an Ionic Bond

Use with pages 210–212.

Solve *Read pages 211–212 in your text.*

You Try It
Problem

Describe the formation of an ionic compound from the elements boron and selenium.

1. Analyze the Problem

Known: the electron configurations of the given elements

_____

Unknown: the number of valence electrons for each neutral atom

_____

_____

2. Solve for the Unknown

Determine how many electrons need to be removed from boron and how many electrons need to be added to selenium to form noble gas configurations.

_____

_____

_____

Determine how many boron atoms and how many selenium atoms must be present for the total number of electrons exchanged between the two elements to be equal.

_____

_____

_____

3. Evaluate the Answer

The overall charge on one unit of this compound is zero.

☐ boron ions (3+/boron ion) + ☐ selenide ions (☐/selenide ion) = ☐ (3+) + ☐ (☐) = 0

Section 2 Ionic Bonds and Ionic Compounds (continued)

⸤Main Idea⸥ ——— ⸤Details⸥ ————————————————————

Properties of Ionic Compounds

Use with pages 212–217.

Analyze *the relationship between the lattice energy of an ionic compound and the force of attraction.*

_____

_____

_____

Describe *the relationship between the size of the ions in a compound and the compound's lattice energy.*

_____

_____

_____

Explain *the relationship between lattice energy and the charge of the ion.*

_____

_____

Organize *the following ionic compounds from those with the least negative lattice energy to those with the most negative lattice energy.*

_____   LiCl

_____   BeS

_____   LiBr

_____   BeO

_____   BeCl$_2$

_____   RbBr

_____   CsI

_____   SrCl$_2$

_____   CsBr

Ionic Compounds and Metals
Section 3 Names and Formulas for Ionic Compounds

Main Idea ————   **Details** ——————————————————————————————

Scan *Section 3 of your text. Use the checklist below as a guide.*

- Read all section titles.

- Read all boldfaced words.

- Read all tables and diagrams.

- Look at all figures and read the captions.

- Study the example problems and note what they are intended to solve.

- Think about what you already know about the formation, formulas, and naming of ions and ionic compounds.

Write *three facts that you discovered about the names and formulas of ionic compounds.*

1. _____

2. _____

3. _____

New Vocabulary   *Use your text to define each term.*

formula unit   _____

monoatomic ion   _____

oxidation number   _____

polyatomic ion   _____

oxyanion   _____

_____

Academic Vocabulary   *Define the following term.*

transfer   _____

Section 3 Names and Formulas for Ionic Compounds (continued)

<table>
<tr><td>⌐Main Idea⌐</td><td>⌐Details⌐</td></tr>
</table>

Formula for an Ionic Compound

Use with Example Problem 1, page 220.

Solve *Read Example Problem 1 in your text.*

You Try It Problem

Calcium can form a cation with a 2+ charge. Write the formula for the ionic compound formed from calcium ion and chlorine.

1. Analyze the Problem

Known: _____the ionic forms of the component elements_____

and _____

Unknown: _____

2. Solve for the Unknown

The smallest number that is divisible by both ionic charges is _____,

so the compound contains _____ calcium ion(s) and _____ chloride

ion(s). The formula for the ionic compound formed is _____.

3. Evaluate the Answer

The overall charge on one formula unit of this compound is zero.

☐ Ca ion(s) (2+/Ca ion) + ☐ Cl ions (1− /Cl ion) = 0

Formula for a Polyatomic Ionic Compound

Use with Example Problem 3, page 222.

Solve *Read Example Problem 3 in your text.*

You Try It Problem

Write the formula for the ionic compound formed from the calcium ion and the bromate ion.

1. Analyze the Problem

Known: _____the ionic forms of the component elements _____

and _____

Unknown: _____

Section 3 Names and Formulas for Ionic Compounds (continued)

⟨Main Idea⟩ ———— ⟨Details⟩ ——————————————————————

	2. Solve for the Unknown The smallest number that is divisible by both ionic charges is _____, so _____ bromate ions combine with _____ calcium ion. The formula for the ionic compound formed is _____. **3. Evaluate the Answer** The overall charge on one formula unit of this compound is zero. 1 Ca ion (2+/Ca ion) + ☐ BrO_3 ions (1−/BrO_3 ion) = 0
Names for Ions and Ionic Compounds *Use with page 222–224.*	**Classify** *the ions listed below as monatomic or polyatomic cations or anions. If the ion is a polyatomic anion, indicate whether it is an oxyanion.* CN^- _____ MnO_4^- _____ Ba^{2+} _____ $Fe(CN)_6^{4-}$ _____ NH_4^+ _____ N^{3-} _____ Hg_2^{2+} _____ $S_2O_3^{2-}$ _____ O^{2-} _____ **Identify** *the ionic compounds listed below.* CaO _____ $KMnO_4$ _____ $Sr(IO_3)_2$ _____ NH_4OH _____ Fe_2S_3 _____ $Sn(NO_3)_4$ _____ $Pb_3(PO_4)_2$ _____ Hg_2SO_4 _____ $PtCl_4$ _____

Ionic Compounds and Metals

Section 4 Metallic Bonds and the Properties of Metals

Main Idea ——— **Details** ————————————————

Skim *Section 4 of your text. Write three questions that come to mind from reading the headings and the illustration captions.*

1. _____

2. _____

3. _____

New Vocabulary *Use your text to define each term.*

electron sea model   _____

_____

_____

delocalized electrons   _____

_____

_____

metallic bond   _____

_____

alloy   _____

_____

Section 4 Metallic Bonds and the Properties of Metals (continued)

⟨**Main Idea**⟩ ——— ⟨**Details**⟩ —————————————————————————

Metallic Bonds

Use with pages 225–226.

Summarize *how the electron sea model accounts for the malleability, high thermal conductivity, and high electrical conductivity of metals.*

_____

_____

_____

_____

_____

Explain *the properties of metals by completing the following sentences.*

The _____ of transition metals increases as the

number of delocalized electrons _____.

Because the _____ in metals are strongly attracted to the delocalized

electrons in the metal, they are not easily _____ from the metal,

causing the metal to be very _____.

Alkali metals are _____ than transition metals because they have only

_____ per atom.

The _____ of metals vary greatly. The melting points are not as

extreme as the _____. It does not take an extreme amount of

energy for _____ to be able to move past each other.

However, during _____, atoms must be separated from a group of

_____, which requires a lot of _____.

Light absorbed and released by the _____ in a metal

accounts for the _____ of the metal.

Section 4 Metallic Bonds and the Properties of Metals (continued)

Main Idea ———— **Details** ——————————————————————

Metal Alloys

Use with pages 227–228.

Match *the alloy composition given in the first column with the common name of the alloy in the second column and the alloy's uses in the third column. Draw lines between the appropriate items. Use Table 13 as a reference.*

45% Cu, 15% Ag, 42% Au	cast iron	tableware, jewelry
75% Fe, 17% Cr, 8% Ni	10-carat gold	dental fillings
97% Fe, 3% C	sterling silver	casting
92.5% Ag, 7.5% Cu	dental amalgam	medals, bells
80% Cu, 15% Zn, 5% Sn	brass	instruments, sinks
85% Cu, 15% Zn	bronze	jewelry
50% Hg, 35% Ag, 15% Sn	stainless steel	hardware, lighting

Contrast *a substitutional alloy with an interstitial alloy. Give an example of each.*

_____

_____

_____

_____

_____

_____

_____

_____

Ionic Compounds and Metals   Chapter Wrap-Up

Now that you have read the chapter, review what you have learned. List three important facts about ionic compounds.

1. _____

2. _____

3. _____

Review   *Use this checklist to help you study.*

☐ Study your Science Notebook for this chapter.

☐ Study the definitions of vocabulary words.

☐ Review daily homework assignments.

☐ Reread the chapter, and review the tables, graphs, and illustrations.

☐ Answer the Section Review questions at the end of each section.

☐ Look over the Study Guide at the end of the chapter.

SUMMARIZE

Explain how the atomic properties of an element determine what sort of ion it will form, and what properties a resulting ionic compound will have.

_____

_____

_____

_____

_____

Covalent Bonding

Before You Read

Review Vocabulary **Define** *the following terms.*

ionic bond _____

_____

octet rule _____

_____

Chapter 4 **Describe** *the structure of an atom.*

_____

_____

Chapter 6 **Explain** *the following concepts: periodic trends and periodic properties of elements.*

_____

_____

_____

Chapter 8 **Identify** *the ions, along with their charges, in the following ionic compounds.*

Li_2S _____

$KMnO_4$ _____

Al_2O_3 _____

Covalent Bonding

Section 1 The Covalent Bond

Main Idea ———————— **Details** ————————————————————

Skim *Section 1 of your text. Write three questions that come to mind from reading the headings and the illustration captions.*

1. _____

2. _____

3. _____

New Vocabulary   *Use your text to define each term.*

covalent bond _____

molecule _____

Lewis structure _____
_____

sigma bond _____
_____

pi bond _____

endothermic reaction _____
_____

exothermic reaction _____
_____

Academic Vocabulary   *Define the following term.*

overlap _____

Section 1 The Covalent Bond (continued)

Main Idea ———— **Details** ————————————————————

Why do atoms bond?

Use with page 240.

Explain *the octet rule by completing the following sentences.*

The _____ rule states that _____

_____

_____. Although exceptions exist, the rule provides a useful framework

for understanding _____.

What is a covalent bond?

Use with page 241.

Complete *the following sentences using words or phrases from your text.*

The force between two atoms is the result of _____

repulsion, nucleus-nucleus _____, and nucleus-electron

_____. At the point of _____, the

_____ forces balance the _____ forces. The most stable

arrangement of atoms exists at the point of _____,

when the atoms bond covalently and a _____ forms.

Solve *Read Example Problem 1 in your text.*

You Try It
Problem

Draw the Lewis structure for hydrogen chloride, HCl.

1. **Analyze the Problem**
   Write the electron-dot structures of each of the two component atoms.

Lewis Structure of a Molecule

Use with Example Problem1, page 244.

   Known: H. and .Cl:

   Unknown: _____ of HCl

   Hydrogen, H, has only one valence electron. Chlorine, Cl, has seven
   valence electrons. Cl needs one electron to complete its octet.

2. **Solve for the Unknown**
   Draw the electron-dot structure for each of the component atoms. Then
   show the sharing of the pairs of electrons.

   +   ⇌

   _____   _____   _____

   hydrogen   *chlorine*   *hydrogen chloride*
                                        molecule

Section 1 The Covalent Bond (continued)

(Main Idea) ——— (Details) ——————————————

> 3. **Evaluate the Answer**
>
> Each atom in the molecule has achieved a _____
>
> configuration and thus is _____.

Multiple covalent Bonds

Use with pages 245–246.

Identify *each bond between the component atoms as sigma bonds (single bonds), one sigma bond and one pi bond (double bonds), or one sigma bond and two pi bonds (triple bonds).*

H–C≡C–H   _____

_____

_____

H–C=O
  |
  H

_____

_____

_____

The Strength of Covalent Bonds

Use with pages 246–247.

Explain *the factors that control the strength of covalent bonds.*

_____

_____

_____

_____

Define *bond dissociation energy.*

_____

REAL-WORLD CONNECTION

Explain how understanding covalent bonding and the chemistry of compounds might help scientists increase food supplies.

_____

_____

_____

Covalent Bonding
Section 2 Naming Molecules

Main Idea ———— **Details** ———————————————————

Scan *Section 2 of your text. Use the checklist below as a guide.*

- Read all section titles.
- Read all boldfaced words.
- Read all tables and graphs.
- Read all formulas.
- Look at all figures and read the captions.
- Think about what you already know about the naming of molecules.

Write *three facts you discovered about the names and formulas of covalent molecules.*

1. _____

2. _____

3. _____

New Vocabulary   *Use your text to define the following term.*

oxyacid   _____

Name _____ Date _____

Section 2 Naming Molecules (continued)

Main Idea — **Details** —

Naming Binary Molecular Compounds

Use with Example Problem 2, pages 249.

Identify *the prefixes for these three binary molecular compounds.*

Ge_3N_2 _____ -germanium _____ -nitride

C_2Cl_4 _____ -carbon _____ -chloride

B_6Si _____ -boron silicide

Solve *Read Example Problem 2 in your text.*

You Try It

Problem

Name the compound N_2O_3.

1. **Analyze the Problem**

   Known: _____

   Unknown: _____

   The formula reveals the elements present and the number of atoms for each element. Only two elements are present, and both are nonmetals, so the compound can be named according to the rules for binary molecular compounds.

2. **Solve for the Unknown**

   The first element present in the compound is _____, _____. The

   second element is _____, _____. The root of this

   name is _____, so the second part of the name is _____. From

   the formula, two _____ atoms and three _____ atoms make

   up a molecule of the compound. The prefix for two is _____ and

   prefix for three is _____. The complete name for the compound is

   _____.

3. **Evaluate the Answer**

   The name _____ shows that a molecule of the

   compound contains _____ atoms and _____

   atoms, which agrees with the chemical formula for the compound,

   N_2O_3.

Chemistry: Matter and Change

100

Science Notebook

Copyright © Glencoe/McGraw-Hill, a division of The McGraw-Hill Companies, Inc.

Section 2 Naming Molecules (continued)

Main Idea ——— **Details** ———————————————

Naming Acids

Use with pages 250–251.

Match *the chemical formulas listed below with the correct acids.*

HF   sulfurous acid

HIO_4 hydrofluoric acid

H_2SO_3 phosphoric acid

H_3PO_4 hypochlorous acid

$HC_2H_3O_2$ periodic acid

H_2CO_3 permanganic acid

HClO   acetic acid

$HMnO_4$ carbonic acid

Writing Formulas from Names

Use with pages 251–252.

Write *the chemical formula for the molecular compound names given below. Use the flow chart in Figure 12 to help you determine the correct formulas.*

_____ dicarbon tetrabromide   _____ tetrasulfur tetranitride

_____ arsenic pentafluoride   _____ arsenic acid

_____ perchloric acid   _____ hydrocyanic acid

SYNTHESIZE

Create questions and answers about naming molecules for your own original quiz game. Include topics such as: prefixes and number of atoms; formulas, common names, and molecular names for covalent binary compounds; and formulas, common names, and molecular names for binary acids and oxyacids.

_____

_____

_____

_____

Covalent Bonding

Section 3 Molecular Structures

Main Idea ——————— **Details** ———————————————————————

Skim *Section 3 of your text. Write three questions that come to mind from reading the headings, illustration captions, and topics for the example problems.*

1. _____

2. _____

3. _____

New Vocabulary

Use your text to define each term.

structural formula _____

_____

resonance _____

_____

coordinate covalent bond _____

_____

Section 3 Molecular Structures (continued)

Main Idea	Details

Structural Formulas

Use with pages 253–257.

List *the steps that should be used to determine Lewis structures.*

1. _____

2. _____

3. _____

_____

4. _____

_____

Lewis Structure for a Covalent Compound with Multiple Bonds

Use with Example Problem 4, page 256.

Solve *Read Example Problem 4 in your text.*

You Try It

Problem

Draw the Lewis structure for FCHO.

1. **Analyze the Problem**

   Known: the compound formula: _____

   Unknown: _____

   Carbon has less attraction for shared electrons, so it is the central atom.

2. **Solve for the Problem**

   Find the total number of valence electrons and the number of bonding pairs.

   ☐ valence electrons/C atom + ☐ valence electrons/F atom

   + 1 valence electron/H atom + ☐ valence electrons/O atom

   = ☐ valence electrons

   ☐ available valence electrons/(2 electrons/pair) = ☐ available pairs

Section 3 Molecular Structures (continued)

(Main Idea) ——— **(Details)** ————————————————————

Draw single bonds, which represent _____ each, from the carbon atom to each terminal atom, and place electron pairs around the _____ and _____ atoms to give them stable

_____.

H—C—Ö:
　　　|
　　 :F:

_____ available pairs – _____ pairs used = 0

Carbon does not have an octet, so one of the lone pairs on the _____ atom must be used to form a _____ bond.

_____

3. Evaluate the Answer

Both carbon and _____ now have an octet, which satisfies the octet rule.

Lewis Structure for a Polyatomic Ion

Use with Example Problem 5, page 257.

Solve *Read Example Problem 5 in your text.*

You Try It

Problem

Draw the Lewis structure for the permanganate ion (MnO_4^-).

1. Analyze the Problem

Known:　　the compound formula: _____

Unknown: _____

Manganese has less attraction for shared electrons, so it is the central atom.

2. Solve for the Unknown

Find the total number of valence electrons and the number of bonding pairs.

1 Mn atom × (☐ valence electrons/Mn atom) + ☐ O atoms

× (6 valence electrons/O atom + ☐ electron(s) from the negative

charge = ☐ valence electrons

Section 3 Molecular Structures (continued)

⬭Main Idea⬭ ——————— ⬭Details⬭ ———————————————

□ available valence electrons/(2 electrons/pair) = □ available pairs

Draw single bonds, which represent an _____, from the

Mn atom to each O atom, and place electron pairs around the

O atoms to give them stable _____.

□ available pairs – □ pairs used = 0

No electron pairs remain available for the Mn atom, so the Lewis

structure for the permanganate ion is:

3. **Evaluate the Answer**
   All atoms now have an octet, and the group of atoms has a net charge
   of _____.

Resonance Structures

Use with page 258.

Explain *resonance structures by completing the following sentences.*
Each actual molecule or ion that undergoes _____ behaves as if it
has only _____ structure. Experimentally measured bond lengths
show that the bonds are _____ to each other.

Exceptions to the Octet Rule

Use with page 258–260.

List *three reasons for exceptions to the octet rule.*

1. _____

   _____

2. _____

   _____

3. _____

   _____

Covalent Bonding

Section 4 Molecular Shapes

◝Main Idea◝ ——— **◝Details◝** ———————————————————

Scan *Section 2 of your text. Use the checklist below as a guide.*

• Read all section titles.

• Read all boldfaced words.

• Read all tables.

• Look at all pictures and read the captions.

• Think about what you already know about the shapes and arrangements of atoms in covalent compounds.

Write *three facts you discovered about the shapes covalent compounds take.*

1. _____

2. _____

3. _____

◝New Vocabulary◝   *Use your text to define each term.*

VSEPR model   _____

_____

_____

_____

hybridization   _____

_____

Section 4 Molecular Shapes (continued)

Main Idea ———— **Details** ——————————————————————

VSEPR Model

Use with pages 261–262.

Match *the molecular shapes listed below with their corresponding bond angles.*

trigonal planar	180°
trigonal pyramidal	120°
bent	109.5°
linear	107.3°
octahedral	104.5°
tetrahedral	90° (out of plane); 120° (in plane)
trigonal bipyramidal	90°

Hybridization

Use with pages 262–263.

Label *the hybrid orbitals in the figures below as sp, sp^2, sp^3 sp^3d, or sp^3d^2.*

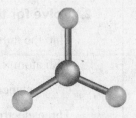

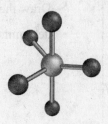

Section 4 Molecular Shapes (continued)

⎛Main Idea⎞ ——— ⎛Details⎞ ———————————————————————

Find the Shape of a Molecule

Use with Example Problem 7, page 264.

Solve *Read Example Problem 7 in your text.*

You Try It
Problem

What is the shape of a SbI_5 molecule? Determine the bond angles, and identify the type of hybrid orbitals that form the molecule's bonds.

1. **Analyze the Problem**

   Known: the compound formula: _____

   Unknown: _____

                 _____

   The molecule contains one central antimony atom bonded to _____ iodine atoms.

2. **Solve for the Unknown**

   Find the number of valence electrons and the number of electron pairs.

   1 Sb atom × (☐ valence electrons/Sb atom) + ☐ I atoms ×

   (☐ valence electrons/I atom) = ☐ valence electrons

   Three electron pairs exist on each iodine atom. This leaves ☐

   available valence electrons for bonding. ☐ available valence

   electrons/(2 electrons/pair) = ☐ available pairs

   Draw the molecule's Lewis structure. From this Lewis structure,

   determine the molecular shape.

   Lewis structure   Molecular shape

   The molecule's shape is _____, with a bond

   angle of _____ in the horizontal plane, and a bond angle of _____

   between the vertical and horizontal bonds. The bonds are made

   up of _____ hybrid orbitals.

3. **Evaluate the Answer**

   Each iodine atom has an octet. The antimony atom has _____ electrons, which is allowed when a d orbital is hybridized.

Covalent Bonding

Section 5 Electronegativity and Polarity

Main Idea ——— **Details** ————————————————

Scan *Section 5 of your text. Use the checklist below as a guide.*

- Read all section titles.

- Read all boldfaced words.

- Read all tables and charts.

- Look at all pictures and read the captions.

- Think about what you already know about the strengths and distribution of charge in covalent bonds.

Write *three facts you discovered about electronegativity.*

1. _____

2. _____

3. _____

Use your text to define the following term.

polar covalent bond _____

_____

Section 5 Electronegativity and Polarity (continued)

Main Idea	Details

Electronegativity and Bond Character

Use with pages 265–265.

Sequence *the following elements from the least electronegative to the most electronegative. Use Figure 20 for reference.*

_____ Au

_____ Y

_____ Ba

_____ P

_____ H

_____ Te

_____ O

_____ I

_____ Co

Polar Covalent Bonds

Use with pages 267–268.

Draw *the Lewis structure for each of the molecular compounds listed below. Analyze the symmetry of the structure to determine whether or not the compound is polar covalent or nonpolar covalent.*

N_2 _____   _____

CO_2 _____   _____

CH_3Cl _____   _____

_____

Section 5 Electronegativity and Polarity (continued)

Main Idea ———— **Details** ————————————————————

Properties of Covalent Compounds

Use with pages 269–270.

Determine *whether each of the properties listed below is characteristic of ionic compounds, covalent compounds, nonpolar covalent compounds, or polar covalent compounds.*

low melting point   _____

very soft solid   _____

high boiling point   _____

weak interaction between formula units   _____

solubility in oil   _____

very hard solid   _____

high melting point   _____

solubility in water   _____

easily vaporized   _____

strong interaction between

formula units   _____

Covalent Network Solids

Use with page 270.

Describe *what the network solid for quartz (SiO_2) molecules is like, and how it has a tetrahedral structure similar to diamond structure.*

_____

_____

_____

_____

_____

_____

_____

_____

_____

_____

_____

Covalent Bonding Chapter Wrap-Up

After reading this chapter, list three key facts about covalent bonding.

1. _____

2. _____

3. _____

Review

Use this checklist to help you study.

☐ Use this Science Notebook to study this chapter.

☐ Study the vocabulary words and scientific definitions.

☐ Review daily homework assignments.

☐ Reread the chapter and review the tables, graphs, and illustrations.

☐ Answer the Section Review questions at the end of each section.

☐ Look over the Study Guide at the end of the chapter.

REAL-WORLD CONNECTION

Explain how covalent bonds in carbon account for the vast number of carbon compounds, including those responsible for living organisms.

_____

_____

_____

_____

_____

Chemical Reactions

Before You Read

Review Vocabulary — *Define the following terms.*

ionic compound

_____

_____

molecular compound

_____

_____

Chapter 7 — **Explain** *how to write formulas for ionic compounds.*

_____

_____

_____

Write *the formula for the following ionic compound.*

aluminum carbonate

_____

Chapter 8 — **Explain** *how to write formulas for molecular compounds.*

_____

_____

Write *the formula for the following molecular compound.*

sulfuric acid

_____

Chemical Reactions

Section 1 Reactions and Equations

Main Idea —— **Details** ————————————————————

Scan *Section 1 of your text. Use the checklist below as a guide.*

- Read all section titles.
- Read all boldfaced words.
- Read all charts and graphs.
- Look at all pictures and read the captions.

Write *three facts about chemical reactions.*

1. _____

2. _____

3. _____

New Vocabulary *In the left column, write the terms defined below.*

_____ *a rearrangement of the atoms in one or more substances to form different substances*

_____ *the starting substances of a chemical reaction*

_____ *the substances formed during a chemical reaction*

_____ *a statement that uses chemical formulas to show the identities and relative amounts of the substances involved in a chemical reaction*

_____ *number written in front of a reactant or product that is used to balance chemical equations*

Academic Vocabulary *Define the following term.*

formula   _____

_____

Section 1 Reactions and Equations (continued)

⊂Main Idea⊃ ———   ⊂Details⊃ —————————————————————

Chemical Reactions

Use with pages 282–283.

Identify *three examples of chemical reactions you have seen, heard, or smelled in the last 24 hours. Think about activities at home, at school, or outside. Include any evidence you had that a chemical reaction was occurring. .*

Reaction	Evidence

Representing Chemical Reactions

Use with pages 283–285.

Organize *types of equations that can express a chemical reaction. In the second column, list the elements (words, coefficients, etc.) that are used to create each equation. In the third column, rank each equation from 1 to 3, giving a 3 to the equation that provides the most information, and a 1 to the equation that provides the least information.*

Type	Elements	Ranking
Word equations		
Chemical equations		
Skeleton equations		

Label *the chemical state each symbol below identifies in a chemical equation.*

(s) _____

(g) _____

(aq) _____

(l) _____

Section 1 Reactions and Equations (continued)

Main Idea ———— **Details** ——————————————————————

Balancing Chemical Equations

Use with pages 285–288.

Solve *Read Example Problem 1 in your text.*

You Try It

Problem

Balance the chemical equation for the reaction in which fluorine reacts with water to produce hydrofluoric acid and oxygen.

1. **Analyze the problem**

   Known: _____

   Unknown: _____

2. **Solve for the Unknown**

   Use the space below to write the skeleton equation:

   _____

   Count the atoms of each element in the reactants.

   ____ F, ____ H, ____ O

   Count the atoms of each element in the products.

   ____ F, ____ H, ____ O

   Insert the coefficient ____ in front of _____ to balance the oxygen atoms.

   Insert the coefficient ____ in front of ____ to balance the _____.

   Insert the coefficient ____ in front of ____ to balance the _____.

   Write the equation after adding the coefficients.

   _____

   Check that the coefficients are at their lowest possible ratio.

   The ratio of the coefficients is _____.

   Write the number of atoms in the balanced equation below:

   Reactants: _____

   Products: _____

3. **Evaluate the Answer**

   The _____ of each element is _____ on both sides of the equation. The _____ are written to the _____ ratio.

Chemical Reactions
Section 2 Classifying Chemical Reactions

Main Idea —————   **Details** ——————————————————

Scan *Section 2 of your text. Use the checklist below as a guide.*

- Read all section titles.
- Read all boldfaced words.
- Read all charts and graphs.
- Look at all pictures and read the captions.
- Think about what you already know about chemical reactions.

Write *three facts you discovered about classifying chemical reactions.*

1. _____

2. _____

3. _____

New Vocabulary   *Use your text to define each term.*

synthesis reaction   _____

_____

combustion reaction   _____

_____

decomposition reaction   _____

_____

single-replacement reaction   _____

_____

double-replacement reaction   _____

_____

precipitate   _____

Chemistry: Matter and Change   117   *Science Notebook*

Section 2 Classifying Chemical Reactions (continued)

Main Idea ———— **Details** ——————————————

Complete *the following diagrams illustrating each classification of chemical reaction. The first one has been completed for you.*

Synthesis Reactions

Use with page 289.

Synthesis reaction

Substance
Substance ————————→ New compound

$A + B \rightarrow$ _____

Combustion Reactions

Use with pages 290–291.

Combustion reactions

Metal, nonmetal, or compound substance ————————→ _____

Decomposition Reactions

Use with page 292.

Decomposition reactions

Compound ————————→ Element or
Element or _____

$AB \rightarrow$ _____

Replacement Reactions

Use with pages 293–298.

Single-replacement reactions

Compound ————————→ Metal or nonmetal

$A + BX \rightarrow$ _____

Double-replacement reactions

Compound with anion ————————→

$AX + BY \rightarrow$ _____

Section 2 Classifying Chemical Reactions (continued)

Main Idea ————— **Details** ————————————————————

Use with pages 289–298.

Organize *types of chemical reactions. The first column in the chart below lists some possible products in a chemical reaction. In the second column, write the type of chemical reaction that is likely to generate each product.*

Products	Possible Chemical Reaction
two different compounds, one of which is often a solid, a gas, or water	
oxide of the metal or a nonmetal or two or more oxides	
two or more elements or compounds	
a new compound and a replaced metal or nonmetal	
one compound	

ANALOGY

Consider the list of metals and halogens and their relative reactivity in Figure 13. Using your own experiences, identify people or things that could be ranked according to how they react in a certain situation.

1. (Example) Rank baseball bats by how likely they are to break.

2. _____

3. _____

4. _____

Chemical Reactions
Section 3 Reactions in Aqueous Solutions

Main Idea ————— **Details** ———————————————————————

Consider *the title and first paragraph in Section 3. Based on what you read, what do you expect to learn in this chapter?*

_____

_____

_____

New Vocabulary    *In the left column, write the terms defined below.*

_____   *the most plentiful substance in a solution*

_____   *substances dissolved in a solution*

_____   *equations that include only particles that participate in a reaction*

_____   *ion that does not participate in a reaction*

_____   *ionic equation that shows all the particles in a solution as they realistically exist*

_____   *a solution in which the most plentiful substance is water*

Section 3 Reactions in Aqueous Solutions (continued)

⟨Main Idea⟩ ———— ⟨Details⟩ ————————————————————————

Aqueous Solutions

Use with pages 299–300.

Connect *English words to their Latin roots. The term aqueous comes from the Latin word for water, aqua. Use a dictionary to find three words that also come from aqua, and list them in the box below together with a brief definition that explains their connection to water.*

Word	Definition

Types of Reactions in Aqueous Solutions

Use with pages 300–301.

Compare *a complete ionic equation and a chemical equation.*

_____

_____

_____

_____

_____

_____

Draw *a circle around the spectator ions in the following equation.*

$$2A^+(aq) + 2B^-(aq) + C^+(aq) + 2D^-(aq) \longrightarrow 2A^+(aq) + 2D^-(aq) + 2BC$$

Identify *whether each of the equations below is a complete ionic equation or a net ionic equation.*

$$A^+(aq) + B^-(aq) + C^+(aq) + D^-(aq) \longrightarrow AD + B^-(aq) + C^+(aq)$$

_____

$$E^+(aq) + F^-(aq) \longrightarrow EF$$

_____

$$G^+(aq) + HI^-(aq) \longrightarrow GI + H(g)$$

_____

Section 3 Reactions in Aqueous Solutions (continued)

(Main Idea)	(Details)

Reactions That Form Water

Use with pages 303–304.

Compare *reactions in aqueous solution that form a precipitate and reactions that form water. Put each of the following characteristics in the corresponding category.*

- can be described with ionic equations
- generates a solid product
- double-replacement reaction
- has no observable evidence

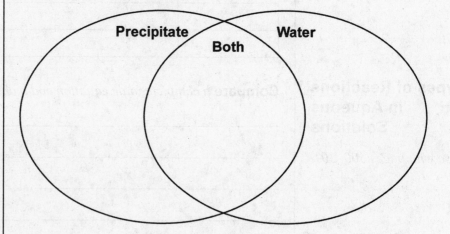

Reactions That Form Gases

Use with pages 305–306.

Identify *three commonly produced gases in reactions in aqueous solutions.*

_____

_____

_____

State *the evidence that would indicate that carbon dioxide gas is escaping from the solution containing sodium hydrogen carbonate shown in Figure 19.*

_____

List *the two reactions that occur when any acidic solution is mixed with sodium hydrogen carbonate.*

_____

_____

Tie-It-All-Together

SYNTHESIZE

Sequence *the steps in writing an overall equation.*

1.

_____

2.

_____

3.

_____

4.

_____

What if *ten years from now, you are a chemist working for a government agency that investigates chemical reactions. Read each of the case studies below, and in the space provided, list the type of chemical reaction that you think is involved and any products or effects that you would expect to discover during or after the chemical reaction.*

1. Owners of an industrial plant plan to mix oxygen with existing chemical substances in order to create a new product.

Type of Reaction	Product or Effect

2. Two vats of chemicals have spilled into a river and created a gelatinous ooze.

Type of Reaction	Product or Effect

Chemical Reactions Chapter Wrap-Up

Now that you have read the chapter, review what you have learned. List three facts you have learned about chemical reactions and the equations that describe them.

_____

_____

_____

Review

Use this checklist to help you study.

☐ Study your Science Notebook for this chapter.

☐ Study the definitions of vocabulary words.

☐ Review daily homework assignments.

☐ Reread the chapter, and review the charts, graphs, and illustrations.

☐ Answer the Section Review questions at the end of each section.

☐ Look over the Study Guide at the end of the chapter.

SYNTHESIZE

Imagine you were asked to give an expert opinion on a magazine article before it is published. The article is on how to make your own household cleansers. You can tell that the author got the ingredients right, and she has amounts in the correct proportion. However, it looks to you like the author mixed up the order in which ingredients should be combined. How would you explain to the author why that matters?

_____

_____

_____

_____

The Mole

Before You Read

Review Vocabulary   *Define the following terms.*

atomic mass _____

atomic mass unit (amu) _____

Chapter 2   **Write** *the following in scientific notation*

0.00582   _____

24, 367   _____

400   _____

Circle *the significant figures in the numbers below.*

75,600,000

0.00033

3.140

The Mole
Section 1 Measuring Matter

Main Idea ———— **Details** ———————————————————————————

Scan *Section 1, using the checklist below to preview your text.*

- Read all section titles.
- Read all boldfaced words.
- Read all tables and graphs.
- Look at all pictures and read the captions.
- Think about what you already know about this subject.

Write three questions that come to mind from your reading.

1. _____

2. _____

3. _____

New Vocabulary   *Use your text to define each term.*

mole _____

Avogadro's number _____

_____

Counting Particles

Use with pages 320–321.

List *three common counting units and their values.*

1. _____

2. _____

3. _____

Section 1 Measuring Matter (continued)

Main Idea —— **Details** ———————————————

Use with page 320–321.

Describe *why chemists needed to invent a new counting unit.*

_____

_____

List *three forms of substances that can be measured using moles.*

1. _____

2. _____

3. _____

Converting Between Moles and Particles

Use with pages 322–323.

Analyze *the usefulness of a conversion factor.*

_____

_____

_____

Write *the equation for finding the number of representative particles in a number of moles.*

Explain *how you would find the number of moles that are represented by a certain number of representative particles.*

Section 1 Measuring Matter (continued)

(Main Idea) —— (Details) _____

Particles-to-Moles Conversion

Use with Example Problem 1, page 324.

Summarize *Fill in the blanks to help you take notes as you read Example Problem 1.*

Problem

Convert 4.50×10^{24} atoms of Zn to find the number of mol of Zn.

1. **Analyze the Problem**

   Known: number of atoms = _____

   1 mole Zn = _____ atoms of Zn

   Unknown: mole Zn = _____

2. **Solve for the Unknown**

   the number of atoms × conversion factor = number of moles

   _____ atoms Zn ×

   = number of moles

   = _____

3. **Evaluate the Answer**

   The answer has _____ significant digits and is less than _____.

REAL-WORLD CONNECTION

Suppose you were given each of the following tasks. Analyze which task(s) the mole would be an effective unit for counting. Explain your answer.

A. Counting the atoms in a single grain of salt.

B. Counting the grains of salt in a very large mine.

C. Counting the grains of salt in the world.

_____

_____

_____

The Mole

Section 2 Mass and the Mole

Main Idea ———   **Details** ————————————————————————

Scan *Section 2, using the checklist below as a guide.*

- Read all section titles.

- Read all boldfaced words.

- Read all tables and graphs.

- Look at all pictures and read the captions.

- Think about what you already know about this subject.

List four things you expect to learn from this section.

1. _____

2. _____

   _____

3. _____

   _____

4. _____

   _____

New Vocabulary

Use your text to define this term.

_____

molar mass

Section 2 Mass and the Mole (continued)

Main Idea	Details

The Mass of a Mole

Use with pages 325–326.

Analyze *molar mass by completing the following statements.*

The mass of one mole of carbon-12 atoms is _____ grams.

The mass of one mole of hydrogen is _____ gram and is _____ the mass of

one mole of _____.

The mass of one mole of helium-4 is _____ the mass of one mole of

_____ and is equal to _____ grams.

One mole of manganese is equal to _____ atoms of Mn.

Using Molar Mass

Use with pages 327–332.

Organize *the following equations by drawing a line from type of conversion to the correct equation.*

mole to mass
$$mass \times \frac{1 \text{ mole}}{\text{number of grams}}$$

mass to mole
$$\begin{cases} mass \times \dfrac{1 \text{ mole}}{\text{number of grams}}, \\ moles \times \dfrac{6.02 \times 10^{23}}{1 \text{ mole}} \end{cases}$$

mass to atoms
$$\text{number of moles} \times \frac{\text{number of grams}}{1 \text{ mole}}$$

atoms to mass
$$\begin{cases} atoms \times \dfrac{1 \text{ mole}}{6.02 \times 10^{23}}, \\ moles \times \dfrac{\text{number of grams}}{1 \text{ mole}} \end{cases}$$

Section 2 Mass and the Mole (continued)

Main Idea ———— **Details** ————————————————————————

| **Using Molar Mass** | **Solve** *Read Example Problem 4.* |

Mass-to-Atoms Conversion

Use with Example Problem 4, page 330.

You Try It

Problem

Determine how many atoms are in 10 g of pure copper (Cu).

1. **Analyze the Problem**

   Known: mass = _____

   Unknown: molar mass

                     number of atoms

2. **Solve for the Unknown**

   Use the periodic table to find the atomic mass of copper and convert it to g/mol.

   _____

   Complete the conversion equations.

   mass Cu × conversion factor = moles Cu

   _____ × _____ g Cu = _____ moles Cu

   moles Cu × conversion factor = atoms Cu

   _____ mol Cu × _____

   atoms Cu

3. **Evaluate the Answer**

   Restate the answer with correct significant digits.

   _____

The Mole
Section 3 Moles of Compounds

(Main Idea) ———— (Details) —————————————————————

Skim *Section 3 of your text. Write three questions that come to mind from your reading.*

1. _____

2. _____

3. _____

Chemical Formulas and the Mole

Use with pages 333–334.

Describe *the relationship between the mole information of a substance and its chemical formula.*

_____

_____

_____

Mole Relationships from a Chemical Formula

Use with Example Problem 6, page 334.

Summarize *Fill in the blanks to help you take notes as you read Problem 6.*

Problem

Determine the number of moles of Al^{3+} ions in 1.25 moles of Al_2O_3.

1. **Analyze the Problem**

   Known: number of moles of alumina = _____

   Unknown: number of moles = _____

2. **Solve for the Unknown**

   Write the conversion factor: ☐ mol Al^{3+} ions / ☐ mol Al_2O_3

   Multiply the known number of moles by the conversion factor.

   ☐ mol Al_2O_3 × ☐ mol Al^{3+} ions/ ☐ mol Al_2O_3

   = ☐ mol Al^{3+} ions

3. **Evaluate the Answer**

   Restate the answer with correct significant digits:

   _____

Section 3 Moles of Compounds (continued)

⟨Main Idea⟩ ———   ⟨Details⟩ ————————————————————

The Molar Mass of Compounds

Use with page 335.

Describe *the molar mass of a compound.*

_____

_____

Investigate *the process of finding molar mass by completing the table below.*

Number of Moles	Molar Mass	=	Number of Grams
mol K	g K/ 1 mol K	=	g
mol Cr	g Cr/ 1 mol Cr	=	g
mol O	g O/ 1 mol O	=	g
mol mass of K_2CrO_4		=	g

The Molar Mass of Compounds

Use with page 336.

Analyze *the process of converting moles of a compound to molar mass by completing the table below. Refer to Example Problem 7.*

Number of Moles	Molar Mass	=	Number of Grams
2 × 3 mol C	g C/ 1 mol C	=	g
2 × 5 mol H	g H/ 1 mol H	=	g
1 mol S	g S/ 1 mol S	=	g
molar mass of $(C_3H_5)_2S$		=	g

Section 3 Moles of Compounds (continued)

Main Idea		**Details**

Converting the Mass of a Compound to Moles

Use with page 337.

Investigate *the process of converting the mass of a compound to moles by completing the following.*

Number of Moles	Molar Mass	=	Number of Grams
1 mol Ca	g Ca/ 1 mol Ca	=	g
2×1 mol O	g O/ 1 mol O	=	g
2×1 mol H	g H/ mol H	=	g
molar mass of $Ca(OH)_2$		=	g

Conversion factor: _____ g of $Ca(OH)_2$/1 mol $Ca(OH)_2$

g $Ca(OH)_2$ × conversion factor = mol $Ca(OH)_2$

_____ × _____ / _____ = _____ mol $Ca(OH)_2$

Converting the Mass of a Compound to Number of Particles

Use with page 338.

Explain *the steps in converting the mass of a compound to number of particles.*

1. Determine the _____.

2. Multiply by the _____ of the molar mass to convert to _____

3. Multiply by _____ to calculate the number of

   _____.

4. Use the ratios from the _____ to calculate the number of

   _____.

5. Calculate the _____ per formula unit.

The Mole

Section 4 Empirical and Molecular Formulas

Main Idea ———— **Details** ——————————————————————————

Skim *Section 4 of your text. Write three questions that come to mind from reading the headings and the illustration captions.*

1. _____

2. _____

3. _____

New Vocabulary   *Use your text to define each term.*

percent composition   _____

empirical formula   _____

molecular formula   _____

_____

Section 4 Empirical and Molecular Formulas (continued)

Main Idea	Details
Percent Composition *Use with pages 341–342.*	**Write** *the equation for determining the percent by mass for any element in a compound.* **Describe** *the general equation for calculating the percent by mass of any element in a compound.*
Empirical Formula *Use with page 344.*	**Explain** *empirical formula by completing the following statements.* To determine the empirical _____ for a compound, you must first determine the smallest _____ of the moles of the elements in the compound. This ratio provides the _____ in the empirical formula. If the empirical formula differs from the molecular formula, the molecular formula will be a _____ multiple of the empirical formula. The data used to determine the chemical formula may be in the form of _____ or it may be the actual masses. When the percent composition is given, you can assume that the total mass of the compound is 100.0 g to simplify calculations. The _____ of elements in a compound must be _____ to whole numbers to be used as _____ in the chemical formula.

Section 4 Empirical and Molecular Formulas (continued)

Main Idea	Details

Molecular Formula

Use with page 346–349.

Explain *how a molecular formula distinguishes two distinct substances sharing the same empirical formula.*

_____

_____

Investigate *molecular formulas by completing the steps below. Refer to Example Problem 12 in your text.*

empirical formula = $C_2H_3O_2$
molar mass = 118.1 g/mol

Identify the molar mass of the compound.

Moles of Element	Mass of Element / 1 Mol of Element	=	Mass of Element
2 mol C	g C/ mol C	=	g C
3 mol H	g H/ mol H	=	g H
2 mol O	g O/ mol O mol C/mol	=	g O
empirical molar mass of $C_2H_3O_2$		=	g

Divide the molar mass of the substance by the molar mass of the compound to determine n.

$$n = \frac{\text{molar mass of substance}}{\text{molar mass of compound}} = \underline{\hspace{2cm}} = \boxed{}$$

Multiply the subscripts in the empirical formula by n. *Write the molecular formula.*

_____

Section 4 Empirical and Molecular Formulas (continued)

⟮**Main Idea**⟯ ——— ⟮**Details**⟯ ————————————————————————

Examine *the flow chart below. Write the steps in determining empirical and molecular formulas from percent composition or mass data next to the relevant boxes in the flow chart.*

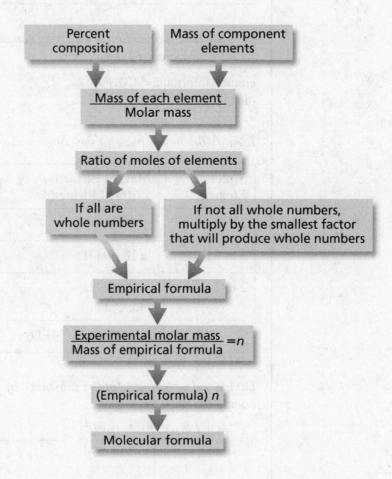

The Mole

Section 5 Formulas of Hydrates

Main Idea ——— **Details** ————————————————————

Skim *Section 5 of your text. Write three questions that come to mind from reading the headings and the illustration captions.*

1. _____

2. _____

3. _____

New Vocabulary *Use your text to define the following term.*

hydrate _____

_____

Naming Hydrates

Use with page 351.

Explain *how hydrates are named by completing the table below.*

Prefix	Molecules of Water
mono-	1
	2
	3
	4
	5
	6
	7
	8
nona-	9
	10

Section 5 Formulas of Hydrates (continued)

Main Idea ———— **Details** ——————————————————

Analyzing a Hydrate

Use with page 352.

Describe *an anyhydrate.*

_____

_____

Determining the Formula of a Hydrate

Use with Example Problem 14, page 353.

Solve *Read Example Problem 14 in your text.*

You Try It

Problem

A 5.00 g sample of barium chloride hydrate was heated in a crucible. After the experiment, the mass of the solid weighed 4.26 g. Determine the number of moles of water that must be attached to $BaCl_2$.

1. **Analyze the Problem**

   Known:　　　 mass of hydrated compound = _____ g $BaCl_2 \cdot x\ H_2O$

   　　　　　　 mass of anhydrous compound = _____ g $BaCl_2$

   　　　　　　 molar mass of H_2O = _____ g/mol

   　　　　　　 molar mass of $BaCl_2$ = 208.23 g/mol

   Unknown:　　 formula for hydrate

   　　　　　　 name of hydrate

Section 5 Formulas of Hydrates (continued)

⟨Main Idea⟩ ————— ⟨Details⟩ ———————————————

2. Solve for the Unknown

Subtract the mass of the anhydrous compound from the hydrated compound.

_____

Calculate the number of moles of H_2O and anhydrous $BaCl_2$ using the conversion factor that relates moles and mass based on the molar mass.

4.26 g $BaCl_2$ x _____ = _____

0.74 g H_2O x _____ = _____

Determine the value of x.

$$x = \frac{\text{moles } H_2O}{\text{moles } BaCl_2} = \qquad\qquad = \underline{\quad}$$

3. Evaluate the Answer

The ratio of H_2O to $BaCl_2$ is _____ so the formula for the hydrate is _____, and the name of the hydrate is _____ _____.

REAL-WORLD CONNECTION

Explain why hydrates are useful in storage and shipping.

_____

_____

_____

The Mole Chapter Wrap-Up

Now that you have read the chapter, review what you have learned and list three things you have learned about moles.

1. _____

2. _____

_____

3. _____

_____

Review

Use this checklist to help you study.

☐ Study your Science Notebook for this chapter.

☐ Study the definitions of vocabulary words.

☐ Review daily homework assignments.

☐ Reread the chapter and review the tables, graphs, and illustrations.

☐ Answer the Section Review questions at the end of each section.

☐ Look over the Study Guide at the end of the chapter.

SUMMARIZE

Summarize the important conversions you have learned in this chapter.

_____

_____

_____

Stoichiometry

Before You Read

Review Vocabulary *Define the following terms.*

mole _____

_____

_____

molar mass _____

conversion factor _____

_____

dimensional analysis _____

_____

_____

law of conservation of mass _____

_____

_____

Chapter 9

Balance *the following equation.*

$$\boxed{}\ Mg\ (s) + \boxed{}\ AlCl_3\ (aq) \rightarrow \boxed{}\ Al\ (s) + \boxed{}\ MgCl_2\ (aq)$$

Chapter 10

Use *the periodic table in the back of your text to complete the chart.*

Pure Substance	Molar Mass
Carbon	12.011
	22.990
	15.999
Sodium carbonate	

Stoichiometry
Section 1 Defining Stoichiometry

Main Idea ——— **Details** ————————————————

Skim *Section 1 of your text. Write three questions that come to mind from reading the headings and the illustration captions.*

1. _____

2. _____

3. _____

New Vocabulary   *Use your text to define each term.*

stoichiometry   _____

_____

_____

mole ratio   _____

_____

_____

Academic Vocabulary   *Define the following term.*

derive   _____

Particle and Mole Relationships

Use with pages 368–369.

Explain *the importance of the law of conservation of mass in chemical reactions.*

_____

_____

_____

_____

_____

Section 1 Defining Stoichiometry (continued)

Main Idea	Details

Interpreting Chemical Equations

Use with Example Problem 1, page 370.

Summarize *Fill in the blanks to help you take notes while you read Example Problem 1.*

Problem

Interpret the equation in terms of _____,

and _____. Show that the law of conservation of mass is _____.

1. **Analyze the Problem**

   Known: _____⟶_____

   Unknown: _____

   _____

   _____

2. **Solve for the Unknown**

   The coefficients indicate the number of _____.

   The coefficients indicate the number of _____.

   Use the space below to calculate the mass of each reactant and each product. Multiply the number of moles by the conversion factor, molar mass.

   $$\text{moles of reactant} \times \frac{\text{grams of reactant}}{1 \text{ mole of reactant}} = \text{grams of} _____$$

   $$\text{moles of product} \times \frac{\text{grams of reactant}}{1 \text{ mole of reactant}} = \text{grams of} _____$$

   Add the masses of the reactants.

   ☐ g C_3H_8 + ☐ g O_2 = ☐ g reactants

   Add the masses of the products.

   ☐ g CO_2 + ☐ g H_2O = ☐ g products

   Determine if the _____ is observed. Does the mass of the reactants equal the mass of the products?

   _____.

3. **Evaluate the Answer**

   Each product or reactant has ☐ significant figures. Your answer must have ☐ significant figures.

Section 1 Defining Stoichiometry (continued)

Main Idea ———   **Details** —————————————————————

Mole ratios

Use with pages 371–372.

Examine *Relationships between coefficients can be used to write conversion factors called _____.*

Example

Given the equation $2KClO_3(s) \longrightarrow 2KCl(s) + 3O_2(g)$
Each substance forms a _____ with the other substances in the reaction.

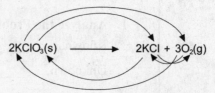

Write *the mole ratios that define the mole relationships in this equation. (Hint: Relate each reactant and each product to each of the other substances.)*

You Try It

Draw *arrows with colored pencils that show the relationships of the substances in this equation.*

$$C_2H_4(g) + 3O_2(g) \longrightarrow 2CO_2(g) + 2H_2O(l)$$

Write *the mole ratios for the above equation.*

Stoichiometry
Section 2 Stoichiometric Calculations

Main Idea ———   **Details** —————————————————————————

Scan *Section 2, using the checklist below to preview your text.*

- Read all section titles.
- Read all boldfaced words.
- Read all tables and graphs.
- Look at all pictures and read the captions.
- Think about what you already know about this subject.

Write *three facts you discovered about stoichiometric calculations.*

1. _____

2. _____

3. _____

Using Stoichiometry

Use with pages 373–374.

Identify *the tools needed for stoichiometric calculations.*

All stoichiometric calculations start with _____ based on a

_____. Finally, _____

are required.

Section 2 Stoichiometric Calculations (continued)

Main Idea	Details

Mole-to-Mass Stoichiometry

Use with Example Problem 3, pages 376.

Solve *Read Example Problem 3 in your text.*

You Try It

Problem

How many grams of solid iron (III) chloride ($FeCl_3$) are produced when 2.00 moles of solid iron (Fe) are combined with chlorine gas (Cl_2)?

1. **Analyze the Problem**

   Known: _____

   Unknown: _____

   You are given the moles of the reactant, Fe, and must determine the mass of the product, $FeCl_3$, therefore, you will do a mole to mass conversion.

2. **Solve for the Unknown**

   Write the balanced chemical equation. Identify the known and unknown substances.

   $\boxed{}$ Fe(s) + $\boxed{}$ Cl_2(g) = $\boxed{}$ $FeCl_3$(s)

   List the mole ratios for this equation. (Hint: *Draw arrows that show the relationships of the substances in this equation.*) Circle the mole ratio that relates moles of Fe to $FeCl_3$.

   Multiply the number of moles of Fe by the mole ratio.

   $\boxed{}$ mol Fe $\times \dfrac{\boxed{}\ \text{mol } FeCl_3}{\boxed{}\ \text{mol Fe}} = \boxed{}$ mol $FeCl_3$

   Multiply the moles of $FeCl_3$ by the molar mass of $FeCl_3$.

   $\boxed{}$ mol $FeCl_3$ $\times \dfrac{\text{g } FeCl_3}{1\ \text{mol } FeCl_3} = \boxed{}$ g $FeCl_3$

3. **Evaluate the Answer**

   The given number of moles has $\boxed{}$ digits, so the mass of $FeCl_3$ must have $\boxed{}$ digits.

Section 2 Stoichiometric Calculations (continued)

⸺Main Idea⸺ ⸺ ⸺Details⸺ _____

Mole-to-Mole Stoichiometry

Use with Example Problem 2, page 375.

Solve *Read Example Problem 2 in your text.*

You Try It

Problem

How many moles of aluminium oxide (Al_2O_3) are produced when 4.0 moles of aluminium (Al) are combined with Oxygen gas (O_2)?

1. Analyze the Problem

Known: _____

Unknown: _____

Both the known and the unknown are in moles, therefore, you will do a mole-to-mole conversion.

2. Solve for the Unknown

Write the balanced chemical equation. Label the known and unknown.

$$\boxed{}\,Al(s) + \boxed{}\,O_2(g) = \boxed{}\,Al_2O_3(s)$$

List the mole ratios for this equation. (Hint: *Draw arrows that show the relationships of the substances in this equation.*)

Circle the mole ratio that converts moles of Al to mol of Al_2O_3.

Multiply the known number of moles Al by the mole ratio to find the moles of unknown Al_2O_3.

$$\boxed{}\ \text{moles Al} \times \frac{\boxed{}\ \text{moles } Al_2O_3}{\boxed{}\ \text{moles of Al}} = \boxed{}\ \text{moles of } Al_2O_3$$

3. Evaluate the Answer

The given number of moles has $\boxed{}$ significant figures. Therefore, the answer must have $\boxed{}$ significant figures.

Section 2 Stoichiometric Calculations (continued)

(Main Idea) ——— **(Details)** ————————————————————————

Mass-to-Mass Stoichiometry

Use with Example Problem 4, page 377.

Solve *Read Example Problem 4 in your text.*

You Try It

Problem

Determine the mass of ammonia (NH_3) produced when 3.75 g of nitrogen gas (N_2) react with hydrogen gas (H_2).

1. **Analyze the Problem**

   Known: _____

   Unknown: _____

   You are given the mass of the reactant, N_2, and must determine the mass of the product NH_3. Do a mass-to-mass conversion.

2. **Solve for the Unknown**

   Write the balanced chemical equation for the reaction.

   $\boxed{} N_2(g) + \boxed{} H_2(g) = \boxed{} NH_3(g)$

   Convert grams of $N_2(g)$ to moles of $N_2(g)$ using the inverse of molar mass as the conversion factor.

   $$\boxed{} \text{g } N_2(g) \times \frac{1 \text{ mol } N_2}{\boxed{} \text{ g } N_2} = \boxed{} \text{ mol } N_2$$

   List the mole ratios for this equation.

   Multiply moles of N_2 by the mole ratio that converts N_2 to NH_3.

   $$\boxed{} \text{ mol } N_2 \times \frac{\boxed{} \text{ mol } NH_3}{\boxed{} \text{ mol } N_2} = \boxed{} \text{ mol } NH_3$$

   Multiply moles of NH_3 by the molar mass.

   $$\boxed{} \text{ mol } NH_3 \times \frac{\boxed{} \text{ g } NH_3}{1 \text{ mol } NH_3} = \boxed{} \text{ g } NH_3$$

3. **Evaluate the Answer**

   The given mass has $\boxed{}$ significant figures, so the mass of NH_3 must have $\boxed{}$ significant figures.

Section 2 Stoichiometric Calculations (continued)

Main Idea ——— **Details** ——————————————————————

Mastering Stoichiometry

Use with page 374.

Sequence *the steps needed to convert from the balanced equation to the mass ofthe unknown.*

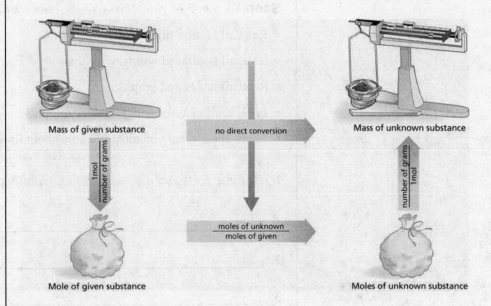

Mass of given substance   no direct conversion   Mass of unknown substance

1mol / number of grams

$\dfrac{\text{moles of unknown}}{\text{moles of given}}$

number of grams / 1mol

Mole of given substance   Moles of unknown substance

Identify *the steps in stoichiometric calculations by completing the summary below.*

1. _____. Interpret the equation in

   terms of _____.

2. _____

   _____. Use the _____

   _____ as the conversion factor.

3. _____

   _____. Use the appropriate mole ratio from

   the _____ as the conversion factor.

4. _____

   _____

   Use _____ as the conversion factor.

Stoichiometry

Section 3 Limiting Reactants

Main Idea —— **Details** _____

Scan *Section 3 of your text. Use the checklist below as a guide.*

- Read all section titles.

- Read all boldfaced words.

- Read all tables and graphs.

- Look at all pictures and read the captions.

- Think about what you already know about limiting reactants.

Write three facts you discovered about limiting reactants.

1. _____

2. _____

3. _____

New Vocabulary *Use your text to define each term.*

limiting reactant _____

_____

excess reactant _____

Section 3 Limiting Reactants (continued)

(Main Idea) ——   **(Details)** ——————————————————

Why do Reactions Stop?

Use with page 379–380.

What if *you have six slices of bread, three tomato slices, and two cheese slices. How many tomato-cheese sandwiches can you make? Which ingredient(s) limit the number of sandwiches you can make?*

_____

Calculating the Amount of a Product When a Reactant is Limiting

Use with page 380–381.

Organize *information about limiting reactants.*

I. _____

   A. Limiting reactant

      1. _____

      2. _____

   B. _____

II. Calculating the product when a reactant is limited

   A. _____

      1. convert the masses to moles

      2. multiply each mass by the inverse of the molar mass

   B. _____

   C. _____

      _____

   D. Determine the amount of product that can be made with the moles of the limiting reactant.

Determining the Limiting Reactant

Use with Example Problem 5, page 382.

Solve *Read Example Problem 5 in your text.*

You Try It

Problem

If 100.0 g of sulfur reacts with 50.0 g of chlorine, what mass of disulfur dichloride is produced?

1. **Analyze the Problem**

   Known: _____

   Unknown: _____

2. **Solve for the Unknown**

   Write the balanced chemical equation.

Section 3 Limiting Reactants (continued)

⟨**Main Idea**⟩ —— ⟨**Details**⟩ ————————————————

List the mole ratios for this equation.

Multiply each mass by the inverse of molar mass.

Calculate the actual ratio of available moles.

Determine the limiting reactant.

_____

_____

Multiply the number of moles of the limiting reactant by the mole ratio of the product to the limiting reactant.

Multiply moles of the product by the molar mass.

Multiply moles of the excess reactant by the molar mass.

Subtract the mass of the excess reactant needed from the mass available.

3. **Evaluate the Answer**

The given mass has ☐ significant figures, so the mass of the unknown must have ☐ significant figures.

Stoichiometry
Section 4 Percent Yield

⟨**Main Idea**⟩ ——— ⟨**Details**⟩ ————————————————————

Skim *Section 4 of your text. Focus on the headings, subheadings, and boldfaced words. Summarize the main ideas of this section.*

_____

_____

_____

⟨**New Vocabulary**⟩   *In the left margin, write the terms defined below.*

_____   *the ratio of actual yield to theoretical yield (from stoichiometric calculations) expressed as a percent*

_____   *in a chemical reaction, the maximum amount of product that can be produced from a given amount of reactant*

_____   *the amount of product actually produced when a chemical reaction is carried out in an experiment*

How much product?

Use with pages 385–386.

Write *the formula for percent yield.*

$$\frac{\underline{\hspace{4cm}} \text{ (from an experiment)}}{\underline{\hspace{4cm}} \text{ (from stoichiometric calculations)}} \times \underline{\hspace{1cm}} = \begin{array}{l} \text{percent} \\ \text{yield} \end{array}$$

Section 4 Percent Yield (continued)

Main Idea —— **Details** ——————————————

Percent Yield

Use with Example Problem 6, page 386.

Solve *Read Example Problem 6 in your text.*

You Try It

Problem

When 100.0 kg sand (SiO_2) are processed with carbon, CO and 51.4 kg SiC are recovered. What is the percent yield of SiC?

1. **Analyze the Problem**
   Known: _____
   Unknown: _____

2. **Solve for the Unknown**
   Write the balanced chemical equation.

   _____ $\longrightarrow$ _____

   Determine the mole ratio that converts _____ to _____.

   Convert kg to g.

   100 kg SiO_2 = _____ g, 51.4 kg SiC = _____ g
   Convert mass to moles using the inverse of molar mass.

   Use the appropriate mole ratio to convert mol SiO_2 to mol SiC.

   Calculate the theoretical yield. Multiply mol SiC by the molar mass.

   Divide the actual yield by the theoretical yield and multiply by 100.

3. **Evaluate the Answer**
   The quantities have ☐ significant figures, so the percent yield must have ☐ significant figures.

Stoichiometry

SYNTHESIZE

Stoichiometry and the Stock Market	*In the left margin, write the stoichiometry concepts that parallel the daily activities of a Wall Street professional.*
_____ _____ _____ _____ _____	1. A stock analyst keeps a close eye on the earnings of corporations. She has determined how much each company should accomplish. 2. The same analyst tracks whether companies meet expectations or fall short. 3. A grain trader wants to be sure to have 100,000 bushels in reserve for the winter selling season. He places an order for 120,000 bushels because he knows spoilage may damage a percentage of the crop. 4. A livestock futures trader knows that one cattle car holds 10 steers averaging 1200 lbs. each. He wants to bid on an identical car full of sheep, which average about 200 lbs. each. He needs to know how many sheep are on the car. 5. A stockbroker learns that a medical supply company has acquired several tons of a rare silver compound that will allow it to make superior dental equipment. The question is whether the company will have enough of the product to meet the demands of the marketplace.

Stoichiometry Chapter Wrap-Up

Now that you have read the chapter, review what you have learned. Write the key equations and relationships.

_____

_____

_____

_____

Review | *Use this checklist to help you study.*

☐ Use this Science Notebook to study this chapter.

☐ Study the vocabulary words and scientific definitions.

☐ Review daily homework assignments.

☐ Reread the chapter, reviewing the tables, graphs, and illustrations.

☐ Answer the Section Review questions at the end of each section.

☐ Look over the Study Guide at the end of the chapter.

REAL-WORLD CONNECTION

Explain how stoichiometry is important to air bags and your safety.

_____

_____

_____

States of Matter

Before You Read

Review Vocabulary

Define the following terms.

gas _____

_____

physical property _____

_____

_____

Chapter 2

Calculate *the density of a sample with a mass of 22.5 g and a volume of 5.0 cm^3. Use the equation: density = mass/volume.*

_____

_____

Chapter 3

Describe *the two essential characteristics that determine the chemical and physical properties of matter.*

_____

_____

_____

_____

Compare and contrast *the chemical and physical properties of gases.*

_____

_____

Name _____   Date _____

States of Matter
Section 1 Gases

⟨Main Idea⟩ —— **⟨Details⟩** ——————————————————

Scan *Section 1, using the checklist below as a guide.*

- Read all section titles.
- Read all boldfaced words.
- Read all tables and graphs.
- Look at all pictures and read the captions.
- Think about what you already know about this subject.

⟨New Vocabulary⟩   *Use your text to define each term.*

kinetic-molecular theory   _____

elastic collision   _____

temperature   _____

diffusion   _____

_____

Graham's law of effusion   _____

_____

pressure   _____

barometer   _____

pascal   _____

atmosphere   _____

Dalton's law of partial pressures   _____

_____

Section 1 Gases (continued)

Main Idea	Details

The Kinetic-Molecular Theory

Use with pages 402–403.

Distinguish *between the three main physical properties of gas particles by completing the passages below.*

1. Size is very _____. It is assumed that there are _____ significant

_____ or _____ forces among gas particles.

2. Motion is _____ moving in a _____ pattern. It is assumed

that gas particles move in a _____ path until they _____.

3. Energy is _____. It is assumed that _____ and

_____ impact the _____ level of a gas_____.

Describe *kinetic energy in equation form by completing the table below.*

$KE = 1/2mv^2$	Variable	Definition
KE		
m		
v		

Explaining the Behavior of Gases

Use with pages 403–405.

Describe *the following concepts as they relate to the behaviors of gases by completing the passages below.*

low density—Gases have low density (_____ per _____) in

comparison to _____. The difference in density is partly due to the mass

of the _____ and also because there is a great deal of

_____between gas particles.

compression and expansion—The large amount of _____

between gas particles allows them to be _____, or pushed, into a

_____ volume. Once the pressure is _____, the particles

_____ to the original _____.

diffusion and effusion—Because there are no _____ forces of

_____ between gas particles, gases _____ past one another. This

_____ motion allows gases to mix until they are_____. The

movement of _____ past one another is called _____. The

process of allowing a gas to escape from a more concentrated container is

called _____.

Section 1 Gases (continued)

Main Idea ——— **Details** ————————————————

Write *Graham's law of effusion as a proportional statement.*

_____

Write *the proportional statement based on Graham's law of effusion that allows you to compare the diffusion rate of two different gases.*

Gas Pressure

Use with pages 406–410.

Describe *pressure as it relates to the behaviors of gases.*

_____

_____

_____

Distinguish *between a barometer and a manometer.*

_____

_____

Explore *the relationship between different units of pressure by filling in the table below.*

Unit Name (unit symbol)	Conversion Ratio: 1 atm = _____	Conversion Ratio: 1 kPa = _____
kilopascal ( )		
millimeters of mercury ( )		
torr		
pounds per square inch (or)		
atmosphere ( )		

States of Matter
Section 2 Forces of Attraction

Main Idea — **Details** _____

Skim *Section 2 of your text. Write three questions that come to mind from reading the headings and the illustration captions.*

1. _____

2. _____

3. _____

New Vocabulary

Use your text to define each term.

dispersion forces _____

_____

dipole-dipole force _____

hydrogen bond _____

_____

_____

Academic Vocabulary

Define the following term.

orient _____

Section 4 Scientific Research (continued)

Main Idea ——— **Details** ————————————————————

Intermolecular Forces

Use with pages 411–414.

Describe *the difference between an intramolecular and an intermolecular force.*

_____

_____

_____

Compare and contrast *intramolecular forces by completing the table below.*

Force	Basis of Attraction	Example
Ionic		
Covalent		
Metallic		

Compare *intermolecular forces by completing the table below.*

Force	Basis of Attraction	Example
Dispersion		
Dipole-dipole		
Hydrogen bond		

States of Matter

Section 3 Liquids and Solids

⟨ **Main Idea** ⟩ ———— ⟨ **Details** ⟩ ————————————————————

Scan *Section 3, using the checklist below as a guide.*

- Read all section titles.
- Read all boldfaced words.
- Read all tables and graphs.
- Look at all pictures and read the captions.
- Think about what you already know about this subject.

⟨ **New Vocabulary** ⟩   *Use your text to define each term.*

viscosity _____

_____

surface tension _____

_____

surfactant _____

_____

crystalline solid _____

_____

unit cell _____

_____

allotrope _____

_____

amorphous solid _____

_____

Section 3 Liquids and Solids (continued)

Main Idea ———— **Details** ——————————————————————

Liquids

Use with pages 415–419.

Compare and contrast *the following paired concepts as they relate to the properties of liquids by completing the following statements.*

Density and compression: A liquid can take the _____, but its volume is _____. The density of a liquid is _____ than the density of the same substance as a _____. Liquids cannot usually be _____ except under _____ pressure.

Fluidity and viscosity: Fluidity is the ability to _____. Liquids flow through each other but at a _____ than _____ do. Viscosity is the measure of the _____ of a liquid to _____. The stronger _____ slow down the ability to flow, which _____ resistance (viscosity).

Viscosity and temperature: Temperature affects the _____ of a _____. Viscosity _____ with temperature.

Analyze *the relationship between viscosity, temperature, and change in kinetic energy by completing the table.*

Temperature	Δ KE	Viscosity	Effect in Liquid
increases			flows faster
decreases		increases	
stays the same	no change		

Section 3 Liquids and Solids (continued)

⊂Main Idea⊃ ——— ⊂Details⊃ ——————————————————————————

Explain *surface tension by completing the web diagram below.*

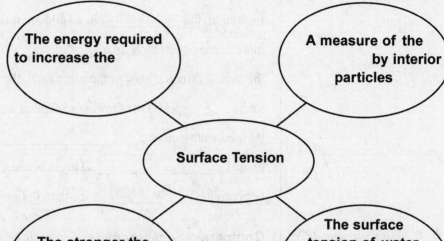

The energy required to increase the

A measure of the by interior particles

Surface Tension

The stronger the between particles, the the surface tension

The surface tension of water is because its molecules form

Use with page 419.

Describe *the following concepts as they relate to the properties of liquids by completing the following passages.*

Capillary action is _____

_____

Cohesion is _____

_____

Adhesion is _____

_____

Section 3 Liquids and Solids (continued)

Main Idea	Details

Solids

Use with pages 420–424.

Contrast *the density of solids and liquids by completing the following paragraph.*

In general, the _____ in a solid are more_____ —that is, more dense—than those in a _____. When liquid and solid states of the same substance exist at the same time, the _____ usually _____ in the _____. One familiar exception is _____. When water is in its solid state as ice, it _____, such as _____ or a(n) _____. This is because there is _____ space between the _____ in ice than in liquid water.

Use with page 422.

Compare *the different types of crystalline solids by completing the following table.*

Type	Unit Particles	Characteristics	Examples
Atomic			
Molecular			
Covalent network			
Ionic			
Metallic			

States of Matter
Section 4 Phase Changes

(Main Idea) ——— **(Details)** ———————————————————

Skim *Section 4 of your text. Write a brief summary of the main topics covered.*

_____

_____

_____

_____

New Vocabulary   *Use your text to define each term.*

vapor pressure   _____

boiling point   _____

_____

condensation   _____

deposition   _____

_____

phase diagram   _____

_____

Compare and contrast *the following terms using your text as a guide.*

melting point, freezing point, and triple point   _____

_____

_____

_____

_____

vaporization and evaporation   _____

_____

Chemistry: Matter and Change

Science Notebook

Section 4 Phase Changes (continued)

Main Idea —— **Details** ————————————————————

Phase Changes That Require Energy

Use with page 425.

Classify *the types of phase changes by completing the table below. Use Figure 23 in your text for reference.*

Phase Transition	Type of Transition
gas to solid	
solid to liquid	
liquid to gas	
liquid to solid	
	condensation
solid to gas	

Use with pages 425–428.

Describe *the phase changes that require energy by completing the following outline.*

  I. Melting

    A. Heat energy disrupts _____.

    B. The amount of energy required depends on

    _____

    C. The melting point is the temperature at which _____

    _____.

    D. The melting point of _____ may be unspecified.

  II. Vaporization

    A. In liquid water, some particles have more _____ .

    B. Particles that escape from liquid enter the _____ .

    C. When vaporization occurs only at a surface it is called

    _____.

    D. The pressure exerted by a vapor over liquid is called _____

    _____.

    E. The temperature at which vapor pressure equals atmospheric pressure is called the _____.

  III. Sublimation

    A. Many solids can become gases without _____

    _____.

    B. Some solids sublime at _____.

    C. The process of _____ is an example of sublimation.

Section 4 Phase Changes (continued)

⟨**Main Idea**⟩ ——— ⟨**Details**⟩ ———————————————

Phase Changes That Release Energy

Use with pages 428–429.

Organize *the phase changes that release energy. Identify the phase, describe the process, and identify the reverse process by completing the table below.*

Phase Change	Process Description	Reverse Process
condensation		vaporization
	process in which a liquid becomes a solid	
deposition		sublimation

Phase Diagrams

Use with pages 429–430.

Explain *how the critical point affects water.*

_____

_____

_____

Identify *normal freezing point, normal boiling point, critical point, and triple point in the phase diagram for H$_2$O below. Use Figure 30 in your text for reference.*

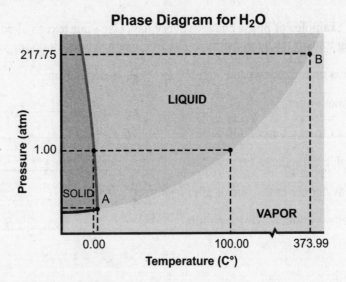

Phase Diagram for H$_2$O

Name _____ Date _____

States of Matter   Chapter Wrap-Up

After reading this chapter, list three key equations and relationships.

1. _____

2. _____

3. _____

Review

Use this checklist to help you study.

☐ Study your Science Notebook for this chapter.

☐ Study the definitions of vocabulary words.

☐ Review daily homework assignments.

☐ Reread the chapter and review the tables, graphs, and illustrations.

☐ Answer the Section Review questions at the end of each section.

☐ Look over the Study Guide at the end of the chapter.

REAL-WORLD CONNECTION

You see examples of phase changes every day. Use your text to identify which phase change each of the following transitions demonstrates. The first one has been done for you.

frost forms on a windowpane   deposition

ice becomes water   _____

steam rises from a cup of coffee   _____

a water pipe bursts on a very cold day   _____

drops of water cover the mirror after a shower _____

snow melts without leaving a puddle   _____

Gases

Before You Read

Review Vocabulary

Define *the following terms.*

density _____

stoichiometry _____

_____

_____

kinetic-molecular theory _____

_____

Chapter 9

Balance *the following equation.*

____Fe + ____H_2SO_4 ⟶ $Fe_2(SO_4)_3$ + ____H_2

Chapter 11

Show *the mole ratios for the following reaction.*

$N_2 + 3H_2$ ⟶ $2NH_3$

1. **a.** mole ratio of N to H_2

2. **b.** mole ratio of NH_3 to H_2

Chapter 12

Explain *how gas particles exert pressure.*

_____

_____

Gases

Section 1 The Gas Laws

Main Idea ———— **Details** ————————————————————

Scan *Section 1 of your text. Use the checklist below as a guide.*

- Read all section titles.
- Read all boldfaced words.
- Read all tables and graphs.
- Look at all pictures and read the captions.
- Think about what you already know about this subject.

Write *three facts you discovered about the gas laws.*

1. _____

2. _____

3. _____

New Vocabulary *Use your text to define each term.*

Boyle's law _____

_____

_____

absolute zero _____

Charles's law _____

_____

Gay-Lussac's law _____

_____

combined gas law _____

_____

Section 1 The Gas Laws (continued)

⟨**Main Idea**⟩ —— ⟨**Details**⟩ ——————————————

Boyle's Law

Use with Example Problem 1, page 443.

Solve *Read Example Problem 1 in your text.*

You Try It
Problem

Helium gas in a balloon is compressed from 4.0 L to 2.5 L at constant temperature. The gas's pressure at 4.0 L is 210 kPa. Determine the pressure at 2.5 L.

1. **Analyze the Problem**

   Known: Unknown:

   $V_1 =$ _____ P_2 _____

   $V_2 =$ _____

   $P_1 =$ _____

   Use the equation for Boyle's law to solve for P_2.

2. **Solve for the Unknown**

   Write the equation for Boyle's law: _____

   To solve for P_2, divide both sides by V_2. $P_2 =$

   Substitute the known values. $P_2 =$

   Solve for P_2. $P_2 =$ _____

3. **Evaluate the Answer**

   When the volume is _____, the pressure is _____. The answer is in _____, a unit of pressure.

Section 1 The Gas Laws (continued)

⟨**Main Idea**⟩ ——— ⟨**Details**⟩ ——————————————————

Charles's Law

Use with Example Problem 2, page 446.

Summarize *Fill in the blanks to help you take notes while you read Example Problem 2.*

Problem

A gas sample at 40.0°C occupies a volume of 2.32 L. Assuming the pressure is constant, if the temperature is raised to 75.0°C, what will the volume be?

1. **Analyze the Problem**

   Known: Unknown:

   $T_1 = $ _____

   $V_1 = $ _____ $V_2 = $ _____

   $T_2 = $ _____

   Use Charles's law and the known values for T_1, V_1, and T_2 to solve for V_2.

2. **Solve for the Unknown**

   Convert the T_1 and T_2 Celsius temperatures to kelvin:

   $T_1 = 273 + 40.0°C = $ _____K $T_2 = 273 + 75.0°C = $ ____ K

   Write the equation for Charles's law:

   $$=$$

   To solve for V_2, multiply both sides by T_2:

   $$V_2 =$$

   Substitute known values:

   $$V_2 =$$

   Solve for V_2.

   $$V_2 = \text{_____}$$

3. **Evaluate the Answer**

   When temperature in kelvin increases by a small amount, the volume _____ by a small amount. The answer is in _____, a unit for volume.

Section 1 The Gas Laws (continued)

| (Main Idea) ——— | (Details) _____ |

Gay-Lussac's Law

Use with Example Problem 3, page 448.

Solve *Read Example Problem 3 in your text.*

You Try It
Problem

The pressure of a gas stored in a refrigerated container is 4.0 atm at 22.0°C. Determine the gas pressure in the tank if the temperature is lowered to 0.0°C.

1. **Analyze the Problem**

   Known: Unknown:

   $P_1 = 4.0$ atm $P_2 = ?$ _____

   $T_1 = $ _____

   $T_2 = $ _____

   Use Gay-Lussac's law and the known values for T_1, V_1, and T_2 to solve for V_2.

2. **Solve for the Unknown**

   Convert the T_1 and T_2 Celsius figures to kelvin.

   $T_1 = $ ___ $+ 22.0$°C $= $ ___ K

   $T_2 = 273 + $ ___ °C $= $ ___ K

   Write the equation for Gay-Lussac's law.

   To solve for P_2, multiply both sides by T_2.

   $P_2 = $

   Substitute known values.

   $P_2 = $

   Solve for P_2.

   $P_2 = 3.7$ atm

3. **Evaluate the Answer**

   The temperature _____ and the pressure _____.

Section 1 The Gas Laws (continued)

⟨Main Idea⟩ ——— **⟨Details⟩** ————————————————————

The Combined Gas Law	**Describe** *the combined gas law.*
Use with page 449.	_____
	_____
	Write *the combined gas law equation.*
	$=$
	Pressure is inversely proportional to _____ and directly proportional to
	_____. Volume also is _____ to temperature.
Use with Example Problem 4, page 450.	**Solve** *Read Example Problem 4 in your text.*

You Try It

Problem

A gas at 100.0 kPa and 30.0°C has an initial volume of 1.00 L. Determine the temperature that could support the gas at 200.0 kPa and a volume of 0.50 L.

1. **Analyze the Problem**

   Known: Unknown:

   $P_1 =$ _____ $T_2 = ?$ °C

   $P_2 =$ _____

   $T_1 =$ _____

   $V_1 =$ _____

   $V_2 =$ _____

   Remember that volume increases as temperature increases, and volume is inversely proportional to pressure.

2. **Solve for the Unknown**

   Convert the T_1 Celsius temperature to kelvin.

   $T_1 =$ _____ $+ 30.0°C =$ _____ K

Section 1 The Gas Laws (continued)

Main Idea ———— **Details** ———————————————————

Write the combined gas law equation.

To solve for T_2, multiply both sides of the equation by T_2.

$$\frac{\quad}{T_1} = P_2 V_2$$

Multiply both sides of the equation by T_1.

$$T_2\, P_1\, V_1 = \underline{\hspace{2cm}}$$

Divide both sides of the equation by $P_1\, V_1$.

$$T_2 =$$

Substitute known values.

$$T_2 = \frac{\quad}{100.0\ \text{kPa} \times 1.00\ \text{L}}$$

Solve for T_2.

$$T_2 = 303\ \text{K} - 273\ \text{K} = 30.0\text{°C}$$

3. **Evaluate the Answer**

As pressure _____ and volume _____ in proportional amounts, the temperature remained constant.

Gases
Section 2 The Ideal Gas Law

Main Idea ———   **Details** ———————————————————————

Skim *Section 2 of your text. Write three questions that come to mind from reading the headings and the illustration captions.*

1. _____

2. _____

3. _____

New Vocabulary   *Use your text to define each term.*

Avogadro's principle   _____

_____

molar volume   _____

_____

standard temperature and pressure (STP)   _____

_____

ideal gas constant (R)   _____

_____

ideal gas law   _____

_____

Section 2 The Ideal Gas Law (continued)

Main Idea	Details
Avogadro's principle *Use with pages 452–453.*	**Explain** *Avogadro's principle by completing the paragraph below.* Avogadro's principle states that _____ _____. The _____ volume for a gas is the volume that one mole occupies at _____ of pressure and a temperature of _____. **Convert** the following volumes of a gas at STP to moles by using 22.4 L/mol as the conversion factor.

$$2.50 \, L \times \frac{1 \, mol}{22.4 \, L} = \underline{\hspace{2cm}}$$

$$7.34 \, L \times \frac{1 \, mol}{22.4 \, L} = \underline{\hspace{2cm}}$$

$$4.7 \, L \times \frac{1 \, mol}{22.4 \, L} = \underline{\hspace{2cm}}$$

Section 2 The Ideal Gas Law (continued)

Main Idea	Details
The Ideal Gas Law *Use with pages 454–455.*	**Analyze** *the ideal gas law.* The equation is written _____ = _____ *P* represents _____ *V* represents _____ *n* represents the number of _____ of gas present R represents the _____ _____ represents temperature The ideal gas law states that _____ _____ _____. The value of R depends on the units used for _____.
Real Versus Ideal Gases *Use with pages 457–459.*	**Describe** *the properties of an ideal gas.* _____ _____ **Describe** *the properties of a real gas.* _____ _____

Section 2 The Ideal Gas Law (continued)

Main Idea	Details

The Ideal Gas Law

Use with Example Problem 6, page 455.

Summarize *Fill in the blanks to help you take notes while you read Example Problem 6.*

Problem

Calculate the number of moles of a gas contained in a 3.0-L vessel at 3.00×10^2 K with a pressure of 1.50 atm.

1. **Analyze the Problem**

   Known: Unknown:

   $V =$ _____ $n = ?$ mol

   $T =$ _____

   $P =$ _____

   $R =$ _____

   Use the known values to find the value of n.

2. **Solve for the Unknown**

   Write the ideal gas law equation.

   _____

   To solve for n, divide both sides by RT.

   $n =$

   Substitute known values into the equation.

   $n =$

   Solve for n.

   $n =$

   $n =$ _____

3. **Evaluate the Answer**

   The answer agrees with the prediction that the number of moles will be _____ one mole. The unit in the answer is the _____.

Gases

Section 3 Gas Stoichiometry

Main Idea ———   **Details** ————————————————————————

Scan *Section 3 of your text. Use the checklist below as a guide.*

- Read all section titles.
- Read all boldfaced words.
- Read all tables and graphs.
- Look at all pictures and read the captions.
- Think about what you already know about this subject.

Write *three facts you discovered about gas stoichiometry.*

1. _____

2. _____

3. _____

Academic Vocabulary   *Define the following terms.*

ratio _____

Section 3 Gas Stoichiometry (continued)

⸺Main Idea⸺ ⸺ ⸺Details⸺ ⸺⸺⸺⸺⸺⸺⸺⸺⸺⸺⸺⸺⸺⸺⸺

Stoichiometry and Volume–Volume Problems

Use with page 460.

Indicate *the moles and volume for the reaction below. Use Figure 10 as a reference.*

$$2C_4H_{10}(g) \quad + \quad 13O_2(g) \longrightarrow 8CO_2(g) \quad + \quad 10H_2O(g)$$

___ moles   ___ moles   ___ moles   ___ moles

___ volumes   ___ volumes   ___ volumes   ___ volumes

The coefficients in the balanced equation represent _____ amounts and

relative _____.

Volume–Volume Problems

Use with Example Problem 7, page 461.

Summarize *Fill in the blanks to help you take notes while you read Example Problem 7.*

Problem

Determine the volume of oxygen gas needed for the complete combustion of 4.00 L of propane gas (C_3H_8).

1. **Analyze the Problem**

   Known: Unknown:

   V of C_3H_8 = _____ V of O_2 = ? L

   Use the known volume of 4.00 L to find the volume needed for the combustion.

2. **Solve for the Unknown**

   Write the balanced equation for the combustion of C_3H_8.

   _____

   Write the volume ratio.

   Multiply the known volume of propane by the volume ratio to find the volume of O_2.

3. **Evaluate the Answer**

   The coefficients of the reactants show that the quantity of _____

   consumed is greater than the amount of propane. The unit of the answer

   is the _____, a unit of volume.

Gases Chapter Wrap-Up

After reading the chapter, review what you have learned. Match each of the gas laws with its equation.

_____ Ideal gas law
1. $\dfrac{V_1}{T_1} = \dfrac{V_2}{T_2}$

_____ Gay-Lussac's law
2. $P_1V_1 = P_2V_2$

_____ Charles's law
3. $\dfrac{P_1}{T_1} = \dfrac{P_2}{T_2}$

_____ Combined gas law
4. $PV = nRT$

_____ Boyle's law
5. $\dfrac{P_1V_1}{T_1} = \dfrac{P_2V_2}{T_2}$

Review *Use this checklist to help you study.*

☐ Study your Science Notebook for this chapter.

☐ Study the vocabulary words and scientific definitions.

☐ Review daily homework assignments.

☐ Reread the chapter and review the tables, graphs, and illustrations.

☐ Answer the Section Review questions at the end of each section.

☐ Look over the Study Guide at the end of the chapter.

REAL-WORLD CONNECTION

Explain why the volume of a balloon increases as you blow into it instead of bursting immediately from the added pressure.

_____

_____

_____

Mixtures and Solutions

Before You Read

Review Vocabulary

Define the following terms.

alloy

_____

_____

solution

_____

Chapter 3

Compare and contrast *a homogeneous mixture with a heterogeneous mixture.*

_____

_____

_____

Chapter 8

Explain *why water is a polar molecule. Include a labeled drawing of a water molecule in your answer.*

_____

_____

_____

Chapter 10

Describe *the relationship between moles and molar mass.*

_____

_____

_____

Mixtures and Solutions

Section 1 Types of Mixtures

⟨**Main Idea**⟩ ——— ⟨**Details**⟩ ————————————————————

Scan *Section 1 of your text, using the checklist below as a guide.*

- Read all section titles.

- Read all boldfaced words.

- Read all tables and graphs.

- Look at all pictures and read the captions.

- Think about what you already know about solutions.

Identify *the unifying theme of this section.*

_____

⟨**New Vocabulary**⟩   *Use your text to define each term.*

suspension _____

colloid _____

_____

Brownian motion _____

_____

Tyndall effect _____

Compare and contrast *soluble and insoluble substances.*

_____

_____

Compare and contrast *miscible and immiscible liquids.*

_____

_____

Section 1 Types of Mixtures (continued)

Main Idea ———— Details ——————————————————

Suspensions

Use with page 476.

List *three properties of a suspension.*

1. _____

2. _____

3. _____

State *three examples of suspensions.*

1. _____

2. _____

3. _____

Colloids

Use with pages 477–479.

Identify *four properties of a colloid.*

1. _____

2. _____

3. _____

4. _____

Section 1 Types of Mixtures (continued)

⟨**Main Idea**⟩ —— ⟨**Details**⟩ ——————————————————

Explain *why particles in Brownian motion do not settle out.*

_____

_____

_____

_____

_____

_____

Identify *each of the following mixtures as a suspension, dilute colloid, or concentrated colloid. Base your answers on the property described.*

Property	Type of Solution
cloudy mixture with particles that move erratically	
large particles with thixotropic behavior	
clear mixture with particles that scatter light	

REAL-WORLD CONNECTION

Describe the properties of fog in terms of being a mixture and why those properties make driving through fog so dangerous.

_____

_____

_____

Mixtures and Solutions
Section 2 Solution Concentration

Main Idea ——— **Details** ——————————————————————————

Scan *Section 2 of your text, using the checklist below as a guide.*

- Read all section titles.
- Read all boldfaced words.
- Read all tables and graphs.
- Look at all pictures and read the captions.
- Think about what you already know about this subject.

Write *three facts you discovered about solutions.*

1. _____

   _____

2. _____

3. _____

New Vocabulary

Use your text to define these terms.

concentration _____

_____

molarity _____

molality _____

_____

mole fraction _____

_____

Academic Vocabulary

Define the following term.

concentrated _____

Section 2 Solution Concentration (continued)

Main Idea ——— **Details** —————————————————

Expressing Concentration

Use with pages 480–481.

Analyze *the similarities in all of the concentration ratios shown in Table 3 in your text.*

_____

_____

_____

Write *the equation for determining percent by mass.*

Percent by mass =

Calculate Percent by Mass

Use with Example Problem 1, page 481.

Summarize *Fill in the blanks to help you take notes as you read Example Problem 1.*

Problem

Determine the percent by mass of 3.6 g NaCl in 100.0 g H_2O.

1. **Analyze the Problem**

   List the knowns and unknowns.

   Known: Unknown:

   mass of solute = _____   percent by mass = ?

   mass of solvent = _____

2. **Solve for the Unknown**

   Find the mass of the solution.

   mass of solution = grams of solute + grams of solvent

   mass of solution = 3.6 g + _____ = _____

   Substitute the known values into the percent by mass equation.

   percent by mass =

3. **Evaluate the Answer**

   The answer should be a small percent, to match the small quantity of _____. The mass of sodium chloride was given in two significant figures, therefore, the answer should have _____ significant figures.

Section 2 Solution Concentration (continued)

Main Idea ——— **Details** ————————————————————————

Molarity

Use with pages 482–485.

Describe *how to calculate the molarity of a solution by completing the following statements.*

To calculate the _____ of a solution, you must know the amount of

dissolved _____ and the volume of _____. The following equation

is used: molarity (M) = _____ of solute/liters of _____.

Explain *why you may need less than one liter of water to prepare a molar solution of one liter.*

_____

_____

Write *the expression that describes the relationship between a stock solution and a dilute solution.*

_____

M_1 = _____

V_1 = _____

M_2 = _____

V_2 = _____

Section 2 Solution Concentration (continued)

Main Idea	Details

Molality and Mole Fraction

Use with pages 487–488.

Explain *how the volume and mass of a solution change with temperature.*

The volume may _____ when heated or _____ when cooled. The mass of the solution _____ change.

Write *the mole fraction equations for a solvent (X_A) and a solute (X_B) below.*

$X_A =$ $X_B =$

Evaluate *the mole fraction for the values given in problem 4 on page 487 of your text. The number of moles for 100 g H_2O is given.*

$n_A = 5.55$ mol H_2O $n_B =$ _____ mol NaCl

$X_{H_2O} =$ = _____

$X_{NaCl} =$ = _____

$X_{H_2O} + X_{NaCl} = 1.000$

_____ + _____ = 1.000

REAL-WORLD CONNECTION

Describe how the mole fractions for a solution are similar to the pieces of a pie.

_____

_____

_____

Mixtures and Solutions
Section 3 Factors Affecting Solvation

Main Idea ———— **Details** ————————————————

Skim *Section 3 of your text. List three main ideas of the section.*

1. _____

2. _____

3. _____

New Vocabulary Use your text to define each term.

solvation _____

_____

heat of solution _____

_____

supersaturated solution _____

_____

_____

Henry's law _____

_____

_____

Compare and contrast *saturated solutions and unsaturated solutions.*

_____

_____

_____

_____

Section 3 Factors Affecting Solvation (continued)

⟨ **Main Idea** ⟩ ——— ⟨ **Details** ⟩ ————————————————

The Solvation Process

Use with pages 489–492.

Describe *solutions by completing the following statements.*

A solution may exist in gas, solid, or liquid form, depending on the state of its

_____ . Some combinations of substances easily form _____ and

others do not. A substance that does not _____ in a solvent is

_____ in that solvent.

Write *the general rule to determine if solvation will occur.*

_____

List *three factors that must be known about component substances to determine if solvation will occur.*

1. _____

2. _____

3. _____

Sequence *the steps required for a sodium chloride crystal to dissolve in water.*

_____ The charged ends of water molecules attract the positive Na ions and the negative Cl ions.

_____ The ions from the crystal break away from the surface.

_____ Water molecules collide with the surface of the crystal.

_____ NaCl crystals are placed in water.

_____ Solvation continues until the entire crystal has dissolved.

_____ The attraction between the dipoles and the ions are stronger than the attractions among the ions in the crystal.

Section 3 Factors Affecting Solvation (continued)

◁Main Idea◁ ——— ◁Details◁ ————————————————————————

Factors That Affect Solvation

Use with page 492.

Organize *the following table on factors that can increase the rate of solvation by increasing the number of collisions.*

Factor	Increase Collisions By
agitating the mixture	
breaking particles into smaller pieces	
increasing temperature of the solvent	

Solubility

Use with pages 493–496.

Explain *how solubility is expressed in units of measurement.*

_____

Review *Table 4 in your text to determine the solubility of the following compounds in water.*

Ca(OH)₂ at 20°C _____

KCl at 60°C _____

Describe *each of these solubility states.*

State	Description
continuing salvation	
dynamic equilibrium	
saturated solution	
unsaturated solution	

Section 3 Factors Affecting Solvation (continued)

(Main Idea) ——— **(Details)** ————————————————————

Describe *how solubility changes with temperature for most substances.*

_____

Explain *why some gases are less soluble as temperature increases.*

_____

_____

Describe *the relationship between solubility and pressure.*

_____

Henry's Law

Use with Example Problem 5, page 497.

Write *the equation for Henry's law.*
Summarize *Fill in the blanks to help you take notes while you read Example Problem 5.*

Problem

Find how much of a gas will dissolve in 1.0 L of water at 1.0 atm, if 0.85 g of that gas will dissolve in 1.0 L of water at 4.0 atm and temperature does not change.

1. **Analyze the Problem**

   List the knowns and unknowns.

   Known: Unknown:

   $S_1 = $ _____

   $P_1 = $ _____ $S_2 = $ _____

   $P_2 = $ _____

2. **Solve for the Unknown**

   Rearrange Henry's Law to solve for S_2.

   $S_2 = $ _____

   Substitute known values and solve.

   $S_2 = $ _____ $\dfrac{(1.0\,\text{atm})}{\text{_____}} = $ _____

3. **Evaluate the Answer**

   The solubility _____ as expected due to the _____ in pressure.

Mixtures and Solutions
Section 4 Colligative Properties of Solutions

Main Idea ——— **Details** ————————————————————————————————

Scan *Section 4 of your text, using the checklist below as a guide.*

- Read all section titles.
- Read all boldfaced words.
- Read all tables and graphs.
- Look at all pictures and read the captions.
- Think about what you already know about solutions.

Write two questions that you would want answers to based on your reading.

1. _____

2. _____

New Vocabulary   *Use your text to define each term.*

colligative property   _____

_____

_____

vapor pressure lowering   _____

_____

boiling point elevation   _____

_____

freezing point depression   _____

_____

osmosis   _____

_____

_____

osmotic pressure   _____

_____

Section 4 Colligative Properties of Solutions (continued)

Main Idea —— **Details** _____

Electrolytes and Colligative Properties

Use with pages 498–499.

Compare and contrast *electrolytes and nonelectrolytes.*

Substances like sodium chloride that _____ in water and conduct an _____ are called _____. Substances like sucrose that dissolve in water but do not _____ and do not conduct an electric current are called _____.

Vapor Pressure Lowering

Use with page 499.

Summarize *why vapor pressure lowering is a colligative property. Include an explanation of vapor pressure.*

_____

_____

_____

_____

_____

_____

_____

Boiling Point Elevation

Use with page 500.

Explain *boiling point elevation by completing the following statements.*

A liquid boils when its _____ equals _____.

Adding a nonvolatile solute lowers the solvent's _____ pressure. More _____ energy must be added to reach the solvent's _____.

The greater the number of _____ particles in the solution, the greater the _____ elevation.

Section 4 Colligative Properties of Solutions (continued)

(Main Idea)	(Details)
Freezing Point Depression *Use with pages 501–502.*	**Describe** *why the freezing point changes when a solute is added to a solution.* _____ _____ _____ _____ _____
Osmotic Pressure *Use with page 504.*	**Evaluate** *the diagram of a semipermeable membrane separating a sucrose-water solution on one side and water on the other side. Draw an arrow to show in which direction more water will flow and circle the side which has the greater osmotic pressure.*

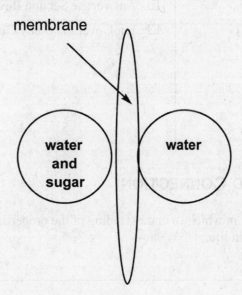

membrane

water and sugar

water

Mixtures and Solutions Chapter Wrap-Up

Now that you have read the chapter, review what you have learned and write the key equations and relationships.

_____

_____

_____

Review

Use this checklist to help you study.

☐ Study your Science Notebook for this chapter.

☐ Study the definitions of vocabulary words.

☐ Review daily homework assignments.

☐ Reread the chapter and review the tables, graphs, and illustrations.

☐ Answer the Section Review questions at the end of each section.

☐ Look over the Study Guide at the end of the chapter.

REAL-WORLD CONNECTION

Identify four ways in which an understanding of the properties of solutions and heterogenous mixtures can be applied to your own life.

1. _____

2. _____

3. _____

4. _____

Energy and Chemical Change

Before You Read

Review Vocabulary   *Define the following terms.*

chemical equation   _____

_____

mole   _____

_____

_____

Chapter 10   **Describe** *the equation you would use to convert mass in grams to moles.*

Chapter 12   **Identify** *the three characteristics of particles about which the kinetic-molecular theory makes assumptions.*

1. _____

2. _____

3. _____

Write the equation that represents the kinetic energy of a particle.

_____

Energy and Chemical Change

Section 1 Energy

Main Idea ———— **Details** ——————————————————————

Skim *Section 1 of your text. Write two facts you discovered about energy.*

1. _____

_____

2. _____

_____

New Vocabulary   *Use your text to define each term.*

energy _____

law of conservation of energy _____

_____

chemical potential energy _____

_____

heat _____

_____

calorie _____

_____

joule _____

specific heat _____

_____

Section 1 Energy (continued)

| Main Idea | Details |

The Nature of Energy

Use with pages 516–518.

Compare *and contrast kinetic energy with potential energy.*

_____

_____

On the curve below that represents the skier on a ski slope on page 516, label the place of greatest kinetic energy A, least kinetic energy B, greatest potential energy C, and least potential energy D.

Describe *the skier above as a function of the law of conservation of energy.*

_____

_____

_____

Explain *chemical potential energy.*

Chemical _____ energy of a substance is a result of the arrangement of its _____ and the strength of the _____ joining the atoms. During some _____ reactions, such as burning _____, much of the potential energy may be released as _____. Some of the energy may be converted to work, which is a form of _____ energy.

Specific Heat

Use with pages 519–520.

Identify *each symbol in the equation for specific heat.*

$q = c \times m \times \Delta T$

_____ represents heat absorbed or released

_____ represents the specific heat of the substance

_____ represents mass of a sample in grams

_____ represents a change in temperature

Section 1 Energy (continued)

Main Idea ———— **Details** ————————————————————————

Calculate Specific Heat

Use with Example Problem 2, page 521.

Summarize *Fill in the blanks to help you take notes while you read Example Problem* 2.

Problem

The temperature of a sample of iron with a mass of 10.0 g changed from 50.4°C to 25.0°C with the release of 114 J heat. Determine the specific heat of iron.

1. Analyze the Problem

Known:

energy released = _____

$\Delta T =$ _____

mass of iron = _____

Unknown:

specific heat of iron = ?

2. Solve for the Unknown

Write the equation for heat absorption.

$q =$ _____

Solve for *c*.

$q =$ $c =$

$c =$

3. Evaluate the Answer

If the values used in the calculations have _____ significant figures, then the answer must also have _____ significant figures. The calculated value matches the value for iron in Table 2.

REAL-WORLD CONNECTION

Describe two potential problems with the use of the Sun as a source of everyday energy.

1. _____

2. _____

Energy and Chemical Change

Section 2 Heat

Main Idea ———— **Details** ————————————————

Skim *Section 2 of your text. Write three questions that come to mind from reading the headings and the illustration captions.*

1. _____

_____

2. _____

3. _____

New Vocabulary — *Use your text to define each term.*

calorimeter _____

_____

thermochemistry _____

_____

system _____

_____

surroundings _____

universe _____

enthalpy _____

enthalpy (heat) _____
of reaction _____

Section 2 Heat (continued)

<table>
<tr><td>Main Idea</td><td>Details</td></tr>
</table>

Calorimetry

Use with pages 523–524.

Describe *how a calorimeter measures heat.*

_____

_____

_____

Using Specific Heat

Use with Example Problem 3, page 525.

Summarize *Fill in the blanks to help you take notes while you read Example Problem 3.*

Problem

Determine the specific heat of a piece of metal with a mass of 4.68 g that _____ 256 J of heat when its temperature increases by 182°C, and explain if the metal could be an _____.

1. **Analyze the problem**

   Known: mass of metal = _____

   quantity of heat absorbed = _____

   _____ = 182°C

   Unknown: specific heat, c = ? J/(g • °C)

2. **Solve for the Unknown**

   Write the equation for absorption of heat.

   $q =$ _____

   Solve for c by dividing both sides of the equation by $m \times \Delta T$.

   $c =$

Section 2 Heat (continued)

Main Idea ———— **Details** ——————————————————

Substitute the known values into the equation.

$c =$ _____ $=$ _____

Table 2 (page 520) indicates the metal could be _____.

3. **Evaluate the Answer**

The quantities used in the calculation have _____ significant

figures, and the answer is correctly stated with _____ significant

figures. The calculation yielded the _____ unit, and the

calculated _____ _____is the same as that for _____.

**Chemical Energy
and the Universe**

Use with pages 525–528.

Compare and contrast *exothermic and endothermic reactions.*

_____

_____

Write *the symbol for enthalpy (heat) of reaction.*

_____

Explain *why chemists prefer to measure change in heat energy, rather
than the total amount of heat energy present.*

_____

_____

_____

_____

Energy and Chemical Change

Section 3 Thermochemical Equations

Main Idea ——— ### Details _____

Skim *Section 3. Focus on the subheadings, boldfaced words, and the main ideas. In the space below, summarize the main idea of this section.*

_____

_____

_____

New Vocabulary

Use your text to define each term.

thermochemical equation

_____

_____

_____

enthalpy (heat) of combustion

_____

_____

molar enthalpy (heat) of vaporization

_____

_____

molar enthalpy (heat) of fusion

_____

_____

Section 3 Thermochemical Equations (continued)

⟨**Main Idea**⟩ ———— ⟨**Details**⟩ ————————————————————

Writing Thermochemical Equations

Use with page 529.

Identify *which of the reactions below is endothermic, and explain how you know.*

1. $4Fe(s) + 3O_2(g) \rightarrow 2Fe_2O_3(s)$ $\Delta H = -1625$ kJ
2. $NH_4NO_3(s) \rightarrow NH_4^+(aq) + NO_3^-(aq)$ $\Delta H = 27$ kJ

_____

_____

_____

Identify *which of the reactions below is exothermic, and explain how you know.*

1. $4Fe(s) + 3O2(g) \rightarrow 2Fe2O3(s)$ $\Delta H = -1625$ kJ
2. $NH_4NO_3(s) \rightarrow NH_4^+(aq) + NO_3^-(aq)$ $\Delta H = 27$ kJ

_____

_____

_____

Changes of State

Use with pages 530–531.

Name *the common states of matter.*

_____

Section 3 Thermochemical Equations (continued)

Main Idea ——— **Details** ————————————————————

Explain *changes in physical states by completing the sentences below.*

During vaporization, a _____ becomes a _____.

Energy must be _____ by the liquid.

During condensation, a _____ becomes a _____.

Energy is _____ by the gas.

During fusion of ice, a _____ becomes a _____.

Energy is _____ by the solid.

Identify *what the following equations represent.*

$\Delta H_{vap} = -\Delta H_{cond}$

_____

_____

$\Delta H_{fus} = -\Delta H_{solid}$

_____

_____

REAL-WORLD CONNECTION

Explain why a farmer would spray his orange trees with water when he knows the overnight temperature will be below 30°C.

_____

_____

_____

Energy and Chemical Change

Section 4 Calculating Enthalpy Change

⟨Main Idea⟩ ——————— **⟨Details⟩** ——————————————————————

Scan *Section 4 of your text. Use the checklist below to preview the section.*

• Read all section titles.

• Read all boldfaced words.

• Read all tables and graphs.

• Look at all pictures and read the captions.

• Think about what you already know about energy and chemical change.

Write *three statements about calculating enthalpy change based on your reading.*

1. _____

2. _____

3. _____

⟨New Vocabulary⟩   *Use your text to define each term.*

Hess's law   _____

_____

_____

standard enthalpy   _____
(heat) of formation
_____

_____

_____

Section 4 Calculating Enthalpy Change (continued)

Main Idea — **Details** _____

Hess's Law

Use with pages 534–536.

Describe *Hess's law by completing the following statement.*

_____ is used to determine the _____ of a system by imagining that each reaction is part of a _____, each of which has a known ΔH.

Examine *Figure 13. Read the caption and follow the arrows. Then apply Hess's law to fill in the blanks below.*

ΔH for equation **c** _____

ΔH for equation **d** _____

sum of ΔH for equations **c** and **d** _____

In other words, the _____ for the conversion of S and

O_2 to SO_3 is _____ .

Standard Enthalpy (Heat) of Formation

Use with pages 537–538.

Explain *standard enthalpy of elements and compounds by completing the following statements.*

An element's _____ is the normal _____ state at one

_____ pressure and _____ . For example, the

standard state for iron is _____ , for mercury is _____ ,

and for oxygen is _____ . Free elements such as these are assigned

a ΔH_f^0, or _____, of exactly

_____ . The ΔH_f^0 of many _____ has been measured

_____ . For example, the standard enthalpies of formation for

the following compounds are:

$NO_2(g)$ _____

$SO_3(g)$ _____

$SF_6(g)$ _____

Section 4 Calculating Enthalpy Change (continued)

Main Idea	Details

The Summation Equation

Use with page 540.

Write *the formula that sums up the procedure for combining standard heats of formation equations to produce the desired equation and its* ΔH^0_{rxn}.

This equation says to _____ the _____ of heats of _____ of the

_____ from the sum of the _____ of formation of the _____ .

Enthalpy Change from Standard Enthalpies of Formation

Use with Example Problem 6, page 540.

Summarize *Fill in the blanks to help you take notes as you work through* *Example Problem 6.*

Problem

Calculate ΔH^0_{rxn} for the combustion of methane.

$$CH_4(g) + 2O_2(g) \rightarrow CO_2(g) + 2H_2O(l)$$

1. **Analyze the Problem**

   Use the formula $\Delta H^0_{rxn} = \Sigma \Delta H^0_f \text{(products)} - \Sigma \Delta H^0_f \text{(reactants)}$ with data from Table R-11 (page 975).

   Known:

   $\Delta H^0_f(CO_2) = $ _____

   $\Delta H^0_f(H_2O) = $ _____

   $\Delta H^0_f(CH_4) = $ _____

   $\Delta H^0_f(O_2) = $ _____

   Unknown:

   $\Delta H^0_{rxn} = ? \text{ kJ}$

Section 4 Calculating Enthalpy Change (continued)

Main Idea ——— **Details** ————————————————————————

2. Solve for the Unknown

Use the formula $\Delta H^0_{rxn} = \Sigma \Delta H^0_f \text{(products)} - \Sigma \Delta H^0_f \text{(reactants)}$

Substitute values in the formula

$\Delta H^0_{rxn} = $ _____

$\Delta H^0_{rxn} = $ _____ = _____

3. Evaluate the Answer

All values are _____ to the stated place. The calculated value matches that in Table R-11 (page 975).

REAL-WORLD CONNECTION

Your family needs to choose a system to heat the new home you are building. From what you have learned so far, write down four questions you will use to evaluate the systems available.

1. _____

2. _____

3. _____

4. _____

Energy and Chemical Change

Section 5 Reaction Spontaneity

Main Idea ———— **Details** ——————————————————

Scan *Section 5, using the checklist below as a guide.*

- Read all section titles.
- Read all boldfaced words.
- Read all tables and graphs.
- Look at all pictures and read the captions.
- Think about what you already know about energy and chemical change.

State *the main concepts of this section.*

_____

_____

_____

_____

New Vocabulary   *Use your text to define each term.*

spontaneous process   _____

entropy   _____

_____

_____

second law of thermodynamics   _____

_____

free energy   _____

Academic Vocabulary   *Define the following term.*

demonstrate   _____

Name _____ Date _____

Section 5 Reaction Spontaneity (continued)

Main Idea	Details
Spontaneous Processes *Use with pages 542–545.*	**Compare and contrast** *spontaneous process and non-spontaneous process.* _____ _____ _____ _____ _____ **Identify** *the parts of the entropy equation.* $\Delta S_{system} = S_{products} - S_{reactants}$ ΔS represents _____. S represents _____. **List** *five reactions or processes in which it is possible to predict change in entropy. For each process, indicate whether entropy will increase or decrease.* 1._____ 2._____ 3._____ 4._____ 5._____

Chemistry: Matter and Change   218   *Science Notebook*

Copyright © Glencoe/McGraw-Hill, a division of The McGraw-Hill Companies, Inc.

Section 5 Reaction Spontaneity (continued)

⟨Main Idea⟩	⟨Details⟩

Entropy, the Universe, and Free Energy

Use with pages 546–548.

Write *the equation for the standard free energy change under standard conditions.*

_____

Predict *whether entropy increases or decreases for the reaction below and explain your reasoning.*

$N_2(g) + 3H_2(g) \rightarrow 2NH_3(g)$

_____

_____

Describe *free energy changes by writing the word positive or negative in the appropriate blank.*

If the sign of the free energy change is _____, the reaction is spontaneous.

If the sign of the free energy system is _____, the reaction is non-spontaneous.

Explain *how* ΔH^0_{system} *and* ΔS^0_{system} *affect reaction spontaneity by completing the following table.*

How ΔH^0_{system} and ΔS^0_{system} Affect Reaction Spontaneity		
	$-\Delta H^0_{system}$	$+\Delta H^0_{system}$
$+\Delta S^0_{system}$		
$-\Delta S^0_{system}$		

Energy and Chemical Change Chapter Wrap-Up

Now that you have read the chapter, review what you have learned and write three key equations or relationships.

1. _____

2. _____

   _____

3. _____

Review

Use this checklist to help you study.

☐ Study your Science Notebook for this chapter.

☐ Study the definitions of vocabulary words.

☐ Review daily homework assignments.

☐ Reread the chapter, reviewing the tables, graphs, and illustrations.

☐ Answer the Section Review questions at the end of each section.

☐ Look over the Chapter Assessment at the end of the chapter.

REAL-WORLD CONNECTION

Explain why the energy that comes from chemical reactions is critical for almost every phase of your daily life.

_____

_____

_____

_____

Reaction Rates

Before You Read

Review Vocabulary *Define the following terms.*

Boyle's law

_____

_____

Charles's law

_____

_____

Gay-Lussac's law

_____

_____

molarity

_____

_____

Chapter 9 **Balance** *the following equation.*

$$\boxed{}\ C_8H_{18}(l) + \boxed{}\ O_2(g) \rightarrow \boxed{}\ CO_2(g) + \boxed{}\ H_2O(l)$$

Reaction Rates
Section 1 A Model for Reaction Rates

Main Idea ——————   **Details** ————————————————

Skim *Section 1 of your text. Preview headings, photos, captions, boldfaced words, problems, and graphs. Write three questions that come to mind.*

1. _____

2. _____

3. _____

New Vocabulary   *Use your text to define each term.*

reaction rate   _____

                     _____

collision theory   _____

activated complex   _____

                     _____

activation energy   _____

                     _____

Academic Vocabulary   *Define the following term.*

investigate   _____

Section 1 A Model for Reaction Rates (continued)

Main Idea ———— **Details** ————————————————————————

Expressing Reaction Rates

Use with pages 560–562.

Identify *what each phrase or symbol represents in this equation.*

$$\text{Average rate} = \frac{\Delta \text{quantity}}{\Delta t}$$

Average rate = the average is used because the rate changes over time

$\Delta = $ _____

$\Delta t = $ _____

Calculate Average Reaction Rates

Use with Example Problem 1, page 562.

Summarize *Fill in the blanks to help you take notes while you read Example Problem 1.*

Problem

Calculate the average reaction rate of the chemical reaction

using the _____ of butyl

chloride in _____.

1. **Analyze the Problem**

   Known: Unknown:

   _____   _____

   _____

   $[C_4H_9Cl]$ at $t_1 = 0.220M$

   _____

2. **Solve for the Unknown**

   Write the equation.

   Average reaction rate =

   Insert known quantities.

   Solve for the average rate   = $\dfrac{\rule{3cm}{0.4pt}}{4.00\ s - 0.00\ s}$

   = _____

   Average reaction rate   = _____

3. **Evaluate the Answer**

   The answer is correctly expressed in _____ significant figures.

Section 1 A Model for Reaction Rates (continued)

⌐Main Idea⌐ ——— ⌐Details⌐ ————————————————

Collision Theory

Use with pages 563–564.

Describe *how each of the items below affects a reaction.*

collision theory

_____

orientation and the activated complex

_____

_____

activation energy and reaction

_____

_____

_____

Analyze *Figure 4. Use colored pencils to draw similar molecules colliding. Be sure to include incorrect orientation, correct orientation, and correct orientation with insufficient energy. Develop a key for your drawings.*

Activation Energy and Reaction Rate

Use with pages 564–565.

Explain *activation energy by completing the following paragraph.*

Some reactions have enough _____ to overcome the _____

_____ of the reaction in order to form products. These are called

_____ . After the _____ is formed,

_____ is released. In other reactions the reactants must absorb energy to

overcome the _____ of the reaction. These reactions are

called _____ .

REAL-WORLD CONNECTION

Describe how the collision theory would apply to a demolition derby.

_____

_____

Reaction Rates

Section 2 Factors Affecting Reaction Rates

Main Idea ———— **Details** ————————————————————————

Scan *Section 2, using the checklist below as a guide.*

- Read all section titles.
- Read all boldfaced words.
- Read all tables and graphs.
- Look at all pictures and read the captions.
- Think about what you already know about this topic.

Write three facts you discovered about reaction rates.

1. _____

2. _____

3. _____

New Vocabulary   *Use your text to define each term.*

catalyst _____

_____

inhibitor _____

heterogeneous catalyst _____

_____

homogeneous catalyst _____

_____

The Nature of Reactants

Use with page 568.

Explain *how reactants influence the rate at which a chemical reaction occurs by completing the following statement.*

As the reactant increases, the _____ increases.

Section 2 Factors Affecting Reaction Rates (continued)

Main Idea —— **Details** —————————————————————

Use with pages 568–573.

Explain *the effect each of the following has on the rate of a reaction.*

reactivity of reactants

_____

concentration

_____

surface area

_____

temperature

_____

catalyst

_____

inhibitors

_____

REAL-WORLD CONNECTION

Compare and contrast the rate at which a sugar cube in cold water and granulated sugar in warm water would dissolve. Include how surface area and the temperature of the water might affect the rate at which each dissolves. Create a statement about which would dissolve faster.

_____

_____

_____

_____

Reaction Rates
Section 3 Reaction Rate Laws

Main Idea ——————— **Details** ————————————————————

Skim *Section 3 of your text. Choose a photograph from this section. Write a question based on what you see and read.*

_____

_____

New Vocabulary

Use your text to define each term.

rate law

_____

_____

specific rate constant

_____

_____

reaction order

_____

_____

method of initial rates

_____

_____

Section 3 Reaction Rate Laws (continued)

(Main Idea) ——— (Details) —————————————————

Writing Reaction Rate Laws

Use with page 574–576.

Explain *what each symbol represents in the following equation.*

Rate = k[A]

k = _____

_____

[A] = _____

Analyze *the rate law reaction for the decomposition of hydrogen peroxide.*

$$2H_2O_2 \longrightarrow 2H_2O + O_2$$

rate law equation: rate = k[A], where [A] = _____

insert the reactant: rate = _____

Express *the rate law reaction for this chemical reaction.*

chemical equation: $2NO(g) + 2H_2(g) \longrightarrow N_2(g) + 2H_2O(g)$

rate law equation: rate = _____, where [A] represents the reactant

_____ and [B] represents the reactant _____

insert the reactants: rate = _____

Section 3 Reaction Rate Laws (continued)

⟨**Main Idea**⟩ —————   ⟨**Details**⟩ —————————————————————————

	Relate *how the reaction rate varies with:*
	concentration
	_____
	_____
	the overall reaction order
	_____
	_____
Determining Reaction Order	**Explain** *reaction order by completing the following sentences.*
Use with page 576.	One of the means of determining reaction order is by comparing
	_____ of a reaction with varying
	_____. This is known as the method of
	_____. This method requires experimentation with differing
	_____ of the reactants and comparing the _____ of
	the reaction at each quantity. While the rate law for a reaction can tell you the
	reaction rate, the rate constant *k*, and the _____,
	actual _____ and _____ of a complex reaction can be
	determined only through experimentation.

REAL-WORLD CONNECTION

Consider whether an average of a student's grades on all chemistry tests is or is not a better way of determining a final grade as compared to using just one test score. Explain which is better and why.

_____

_____

Reaction Rates

Section 4 Instantaneous Reaction Rates and Reaction Mechanisms

Main Idea ———   **Details** ————————————————————————

Skim *Section 4 of your text. Preview the headings, photos, captions, boldfaced words, problems, and graphs. Write three questions that come to mind.*

1. _____

2. _____

3. _____

New Vocabulary   *Use your text to define each term.*

instantaneous rate _____

_____

complex reaction _____

_____

reaction mechanism _____

_____

intermediate _____

_____

rate-determining step _____

_____

Section 4 Instantaneous Reaction Rates and Reaction Mechanisms (continued)

Main Idea	Details

Calculate Instantaneous Reaction Rates

Use with Example Problem 2, page 579.

Summarize *Fill in the blanks to help you take notes while you read Example Problem 2.*

Problem

Calculate the instantaneous rate for this reaction, given the quantities for NO and H_2.

$$2NO(g) + H_2(g) \longrightarrow N_2O(g) + H_2O(g)$$

1. **Analyze the Problem**

   Known: Unknown:

   quantity of [NO] = 0.00200M rate = ? mol/(L · s)

   quantity of [H_2] = _____

   k = _____

2. **Solve for the Unknown**

   Insert the known quantities into the rate law equation.

   rate = _____

   rate = _____

   rate = _____

3. **Evaluate the Answer**

   Are your units correct? Is your magnitude reasonable?

Reaction Mechanisms

Use with pages 580–582.

Compare *the reaction mechanism using the terms complex, intermediate, and rate-determining step to the process of building a car. Show that you understand the vocabulary.*

_____

_____

_____

_____

_____

_____

Reaction Rates Chapter Wrap-Up

Now that you have read the chapter, list three facts you learned about reaction rates:

1. _____

2. _____

3. _____

Review

Use this checklist to help you study.

☐ Study your Science Notebook for this chapter.

☐ Study the definitions of vocabulary words.

☐ Review daily homework assignments.

☐ Reread the chapter and review the tables, graphs, and illustrations.

☐ Answer the Section Review questions at the end of each section.

☐ Look over the Chapter Assessment at the end of the chapter.

REAL-WORLD CONNECTION

Suppose you obtain a part-time job working for a lawn care business. Your new boss wants you to help her choose the right fertilizer for most of the lawns you will see. Use the terms from this chapter to explain to your boss what she should look for in a fertilizer.

_____

_____

_____

Chemical Equilibrium

Before You Read

Review Vocabulary

Define the following terms.

chemical equation

_____

_____

_____

_____

reaction rate

_____

_____

rate law

_____

_____

_____

Chapter 9

Balance *the chemical equation below.*

$$NO(g) + H_2(g) \qquad\qquad N_2O(g) + H_2O(g)$$

Chapter 16

Write *the rate law for the reaction below.*

$$H_2(g) + I_2(g) \longrightarrow 2\,HI(g)$$

Rate = _____

Chemical Equilibrium

Section 1 A State of Dynamic Balance

Main Idea ——— **Details** ———————————————————————

Skim *Section 1 of your text. Write a statement that describes the nature of equilibrium from your reading of the headings, boldface terms, and illustration captions.*

_____

_____

_____

New Vocabulary

Use your text to define each term.

reversible reaction _____

chemical equilibrium _____

_____

law of chemical equilibrium _____

_____

_____

equilibrium constant _____

_____

_____

homogeneous equilibrium _____

_____

heterogeneous equilibrium _____

_____

Section 1 A State of Dynamic Balance (continued)

Main Idea ——— **Details** ———————————————————

What is equilibrium?

Use with pages 594–598.

Explain *reversible reactions by inserting the words left and right in the following statements.*

The reactants for the forward reaction are on the _____. The products are

on the _____. The reactants for the reverse reaction are on the _____.

The products are on the _____.

List *the reactants and products of the following reversible reaction.*

$N_2(g) + 3H_2(g) \rightleftharpoons 2NH_3(g)$

	Reactants	Products
Forward reaction		
Reverse reaction		

Complete *the following statement.*

The state in which forward and reverse reactions balance each other because

they take place at equal rates is called _____. Although a

chemical reaction may be in equilibrium, the _____ and _____

may continually be _____ because chemical equilibrium is a dynamic

process.

Equilibrium Expressions and Constants

Use with pages 599–604.

Identify *the parts of the equilibrium constant expression.*

$$K_{eq} = \frac{[C]^c[D]^d}{[A]^a[B]^b}$$

K_{eq} = _____

[C][D]   = _____

[A][B]   = _____

a, b, c, and d   = _____

Section 1 A State of Dynamic Balance (continued)

⟨**Main Idea**⟩ ——— ⟨**Details**⟩ ———————————————

Write *the equilibrium constant expression for the following balanced chemical equation.*

$$N_2(g) + 3H_2(g) \rightleftharpoons 2NH_3(g)$$

$K_{eq} =$ _____

_____

Compare and contrast *homogeneous equilibrium and heterogeneous equilibrium by completing the following sentences.*

Homogeneous equilibrium occurs when _____ and _____ of a

reaction are in the _____ physical state. Heterogeneous equilibrium

occurs when _____ and _____ of a reaction are

in more than _____ physical state. Equilibrium depends on the

_____ in the system.

Write *the equilibrium expression for this reaction.*

$$I_2(s) \rightleftharpoons I_2(g)$$

_____

REAL-WORLD CONNECTION

Discuss why sodium hydrogen carbonate is valuable in baking.

_____

_____

_____

Section 1 A State of Dynamic Balance (continued)

Main Idea	Details

The Value of Equilibrium Constants

Use with Example Problem 3, page 605.

Summarize *Fill in the blanks to help you take notes while you read Example Problem 3.*

Problem

Calculate the value of K_{eq} for the equilibrium constant expression.

$$K_{eq} = \frac{[NH_3]^2}{[N_2][H_2]^3}$$

1. Analyze the Problem

List the knowns and unknowns.

Known: the equilibrium constant expression:

_____

Known: the concentration of each reactant and product:

$[NH_3] = $ _____

$[N_2] \ = $ _____

$[H_2] \ = $ _____

Unknown: the value of the equilibrium constant

2. Solve for the Unknown

Substitute the _____ into the equilibrium

_____ and calculate its value.

$$K_{eq} = \frac{\rule{3cm}{0.4pt}}{[0.533]} = \rule{2cm}{0.4pt}$$

3. Evaluate the Answer

The given concentrations have _____ significant figures, therefore

the answer must have _____ significant figures.

Chemical Equilibrium
Section 2 Factors Affecting Chemical Equilibrium

Main Idea ———— **Details** ——————————————————

Scan *Section 2 of your text. Use the checklist below as a guide.*

- Read all section heads.
- Read all boldfaced words.
- Read all the tables and graphs.
- Look at all pictures and read the captions.
- Think about what you already know about chemical equilibrium.

Write *four facts you discovered about chemical equilibrium.*

1. _____

2. _____

3. _____

4. _____

New Vocabulary   *Use your text to define the following term.*

Le Châtelier's principle _____

_____

Section 2 Factors Affecting Chemical Equilibrium (continued)

(Main Idea) ———— (Details) ————————————————————————

Applying Le Châtelier's Principle

Use with pages 607–610.

Determine *how each of the following changes affects a system in equilibrium. Write a sentence that includes the term(s) in parentheses.*

changes in concentration (collisions)

_____

_____

changes in volume (pressure, products)

_____

_____

changes in temperature (endothermic, exothermic)

_____

_____

_____

REAL-WORLD CONNECTION

Describe how your body would relieve the stress placed on it by climbing to a high altitude.

_____

_____

_____

Name _____ Date _____

Chemical Equilibrium
Section 3 Using Equilibrium Constants

Main Idea ————— **Details** ——————————————————

Scan *Section 3 of your text. Use the checklist below as a guide.*

- Read all section heads.
- Read all boldfaced words.
- Read all the tables and graphs.
- Look at all pictures and read the captions.
- Think about what you already know about equilibrium constants.

Write *three facts you discovered about using equilibrium constants.*

1. _____

2. _____

3. _____

New Vocabulary *Use your text to define each term.*

solubility product
constant   _____

                  _____

common ion   _____

common ion effect _____

                  _____

Chemistry: Matter and Change

Science Notebook

240

Section 3 Using Equilibrium Constants (continued)

⌐Main Idea⌐ ——— ⌐Details⌐ ————————————————————————

Calculating Equilibrium Concentrations

Use with Example Problem 4, page 613.

Summarize *Fill in the blanks to help you take notes while you read Example Problem 4.*

Problem

At 1405 K, hydrogen sulfide _____ to form

_____ and a diatomic _____ molecule, S_2. The

_____ for the reaction is 2.27×10^{-3}.

$$2H_2S(g) \rightleftharpoons 2H_2(g) + S_2(g)$$

What is the concentration of $H_2(g)$ if

$[S_2] = 0.0540$ mol/L and $[H_2S] = 0.184$ mol/L?

1. **Analyze the Problem**

   List the knowns and unknowns.

   Known: Unknown:

   $K_{eq} = $ _____ $[H_2] = $ _____

   $[S_2] = $ _____

   $[H_2S] = $ _____

2. **Solve for the Unknown**

   Write the equilibrium constant expression.

   $K_{eq} = $

   _____

   Substitute known quantities.

   _____

   Solve for the unknown.

   _____

   _____

3. **Evaluate the Answer**

   The number of significant figures in the data is _____ Therefore,

   the number of significant figures in the answer must be _____.

Section 3 Using Equilibrium Constants (continued)

⟨ **Main Idea** ⟩ ———— ⟨ **Details** ⟩ ————————————————————

The Solubility Product Constant

Use with pages 614–619.

Describe *solubility equilibrium.*

_____

_____

Identify *the part of the equation that shows equilibrium and circle it.*

$$BaSO_4(s) \rightleftharpoons Ba^{2+}(aq) + SO_4^{2-}(aq)$$

Explain *solubility by completing the following statements.*

_____ is the amount of a substance that will _____ in a given

volume of _____.

K_{sp} represents the _____.

K_{sp} is the _____ of the concentration _____ each raised to the

power equal to the _____ of the ion in the _____.

K_{sp} depends only on the _____ of the _____ in a

saturated _____.

Explain *why it benefits doctors to understand solubility.*

_____

Calculating Molar Solubility

Use with Example Problem 5, page 616.

Summarize *Fill in the blanks to help you take notes while you read Example Problem 5.*

Problem

Calculate the solubility in mol/L of copper(II) carbonate ($CuCO_3$) at 298 K.

1. **Analyze the Problem**

   List the knowns and unknowns.

   Known: Unknown:

   K_{sp} ($CuCO_3$) = _____   solubility ($CuCO_3$) = _____

Section 3 Using Equilibrium Constants (continued)

Main Idea ———— **Details** ————————————————————————

2. **Solve for the Unknown**

Write the balanced chemical equation.

_____ _____

Write the solubility constant expression (remember only the ions are used).

_____

$s = [Cu^{2+}] = $ _____

Substitute s for $[Cu^{2+}]$ and _____

_____

_____

3. **Evaluate the Answer**

K_{sp} has _____ significant figures so the answer must be expressed with _____ significant figures.

Describe *conditions in which precipitates are likely to form.*

1. _____

2. _____

3. _____

_____

The Common Ion Effect

Use with pages 620–621.

Discuss *the common ion effect by completing the following paragraph.*

An ion that is common to two or more ionic compounds is known as a

_____. The lowering of the solubility of a substance by the

presence of a common ion is called the _____.

Chemical Equilibrium Chapter Wrap-Up

Now that you have read the chapter, review what you have learned.

Describe *chemical equilibrium.*

_____

_____

_____

Explain *Le Châtelier's principle.*

_____

_____

Review

Use this checklist to help you study.

☐ Study your Science Notebook for this chapter.

☐ Study the vocabulary words and scientific definitions.

☐ Review daily homework assignments.

☐ Reread the chapter and review the tables, graphs, and illustrations.

☐ Answer the Section Review questions at the end of each section.

☐ Look over the Study Guide at the end of the chapter.

REAL-WORLD CONNECTION

Describe several uses of solubility in your home.

_____

_____

_____

Acids and Bases

Before You Read

Review Vocabulary   *Define the following terms.*

chemical equilibrium

_____

_____

Chapter 9   **Write** *the equation for hydrogen chloride dissolving in water to form hydrogen ions and chloride ions.*

_____

Explain *what type of compound hydrogen chloride is since it produces hydrogen ions in aqueous solution.*

_____

Chapter 16   **Identify** *five factors that influence reaction rate.*

1. _____

2. _____

3. _____

4. _____

5. _____

Acids and Bases
Section 1 Introduction to Acids and Bases

Main Idea ————— **Details** ————————————————————

Skim *Section 1 of your text. Write two questions that come to mind from reading the headings and the illustration captions.*

1. _____

2. _____

New Vocabulary *Use your text to define each term.*

acidic solution _____

basic solution _____

Arrhenius model _____

_____

_____

_____

Brønsted-Lowry model _____

_____

conjugate acid _____

conjugate base _____

conjugate acid-base pair _____

_____

amphoteric _____

Lewis model _____

Section 1 Introduction to Acids and Bases (continued)

⟮ **Main Idea** ⟯ ——— ⟮ **Details** ⟯ ———————————————

Properties of Acids and Bases

Use with pages 634–636.

Compare and contrast *the properties of an acid and a base by placing an X in the Acid column if the property applies to an acid and in the Base column if the property applies to a base.*

Acid	Properties	Base
	tastes sour	
	tastes bitter	
	feels slippery	
	affects color	
	reacts with metal	
	conducts electricity	
	has more hydrogen ions than hydroxide ions	
	has more hydroxide ions than hydrogen ions	

Write *the chemical equation for the self-ionization of water.*

The Arrhenius and Brønsted-Lowry Models

Use with pages 637–639.

Analyze *why the Arrhenius model of acids and bases does NOT include ammonia (NH₃) in solution as a base.*

_____

_____

_____

Identify *which of the following statements describes the Arrhenius model and which describes the Brønsted-Lowry model by filling in the blanks.*

The _____ model is based on the dissociation of compounds, while the

_____ model is based on the donation and acceptance of

hydrogen ions. Conjugate acid-base pairs are a component of the

_____ model and are NOT a component of the _____

model.

Section 1 Introduction to Acids and Bases (continued)

Main Idea ———— **Details** ——————————————————————

Describe *what happens in the forward and reverse reactions when ammonia is dissolved in water. Identify the conjugate acid, the conjugate base, and the two conjugate acid-base pairs.*

_____

_____

_____

_____

_____

_____

_____

_____

Monoprotic and Polyprotic Acids

Use with pages 640–641.

Explain *what a polyprotic acid is.*

_____

Sequence *the following equations in the steps of the ionization of phosphoric acid in the correct order.*

____ $HPO_4^{2-}(aq) + H_2O(l) \rightleftarrows H_3O^+(aq) \lessgtr PO_4^{3-}(aq)$

____ $H_3PO_4(aq) + H_2O(l) \rightleftarrows H_3O^+(aq) \lessgtr H_2PO_4^{2-}(aq)$

____ $H_2PO_4^-(aq) + H_2O(l) \rightleftarrows H_3O^+(aq) \lessgtr HPO_4^-(aq)$

The Lewis Model

Use with pages 641–643.

Define and give examples of an anhydride, distinguishing between those that produce an acid and those that produce a base.

_____

_____

_____

_____

Acids and Bases
Section 2 Strengths of Acids and Bases

Main Idea ———— **Details** ——————————————————————————

Skim *Section 2 of your text. Focus on the headings, subheadings, boldfaced words, and the main ideas. Write three questions about strengths of acids and basis based on what you have read.*

1. _____

2. _____

3. _____

New Vocabulary *Use your text to define each term.*

strong acid _____

weak acid _____

acid ionization constant _____

_____

strong base _____

weak base _____

_____

base ionization constant _____

_____

Section 2 Strengths of Acids and Bases (continued)

⟨Main Idea⟩ ——— ⟨Details⟩ ——————————————————————————

Strengths of Acids

Use with pages 644–647.

Explain *why all acids are not equal in strength.*

_____

_____

Identify *the acids in the following table as strong or weak.*

Acid	Strong or Weak	Acid	Strong or Weak
acetic		hydroiodic	
carbonic		hydrosulfuric	
		hypochlorous	
hydrochloric		nitric	
hydrofluoric		sulfuric	

Describe *the difference in conductivity between strong and weak acids.*

_____

_____

_____

_____

Analyze *equilibrium constant expressions by completing the following statements.*

The concentration of liquid water in the denominator of an equilibrium

constant expression is considered to be _____ in dilute aqueous

solutions. Therefore, liquid water can be _____ K_{eq} to give a

new equilibrium constant, K_a. For weak acids, the equilibrium

_____ of the _____ in the numerator tends to be small

compared to the equilibrium _____ of the _____ in the

denominator. The weakest acids have the _____ K_a values because

their solutions have the highest concentrations of _____ acid

molecules.

Section 2 Strengths of Acids and Bases (continued)

⟨Main Idea⟩ ——— **⟨Details⟩** ———————————————————————

Strengths of Bases

Use with pages 648–649.

Compare and contrast *the strengths of acids and bases by completing this concept map using the terms ionize, ionization constant, strong, stronger, weak, and weaker.*

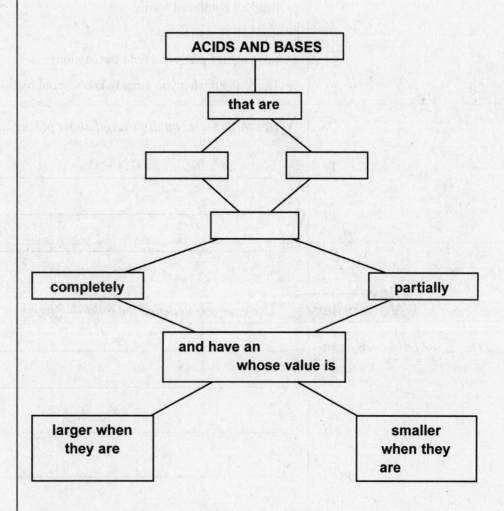

ACIDS AND BASES

that are

completely   partially

and have an
whose value is

larger when
they are

smaller
when they
are

Describe *the differences between the strength and the concentration of acids and bases by completing the following statements.*

The number of the acid or base molecules dissolved is described by the terms

_____ and _____. The degree to which an acid or base separates

into ions is described by the terms _____ and _____. A strong acid

can be a _____ solution and a _____ acid can be a concentrated

solution.

Acids and Bases
Section 3 Hydrogen Ions and pH

Main Idea ——— **Details** ——————————————————

Scan *Section 3 of your text. Use the checklist below as a guide.*

- Read all section titles.
- Read all boldfaced words.
- Read all formulas.
- Look at all figures and read the captions.
- Think about what you already know about hydrogen ions and pH

Write *three facts you discovered about pH as you scanned the section.*

1. _____

2. _____

3. _____

New Vocabulary   *Use your text to define the following terms.*

ion product constant   _____
for water
                              _____

pH   _____

pOH   _____

Section 3 Hydrogen Ions and pH (continued)

Main Idea ———— **Details** ————————————————————————

Ion Product Constant for Water

Use with pages 650–651.

Describe *how the ion product constant for water is derived from the self-ionization equation.*

$$H_2O(l) \rightleftharpoons \underline{\hspace{3cm}}$$

$$K_{eq} = \underline{\hspace{3cm}}$$

$$K_{eq}\,[H_2O] = \underline{\hspace{3cm}}$$

$$K_w = [H^+][OH^-] = \underline{\hspace{5cm}}$$

Calculate [H⁺] and [OH⁻] Using K_w

Use with Example Problem 1, page 651.

Summarize *Fill in the blanks to help you take notes while you read Example Problem 1.*

Problem

Calculate [OH⁻] using _____ and the concentration of _____, and determine if the solution is acidic, basic, or neutral.

Step 1: Analyze the Problem

Known: Unknown:

[H⁺] = _____   [OH⁻] = ? mol/L

K_w = _____

Write what you can predict about [OH⁻]:

_____

Step 2: Solve for the Unknown

Write the ion product constant expression

K_w = _____

Solve for [OH⁻] by _____

[OH⁻] = _____

[OH⁻] = _____

Since [H⁺] > [OH⁻], _____

Section 3 Hydrogen Ions and pH (continued)

Main Idea ———— **Details** ———————————————————————

Step 3: Evaluate the Answer

The answer is correctly stated with _____ significant figures because [H⁺] and [OH⁻] each have two. The hydroxide ion concentration _____ the prediction.

pH and pOH

Use with pages 652–658.

Compare and contrast *pH and pOH by completing the following table.*

Solution Type	Scale Measure	Relationship (Equation)
acid	pH	
base		
acid and base		

Analyze *the process of calculating pH and pOH from the hydroxide concentration.*

_____

_____

_____

Describe *the process of calculating the hydrogen ion and hydroxide ion concentrations from pH.*

_____

_____

_____

Describe *the process of calculating K_a from pH for a 0.100M weak acid.*

_____

_____

_____

_____

_____

Acids and Bases
Section 4 Neutralization

Main Idea ———— **Details** ————————————————————

Skim *Section 4 of your text. Focus on the headings, subheadings, boldfaced words, and the main ideas. Write three questions about strengths of acids and bases based on what you have read.*

1. _____

2. _____

3. _____

New Vocabulary *Define the following term.*

neutralization reaction _____

_____

salt _____

titration _____

_____

titrant _____

equivalence point _____

_____

acid-base indicator _____

end point _____

_____

salt hydrolysis _____

_____

buffer _____

buffer capacity _____

_____

Section 4 Neutralization (continued)

(Main Idea) ———   (Details) ————————————————————

**Reactions
Between Acids
and Bases**

Use with pages 659–664.

Write *the full equation of the neutralization reaction for magnesium
hydroxide and hydrochloric acid.*

_____

Draw *the titration curve for
50.0 mL 0.100M HCl titrated
with 0.100M NaOH. Label
the pH and volume vectors,
as well as the equivalence
point.*

Describe *the indicator that matches each of the following pH levels. Use
Figure 24 as a guide.*

pH	Indicator
7.2	
4.2	
1.8	
1–12	

Explain *the process for calculating the molarity of an unknown HCOOH
solution by completing the equations below.*

Balanced equation:
$HCOOH(aq) + NaOH(aq) \rightarrow HCOONa(aq) + H_2O(l)$

18.28 mL NaOH × _____ = _____ L NaOH

0.01828 L NaOH × _____

= _____ mol NaOH

1.828×10^{-3} mol NaOH × _____

= _____ mol HCOOH

1.828×10^{-3} mol HCOOH / _____

= _____ *M* HCOOH

Section 4 Neutralization (continued)

◖Main Idea◗ ——————	◖Details◗ —————————————————

Salt Hydrolysis

Use with page 665.

Describe *salt hydrolysis by completing the following statements.*

Some aqueous salt solutions are neutral, some are basic, and some are

_____. The reason for this is a process known as _____. In this

process, the anions of the dissociated salt donate _____ to

water. Salts that will hydrolyze have a weak acid and a _____ or a

strong acid and a _____. A salt formed from a strong acid and a

weak base will form an _____. A salt formed from a strong base

and a weak acid will form a _____. Salts formed from weak acids

and bases or from strong acids and bases will not hydrolyze and form

_____.

**Buffered
Solutions**

Use with pages 666–667.

Explain *how a buffer works by completing the table below.*

The equation at equilibrium	$HF(aq) \rightleftharpoons H^+(aq) + F^-(aq)$	
Δ Condition	**Equilibrium Shift**	**The Process**
add acid	left	The H^+ ions react with F^- ions to form
add base	right	The OH^- ions react with H^+ ions to form water. This decreases the concentration of the H^+ ions so that
A greater of the buffering molecules and ions in the solution leads to a of the solution.		
A buffer has of an acid and its or a base with its		

Acids and Bases Chapter Wrap-Up

Now that you have read the chapter, review what you have learned; write out three key equations and relationships.

1. _____

   _____

2. _____

3. _____

Review

Use this checklist to help you study.

☐ Study your Science Notebook for this chapter.

☐ Study the definitions of vocabulary words.

☐ Review daily homework assignments.

☐ Reread the chapter and review the tables, graphs, and illustrations.

☐ Answer the Section Review questions at the end of each section.

☐ Look over the Study Guide at the end of the chapter.

REAL-WORLD CONNECTION

Suppose you are on the bench for your school's soccer team when one of the players comes out of the game with a cramp. A teammate suggests that she start breathing into a paper bag to recover sooner. Explain whether or not this is good advice.

_____

_____

_____

Redox Reactions

Before You Read

Review Vocabulary *Define the following terms.*

electronegativity _____
_____
_____

chemical reactions _____
_____
_____

Chapter 7 **Compare and contrast** *monatomic ions and polyatomic ions.*

_____
_____
_____

Chapter 9 **List** *five types of chemical reactions.*

1. _____

2. _____

3. _____

4. _____

5. _____

Redox Reactions

Section 1 Oxidation and Reduction

Main Idea —————— **Details** ———————————————————

Skim *Section 1 of your text. Write three questions that come to mind from reading the headings and the illustration captions.*

1. _____

2. _____

3. _____

New Vocabulary   *Use your text to define each term.*

oxidation-reduction reaction   _____

redox reaction   _____

oxidation   _____

reduction   _____

oxidizing agent   _____

reducing agent   _____

Electron Transfer and Redox Reactions

Use with pages 680–682.

Describe *redox reactions by completing the statement below. Use Figure 1 in your text as reference.*

A redox reaction consists of two complimentary processes. Oxidation results in a _____ and an increased _____ .

Reduction results in a _____ and a _____

oxidation number.

Section 1 Oxidation and Reduction (continued)

(Main Idea) ———   **(Details)** ——————————————————————

Oxidizing and Reducing Agents

Use with page 683.

Compare and contrast *an oxidizing agent and a reducing agent.*

_____

_____

_____

_____

_____

Identify Oxidation–Reduction Reactions

Use with Example Problem 1, page 685.

Summarize *Fill in the blanks to help you take notes while you read Example Problem 1.*

Problem

Write the equation for the redox reaction:

Identify what is _____ and what is _____ in the redox reaction of aluminum and iron. Identify the _____ and the _____.

1. **Analyze the Problem**

   Known: _____

   Unknown: _____

2. **Solve for the Unknown**

   Al becomes Al^{3+} and _____ electrons.

   Fe^{3+} becomes Fe and gains _____ electrons.

3. **Evaluate the Answer**

   Aluminum _____ electrons and is _____. It is the _____ agent. Iron _____ electrons and is _____. It is the _____ agent.

Section 1 Oxidation and Reduction (continued)

Main Idea	Details

Determining Oxidation Numbers

Use with page 686.

Describe *the rules for determining oxidation numbers by completing these statements.*

1. The oxidation number of an uncombined atom is _____.

2. The oxidation number of a monatomic ion is equal to
_____.

3. The oxidation number of the more electronegative atom in a molecule or a complex ion is the same as _____
_____.

4. The oxidation number of fluorine, the most electronegative element, when it is bonded to another element is _____.

5. The oxidation number of oxygen in compounds is _____, except in peroxides where it is _____. The oxidation number of oxygen when it bonds to fluorine is _____.

6. The oxidation number of hydrogen in most of its compounds is _____.

7. The oxidation numbers of the metal atom in the compounds formed by the metals of groups 1 and 2 and aluminum in group 13 are
_____, respectively. These oxidation numbers are equal to
_____.

8. The sum of the oxidation numbers in a neutral compound is _____.

9. The sum of the oxidation numbers of the atoms in a polyatomic ion is equal to _____.

Oxidation Numbers in Redox Reactions

Use with page 688.

Describe *the redox reaction for the equation listed below. Use the example on page 688 of your text to complete the table, then label the oxidation numbers of the elements in the equation and indicate the change in each.*

$2Al + Fe_2O_3 \rightarrow 2Fe + Al_2O_3$

Element	Oxidation Number	Rule
Al		
Fe in Fe_2O_3		
O in Fe_2O_3		
Fe		
Al in Al_2O_3		
O in Al_2O_3		

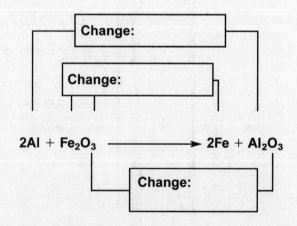

Redox Reactions

Section 2 Balancing Redox Equations

Main Idea ———— **Details** ————————————————————

Scan *Section 2 of your text, using the checklist below as a guide.*

- Read all section titles.
- Read all boldfaced words.
- Read all formulas.
- Look at all figures and read the captions.
- Think about what you already know about redox reactions.

Write *three facts you discovered about balancing redox reactions.*

1. _____

2. _____

3. _____

New Vocabulary

Use your text to define this term.

oxidation-number method _____

_____

_____

species _____

half-reaction _____

_____

The Oxidation-Number Method

Use with page 689.

Sequence *the steps for balancing redox reactions by the oxidation-number method.*

_____ Identify the atoms that are oxidized and the atoms that are reduced.

_____ Assign oxidation numbers to all atoms in the equation.

_____ Make the change in oxidation numbers equal in magnitude by adjusting coefficients in the equation.

_____ If necessary, use the conventional method to balance the remainder of the equation.

_____ Determine the change in oxidation number for the atoms that are oxidized and for the atoms that are reduced.

Section 2 Balancing Redox Equations (continued)

(Main Idea) ———— **(Details)** —————————————————

The Oxidation-Number Method

Use with Example Problem 3, page 690.

Summarize *Fill in the blanks to help you take notes while you read Example Problem 3.*

Problem

Balance the _____ equation for the _____ that produces

_____.

$$Cu + HNO_3 \rightarrow Cu(NO_3)_2 + NO_2 + H_2O$$

1. **Analyze the Problem**

Known:

The formulas for the reactants and _____; the rules for

determining _____; and the fact that the increase in

the oxidation number of the _____ must equal the

_____ of the reduced atoms.

Unknown: _____

2. **Solve for the Unknown**

Step 1 Assign oxidation numbers to all the atoms in the equation.

_____

$$Cu + H\,N\,O_3 \rightarrow Cu(N\,O_3)_2 + N\,O_2 + H_2\,O$$

Step 2 Identify which atoms are oxidized (using thin arrows) and which are reduced (using thick arrows).

_____

$$Cu + H\,N\,O_3 \rightarrow Cu\,(N\,O_3)_2 + N\,O_2 + H_2\,O$$

Step 3 Determine the change in oxidation number for the atoms that are oxidized and for the atoms that are reduced. Complete the following tables.

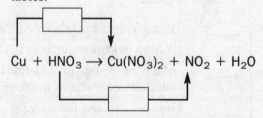

Step 4 To make the net changes in oxidation number have the same magnitude, HNO_3 on the left and NO_2 on the right must be multiplied by _____.

Section 2 Balancing Redox Equations (continued)

Main Idea	Details

Step 5 Increase the coefficient of HNO_3 from 2 to _____ to balance the nitrogen atoms in the products. Add a coefficient of _____ to H_2O to balance the number of hydrogen atoms on the left.

3. Evaluate the Answer

The number of atoms of each element is _____ on both sides of the equation. No subscripts have been _____ .

Balancing Net Ionic Redox Equations

Use with page 691.

Describe *the difference in the way each of the equations for the oxidation of copper by nitric acid are written.*

$Cu(s) + 4HNO_3(aq) \rightarrow Cu(NO_3)_2(aq) + 2NO_2(g) + 2H_2O(l)$

$Cu(s) + 4H^+(aq) + 4NO_3^-(aq) \rightarrow Cu^{2+}(aq) + 2NO_3^-(aq) + 2NO_2(g) + 2H_2O(l)$

_____

_____

_____

Balance a Net Ionic Redox Equation

Use with Example Problem 4, page 692.

Solve *Read Example Problem 4 in your text.*

You Try It

Problem

Balance the net ionic redox equation for the reaction between the perchlorate ion and the iodide ion in acid solution.

$ClO_3^-(aq) + I^-(aq) \rightarrow Cl^-(aq) + I_2(s)$ (in acid solution)

1. Analyze the Problem

Known: _____

_____

Unknown: _____

Section 2 Balancing Redox Equations (continued)

⟨Main Idea⟩ —— ⟨Details⟩ ————————————————————

2. **Solve for the Unknown**

Step 1 Assign oxidation numbers to all the atoms in the equation.

_____

$Cl\,O_3^-$ (aq) $+ I^-$(aq) $\rightarrow Cl^-$(aq) $+ I_2$(s) (in acid solution)

Step 2 Identify which atoms are oxidized (using thin arrows) and which are reduced (using thick arrows).

_____

$Cl\,O_3^-$ (aq) $+ I^-$(aq) $\rightarrow Cl^-$(aq) $+ I_2$(s) (in acid solution)

Step 3 Determine the change in oxidation number for the atoms that are oxidized and for the atoms that are reduced. Complete the following tables.

ClO_3^- (aq) $+ I^-$(aq) $\rightarrow Cl^-$(aq) $+ I_2$(s) (in acid solution)

Step 4 To make the net changes in oxidation number have the same magnitude, place the appropriate coefficients in front of the formulas in the equation.

ClO_3^- (aq) $+ 6I^-$(aq) $\rightarrow Cl^-$(aq) $+ 3I_2$(s) (in acid solution)

Step 5 Write an equation that adds enough hydrogen ions and water molecules to balance the oxygen atoms on both sides.

_____

3. **Evaluate the Answer**

The number of atoms of each element is _____ on both sides of the equation. The net charge on the right _____ the net charge on the left. No subscripts have been _____.

Section 2 Balancing Redox Equations (continued)

Main Idea	Details

Balancing Redox Equations Using Half-Reactions

Use with pages 693–694.

Identify *the number of species in each reaction. Then, show the oxidation half-reaction and the reduction half-reaction for each equation.*

Reaction	No. of Species	Half-Reaction	
		Oxidation	Reduction
$4Fe + 3O_2 \rightarrow 2Fe_2O_3$			
$4Fe + 3Cl_2 \rightarrow 2Fe_2Cl_3$			

Sequence *the steps for balancing by half-reactions.*

_____ Adjust the coefficients so that the number of electrons lost in oxidation equals the number of electrons gained in reduction.

_____ Write the net ionic equation for the reaction, omitting spectator ions.

_____ Add the balanced half-reactions and return spectator ions.

_____ Write the oxidation and reduction half-reactions for the net ionic equation.

_____ Balance the atoms and charges in each half-reaction.

Section 2 Balancing Redox Equations (continued)

Main Idea ——————— **Details** ———————————————

Balance a Redox Equation by Using Half-Reactions

Use with Example Problem 5, page 695 .

Summarize *Fill in the blanks to help you take notes while you read Example Problem 5.*

Problem

Balance the redox equation for the _____ of permanganate and sulfur dioxide when sulfur dioxide _____ is bubbled into an _____ solution of _____ .

$KMnO_4(aq) + SO_2(g) \rightarrow MnSO_4(aq) + K_2SO_4(aq)$

1. **Analyze the problem**

   Known: _____

   _____

   Unknown: _____

   _____

2. **Solve for the Unknown**

   Step 1: Write the net ionic equation for the reaction:

   _____

   Step 2: Using rule number 5, the oxidation number for Mn in MnO_4^- is

   _____ . Using rule number 2, the oxidation number for Mn^{2+} is

   _____ . The reduction half-reaction is _____ .

   Step 3(a): Balance the atoms and charges in the half-reaction.

   _____ .

Section 2 Balancing Redox Equations (continued)

Main Idea ———— **Details** —————————————————————

Step 3(b): The _____ ions are readily available and can be used to balance the charge in half-reactions in acid solutions. The number of H+ ions added to the right side of the oxidation half-reaction is _____. The number of H+ ions added to the left side of the reduction half-reaction is _____.

Write the oxidation half-reaction: _____.

Write the reduction half-reaction: _____.

Step 4: The number of electrons lost in oxidation is _____. The number of electrons gained in reduction is _____. The least common multiple of these numbers is _____. To balance the half-reactions, the atoms in the oxidation half-reaction must be multiplied by _____ and the atoms in the reduction half-reaction must be multiplied by _____. The oxidation half-reaction is now

_____

The reduction half-reaction is now

_____

Step 5 After adding the balanced half-reactions, write the redox reaction equation:

_____

Cancel or reduce like terms on both sides of the equation, then write the simplified equation:

_____

Return spectator ions _____ and restore the state descriptions.

_____

3. **Evaluate the Answer**

The number of _____ for each element is _____ on both sides of the equation and none of the subscripts have been changed.

Redox Reactions Chapter Wrap-Up

After reading this chapter, summarize the processes that occur in a redox reaction.

_____

_____

_____

_____

_____

Review

Use this checklist to help you study.

☐ Study your Science Notebook for this chapter.

☐ Study the definitions of vocabulary words.

☐ Review daily homework assignments.

☐ Reread the chapter and review the tables, graphs, and illustrations.

☐ Answer the Section Review questions at the end of each section.

☐ Look over the Study Guide at the end of the chapter.

REAL-WORLD CONNECTION

Photosynthesis is an example of a series of naturally occurring redox reactions. In this context, discuss the importance of redox reactions to life on Earth.

_____

_____

_____

_____

Electrochemistry

Before You Read

Review Vocabulary

Define the following terms.

energy _____

chemical potential energy _____

spontaneous process _____

oxidation _____

reduction _____

half-reaction _____

_____

_____

_____

_____

Chapter 9

Identify three types of reactions.

1. _____

2. _____

3. _____

Organize *the following elements from least active to most active.*

Refer to the activity series in Figure 13.

aluminum, copper, calcium, gold, rubidium, iron, lead, potassium

_____

Electrochemistry
Section 1 Voltaic Cells

Main Idea — **Details**

Skim *Section 1 of your text. Focus on the headings, subheadings, boldfaced words, and the main ideas. Summarize three main ideas of this section.*

1. _____

2. _____

3. _____

New Vocabulary   *Use your text to define each term.*

salt bridge _____

electrochemical cell _____

_____

voltaic cell _____

_____

half-cell _____

_____

anode _____

cathode _____

reduction potential _____

standard hydrogen
electrode _____

_____

Academic Vocabulary   *Define the following terms.*

correspond _____

Section 1 Voltaic Cells (continued)

Main Idea	Details

Redox in Electrochemistry

Use with pages 708–709.

Explain *the branch of chemistry called electrochemistry.*

_____

_____

Write *the half-reactions of copper and zinc as indicated in Figure 2.*

_____ (reduction half-reaction: electrons _____)

_____ (oxidation half-reaction: electrons _____)

Explain *how an electrochemical cell uses a redox reaction.*

_____

_____

Chemistry of Voltaic Cells

Use with pages 710–711.

Complete *each of the following statements.*

1. The electrode where oxidation takes place is called the _____.

2. The electrode where reduction takes place is called the _____.

3. An object's potential energy is _____.

4. In electrochemistry, _____ is a measure of

   the amount of _____ that can be generated from a

   _____ to do work.

Sequence *the steps of the electrochemical process that occur in a zinc-copper voltaic cell. The first one has been done for you.*

_____ To complete the circuit, both positive and negative ions move through the salt bridge. The two half-reactions can be summed to show the overall cell reaction.

_____ The electrons flow from the zinc strip and pass through the external circuit to the copper strip.

___1___ Electrons are produced in the oxidation half-cell according to this half-reaction: $Zn(s) \rightarrow Zn^{2+}(aq) + 2e^-$.

_____ Electrons enter the reduction half-cell where the following half-reaction occurs: $Cu^{2+}(aq) + 2e^- \rightarrow Cu(s)$.

Section 1 Voltaic Cells (continued)

⟨Main Idea⟩ ———— **⟨Details⟩** ————————————————————

Calculating Electrochemical Cell Potential

Use with pages 711–714.

Describe *reduction potential in relation to an electrode.*

_____

_____

_____

Analyze *Table 1. Some of the E^0 values are positive, some are negative. Explain the difference.*

_____

_____

_____

Write *the abbreviated E^0 and half-reaction for each of the following:*

Element	Half-Reaction	E^0 (V)
Li		
Au		
PbSO$_4$		
Na		

Calculate a Cell Potential

Use with Example Problem 1, pages 715.

Summarize *Fill the blanks to help you take notes while you read Example Problem 1.*

Problem

Calculate the overall cell reaction and the standard potential for the half-cells of a voltaic cell.

$$I_2(s) + 2e^- \rightarrow 2I^-(aq)$$

$$Fe^{2+}(aq) + 2e^- \rightarrow Fe(s)$$

1. Analyze the Problem

List the known and the unknown.

Known: Standard reduction potentials for the half-cells

_____

Unknown: _____

Section 1 Voltaic Cells (continued)

Main Idea ———— **Details** ————————————————————

2. **Solve for the Unknown**

Find the standard reduction potentials for half-reactions.

$E^0_{I_2|I^-}$ = _____

$E^0_{Fe^{2+}|Fe}$ = _____

Rewrite the half–reactions in the correct direction.

reduction half–cell reaction: _____

oxidation half–cell reaction: _____

overall cell reaction: _____

Balance the reaction if necessary.

_____

Calculate the cells standard potential.

$E^0_{cell} = E^0_{reduction} - E^0_{oxidation}$

$E^0_{cell} = +0.536 \text{ V} -$ _____

$E^0_{cell} = +$ _____

Write the reaction using cell notation.

_____

3. **Evaluate the Answer**

The answer seems reasonable given the _____ of

the _____ that comprise it.

Using Standard Reduction Potentials

Use with page 716.

Write *the steps for the process of predicting whether any proposed redox reaction will occur spontaneously.*

1. _____

2. _____

3. _____

_____

4. _____

5. _____

Electrochemistry
Section 2 Batteries

Main Idea — **Details** —————————————————————

Skim *Section 2 of your text. Write three questions that come to mind after reading the headings and the illustration captions.*

1. _____

2. _____

3. _____

New Vocabulary

Use your text to define each term.

battery

_____

dry cell

_____

primary battery

_____

secondary battery

_____

fuel cell

_____

corrosion

_____

galvanization

_____

Section 2 Batteries (continued)

⟨**Main Idea**⟩ ————— ⟨**Details**⟩ —————————————————————————

Dry Cells

Use with pages 718–720.

Write *the oxidation half-reaction for the dry cell of the most commonly used voltaic cell.*

_____

List *the paste and cathode type for each of the following batteries. So-called dry cell batteries contain different moist pastes in which the cathode half-reaction takes place.*

Zinc-carbon battery

Paste   _____

Cathode type   _____

Alkaline battery

Paste   _____

Cathode type   _____

Mercury battery

Paste   _____

Cathode type   _____

Compare and contrast *primary and secondary batteries.*

_____

_____

_____

_____

Explain *how NiCad batteries, often found in cordless tools and phones, are recharged.*

_____

_____

_____

Section 2 Batteries (continued)

⟨**Main Idea**⟩ ———— ⟨**Details**⟩ ———————————————————

**Lead-Acid
Storage Battery**

Use with pages 720–721.

Explain *how the following overall reaction of lead-acid batteries is different from traditional redox reactions.*

$Pb(s) + PbO_2(s) + 4H^+ (aq) + 2SO_4^{2-} (aq) \rightarrow 2PbSO_4(s) + 2H_2O(l)$

_____

_____

_____

_____

Lithium Batteries

Use with pages 721–722.

List *two reasons that scientists and engineers have focused a lot of attention on the element lithium to make batteries.*

1. _____

2. _____

Describe *two applications of lightweight lithium batteries.*

_____

_____

Fuel Cells

Use with pages 722–723.

Explain *the makeup of a fuel cell by completing the following paragraph and accompanying reactions.*

In a fuel cell, each electrode _____

that allows contact between the _____

_____. The walls of the chamber also contain _____,

such as powdered platinum or palladium, which _____.

oxidation half-reaction: _____

reduction half-reaction: _____

overall cell reaction: _____

The overall cell reaction is the same as the equation for the _____

_____.

List *three reasons why PEMs are used instead of a liquid electrode.*

1. _____

2. _____

3. _____

Section 2 Batteries (continued)

Main Idea ——— **Details**

Corrosion

Use with pages 724–727.

Compare *rusting of metal to redox reactions in voltaic cells.*

_____

_____

_____

Draw *and label the parts of the corrosion reaction in Figure 15. Be sure to identify the anode and cathode.*

Explain *why rusting is a slow process. List a way that it might be sped up in certain areas.*

_____

_____

_____

_____

Explain *the two ways galvanizing helps prevent corrosion.*

1. _____

_____

_____

2. _____

_____

_____

Electrochemistry
Section 3 Electrolysis

Main Idea ——— **Details** —————————————————————

Scan *Section 3 of your text. Use the checklist below as a guide.*

- Read all section titles.

- Read all boldfaced words.

- Read all formulas.

- Look at all figures and read the captions.

- Think about what you already know about electrolysis.

Write *three facts you discovered about electrolysis as you scanned the section.*

1. _____

2. _____

3. _____

New Vocabulary *Use your text to define each term.*

electrolysis _____

_____

electrolytic cell _____

_____

Section 3 Electrolysis (continued)

Main Idea	Details

Reversing Redox Reactions

Use with page 728.

Describe *how it is possible to reverse a spontaneous redox reaction in an electrochemical cell.*

_____

_____

_____

Applications of Electrolysis

Use with pages 729–732.

Compare *the reactions involved in sodium chloride to those in the electrolysis of brine.*

_____

_____

_____

Explain *the importance of electrolysis in the purification of metals.*

_____

_____

_____

_____

_____

_____

Electrochemistry Chapter Wrap-Up

After reading this chapter, list three important facts you have learned about electrochemistry.

1. _____

_____

2. _____

_____

3. _____

Review

Use this checklist to help you study.

☐ Study your Science Notebook for this chapter.

☐ Study the definitions of vocabulary words.

☐ Review daily homework assignments.

☐ Reread the chapter and review the tables, graphs, and illustrations.

☐ Review the Section Assessment questions at the end of each section.

☐ Look over the Study Guide at the end of the chapter.

REAL-WORLD CONNECTION

Describe how electrochemistry is involved in producing energy in batteries.

_____

_____

_____

Hydrocarbons

Before You Read

Review Vocabulary	**Define** *each term.*
covalent bond	_____
	_____
	_____
Lewis structure	_____
	_____
Chapter 8	**Draw** *the Lewis structure for NH₃.*
Chapter 12	**Compare and contrast** *melting and boiling.*
	_____
	_____
	_____

Hydrocarbons
Section 1 Introduction to Hydrocarbons

Main Idea ———— **Details** ————————————————

Scan *Section 1 of your text. Use the checklist below as a guide.*

- Read all section titles.
- Read all boldfaced words.
- Look at all pictures and read the captions.
- Think about what you already know about this subject.

Write *three facts you discovered about hydrocarbons.*

1. _____

2. _____

_____

3. _____

New Vocabulary *Use your text to define each term.*

organic compound _____

_____

hydrocarbon _____

_____

saturated hydrocarbon _____

unsaturated hydrocarbon _____

_____

fractional distillation _____

cracking _____

_____

Section 1 Introduction to Hydrocarbons (continued)

⟨ **Main Idea** ⟩ ——— ⟨ **Details** ⟩ ——————————————————————

**Organic
Compounds**

Use with pages 744–745.

Explain *the evolution of the contemporary understanding of the term
organic compound.*

> In the early nineteenth century, chemists referred to the variety of
> carbon compounds produced by living things as **organic compounds.**

↓

> _____
>
>
>
>

↓

> Today the term **organic compound** is applied to all carbon-containing
> compounds with the primary exceptions of carbon oxides, carbides, and
> carbonates, which are considered inorganic.

Explain *why many compounds contain carbon by completing the
following statements.*

Carbon's _____ allows it to make four covalent bonds.

In organic compounds, carbon atoms bond to _____ or other

elements near carbon on the periodic table. Carbon atoms also bond to

_____ and can form long _____.

Hydrocarbons

Use with pages 745–746.

Label *the web below with the correct name for each model of methane.*

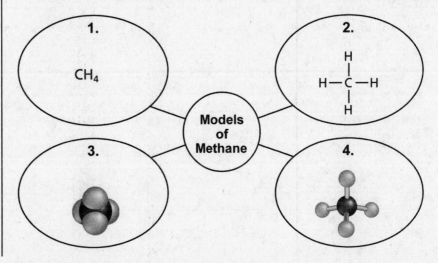

Section 1 Introduction to Hydrocarbons (continued)

Main Idea	Details

Multiple Carbon-Carbon Bonds

Use with page 746.

Organize *the outline below.*

I. Ways that carbon atoms bond to each other

    A. _____

        1. share _____

        2. also called _____

    B. _____

        1. share _____

        2. also called _____

    C. _____

        1. share _____

        2. also called _____

Draw *models of each carbon-carbon bond and label them appropriately.*

Single Covalent Bond	Double Covalent Bond	Triple Covalent Bond

Section 1 Introduction to Hydrocarbons (continued)

| Main Idea | Details |

Refining Hydrocarbons

Use with pages 747–748.

Identify *natural sources of hydrocarbons by completing the following statements.*

The main natural source of hydrocarbons is _____, a complex

mixture containing more than a thousand _____.

Petroleum is more useful to humans when _____

_____, called _____. Separation is carried out by

_____

_____, a process called fractional distillation.

Sequence *the process of fractional distillation.*

_____ Vapors travel up through the column.

_____ Temperature is controlled to remain near 400° at the bottom of the fractionating tower.

_____ Hydrocarbons with fewer carbon atoms remain in the vapor phase until they reach regions of cooler temperatures farther up the column.

_____ Hydrocarbons with more carbon atoms condense closer to the bottom or the tower and are drawn off.

_____ Petroleum boils and gradually moves toward the top.

Match the names of these two processes with their definitions.

1. fractional distillation   **2.** cracking

_____ is done to break the larger molecules of petroleum

components into smaller molecules.

_____ separates petroleum into simpler components.

Rating Gasoline

Use with pages 748–749.

Explain *why branched-chain alkanes make better gasolines than straight-chain hydrocarbons.*

_____

_____

_____

_____

Hydrocarbons

Section 2 Alkanes

Main Idea ———— **Details** ————————————————

Skim *Section 2 of your text. Write three questions that come to mind from reading the headings and the illustration captions.*

1. _____

2. _____

3. _____

_____

New Vocabulary *Use your text to define each term.*

alkane _____

homologous series _____

parent chain _____

substituent group _____

_____

cyclic hydrocarbon _____

cycloalkane _____

Academic Vocabulary *Define the following term.*

substitute _____

Section 2 Alkanes (continued)

Main Idea ——— **Details** ————————————————————

Straight-Chain Alkanes

Use with pages 750–751.

Compare and contrast *the models in the table below.*

Type of Model	Description of Model
1. Molecular formula	
2. Structural formula	
3. Space-filling model	
4. Ball-and-stick model	

Describe *straight-chain alkanes by completing the following sentences.*

The first four compounds in the straight-chain series of alkanes are

_____. The names of all alkanes

end in _____. Because the first four alkanes were named before there was a

complete understanding of alkane structures, their names do not have

_____ as do the alkanes with _____ in a

chain. Chemists use _____ to save space.

Explain *the structural formula of the following hydrocarbons. The first has been done for you.*

1. Methane is formed from one atom of carbon and four atoms of hydrogen.

2. Butane is formed _____.

3. Octane is formed _____

_____.

4. Decane is formed _____

_____.

Analyze *how the function of a homologous series is evidenced in the condensed structural formula of nonane.*

_____

_____

_____

Section 2 Alkanes (continued)

⟨ **Main Idea** ⟩ ———— ⟨ **Details** ⟩ ————————————————

Branched-Chain Alkanes	**Compare** *three characteristics of butane and isobutane.*
Use with pages 752–753.	_____
	_____
	_____
	_____
	_____
	_____
	_____

Naming Branched-Chain Alkanes	**Describe** *naming branched-chain alkanes.*
Use with page 753.	

> A straight-chain and a branched-chain alkane can have the same molecular formula.

> **PRINCIPLE**
> Therefore, the name of an organic compound also must describe

> **NAMING PROCESS**
> Branched-chain alkanes are viewed as consisting of a

> **NAMING, PART 1**
> The longest continuous chain of carbon atoms is called
> .

> **NAMING, PART 2**
> All side branches are called   because they appear to substitute for a hydrogen atom in the straight chain.

> **NAMING, PART 3**
> Each alkane-based substituent group branching from the parent chain is named

Section 2 Alkanes (continued)

⟨Main Idea⟩ ──── ⟨Details⟩ ────────────────────────────

Cycloalkanes

Use with pages 755–756.

Organize *the concept web below.*

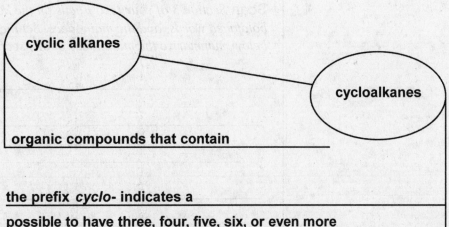

organic compounds that contain _____

the prefix *cyclo-* indicates a _____

possible to have three, four, five, six, or even more ____

represented by condensed, skeletal, _____

can have _____ groups

Properties of Alkanes

Use with pages 757–758.

Classify *the properties of alkanes into categories.*

General Properties (3)	Physical Properties (4)	Chemical Properties (2)

Hydrocarbons
Section 3 Alkenes and Alkynes

Main Idea ——   **Details** ——————————————————

Scan *Section 3 of your text. Focus on the headings, subheadings, boldfaced words, and the main ideas. Set the book aside and, in the space below, summarize the main ideas of this section.*

_____

_____

_____

_____

_____

New Vocabulary   *Use your text to define each term.*

alkene   _____

_____

alkyne   _____

_____

Section 3 Alkenes and Alkynes (continued)

Main Idea ——— **Details** ——————————————————————

Alkenes

Use with pages 759–760.

Identify *five facts about alkenes as discussed in your text.*

1. _____

   _____

2. _____

3. _____

   _____

4. _____

5. _____

Sequence *the factors involved in naming an alkene with four or more carbons in the chain using the web below and number the steps.*

1. Change the
   –*ane* ending of the
   corresponding alkane
   to

2. Specify the
   location of the

Naming Alkenes

3. Number the
   carbons in the parent
   chain starting

4. Use only that
   number

Naming Branched-Chain Alkenes

Use with Example Problem 3, page 761.

Summarize *Use the following to help you take notes as you read Example Problem 3 in your text.*

Problem

Name the following alkene. $CH_3CH=CHCHCH_2CHCH_3$

$\quad\quad\quad\quad\quad\quad\quad\quad\quad\quad\quad\quad | \quad\quad |$

$\quad\quad\quad\quad\quad\quad\quad\quad\quad\quad\quad CH_3 \quad CH_3$

Section 3 Alkenes and Alkynes (continued)

(Main Idea)———— (Details)————————————————————————

1. **Analyze the Problem**

   You are given a branch-chained alkene that contains one double bond and two alkyl groups. Follow the IUPAC rules to name the organic compound.

2. **Solve for the Unknown**

   a. The longest continuous carbon chain that includes the double bond contains _____ carbons. The _____ alkane is heptane, but the name is changed to _____ because a double bond is present. *Write the 2-heptene parent chain.*

   b. and c. Number the chain to give the lowest number to the double bond and name each substituent.

   d. Determine how many of each substituent is present, and assign the correct prefix to represent that number. Then, include the position numbers to get the complete prefix.

   e. The names of substituents _____

   _____.

   f. Apply the complete prefix to the name of the parent alkene chain. Use commas to separate numbers and hyphens between numbers and words. Write the name _____.

3. **Evaluate the Answer**

   The longest carbon chain includes the _____, and the position of the double bond has the _____.
   Correct prefixes and alkyl-group names _____.

Alkynes

Use with pages 762–764.

Compare and contrast *alkenes and alkynes.*

_____

_____

Hydrocarbons
Section 4 Hydrocarbon Isomers

Main Idea ——— **Details**

Skim *Section 4 of your text. Write two questions that come to mind from reading the headings and the illustration captions.*

1. _____

2. _____

New Vocabulary   *Use your text to define each term.*

isomer   _____
_____

structural isomer   _____
_____

stereoisomer   _____
_____

geometric isomer   _____
_____

chirality   _____

asymmetric carbon   _____
_____

optical isomer   _____
_____
_____

optical rotation   _____
_____
_____

Section 4 Hydrocarbon Isomers (continued)

Main Idea ———— **Details** ———————————————————————

Structural Isomers	**Organize** *the outline below.*
Use with page 765.	I. _____: Two or more compounds that have the same molecular formula but different molecular structures.
	A. Two types of isomers
	1. Structural isomers
	a. _____
	b. _____
	_____
	i. Examples include _____
	_____
Stereoisomers	2. Stereoisomers
Use with page 766.	a. _____
	_____
	i. _____
	ii. _____
	_____
	b. _____
	i. Result from different arrangements of groups around a double bond
	1. Possible _____ with *trans*-fatty acids.
	2. The _____ seem not to be as harmful.

Chiralty	**Describe** *chirality by completing the flow chart below.*
Use with page 767.	

Chirality occurs whenever → a compound contains an → which has ___ or ___ attached to it.

These isomers are called ← The molecules are ← The four groups can be

Section 4 Hydrocarbon Isomers (continued)

⟨Main Idea⟩ ———— ⟨Details⟩ ————————————————————————

Optical Isomers

Use with pages 768–769.

Identify *the types of isomers shown below. Which pair are optical isomers?*

D-glyceraldehyde　　　**L-glyceraldehyde**

```
      CHO                    CHO
       |                      |
  H — C — OH           HO — C — H
       |                      |
     CH₂OH                  CH₂OH
```

ethanol　　　　　　　**methoxymethane**

```
    H   H                  H        H
    |   |     H            |        |
H — C = C — O         H — C — O — C — H
    |   |                  |        |
    H   H                  H        H
```

trans-1,2-dichloroethene　　**cis-1,2-dichloroethene**

```
  H       Cl              H       H
    \\    /                 \\    /
     C = C                   C = C
    /     \\                 /     \\
  Cl       H              Cl       Cl
```

COMPARE

Explain what a pair of shoes and crystals of the organic compound tartaric acid have in common.

_____

_____

_____

Hydrocarbons
Section 5 Aromatic Hydrocarbons

Main Idea ——— **Details** ——————————————————————

Skim *Section 5 of your text. Focus on the headings, subheadings, boldfaced words, and the main ideas. Summarize the main ideas of this section.*

_____

_____

_____

_____

_____

New Vocabulary   *Use your text to define each term.*

aromatic compound

_____

_____

aliphatic compound

_____

_____

Section 5 Aromatic Hydrocarbons (continued)

Main Idea ———— **Details** ——————————————————

Aromatic Compounds

Use with pages 771–774.

Classify *the properties of aromatic and aliphatic compounds.*

	Structural Characteristics	Reactivity
Aromatic Compounds		
Aliphatic Compounds		

Model *Draw a model of a fused ring system.*

Explain *how substituted benzene rings are numbered.*

_____

_____

Number *the substituted benzene ring in the structure below, then name the structure.*

CH$_3$
CH$_2$CH$_3$
CH$_3$

Hydrocarbons Chapter Wrap-Up

Now that you have read the chapter, review what you have learned; list the types of models used to represent chemical compounds and name the different categories of hydrocarbons.

Hydrocarbons: **Models:**

Alkanes

_____   _____

_____   _____

_____   _____

Alkenes

_____

Alkynes

Isomers

_____   _____

_____   _____

Aromatic _____   Aliphatic _____

Review   *Use this checklist to help you study.*

☐ Study your Science Notebook for this chapter.

☐ Study the definitions of vocabulary words.

☐ Review daily homework assignments.

☐ Reread the chapter and review the tables, graphs, and illustrations.

☐ Review the Section Assessment questions at the end of each section.

☐ Look over the Study Guide at the end of the chapter.

SUMMARIZE

Explain how hydrocarbons have contributed to space exploration.

_____

_____

_____

Substituted Hydrocarbons and Their Reactions

Before You Read

Review Vocabulary

Define the following terms.

periodic table

_____

_____

compound

_____

_____

_____

halogens

_____

_____

chemical bond

_____

_____

_____

catalyst

_____

_____

Chapter 21 **Compare and contrast** *stereoisomers with structural isomers.*

_____

_____

_____

_____

_____

_____

Substituted Hydrocarbons and Their Reactions

Section 1 Alkyl Halides and Aryl Halides

Main Idea ──────────── **Details** ────────────────────────────

Skim *Section 1 of your text. Write three questions that come to mind from reading the headings and the illustration captions.*

1. _____

2. _____

3. _____

New Vocabulary   *Use your text to define each term.*

functional group _____

_____

halocarbon _____

_____

alkyl halide _____

_____

aryl halide _____

_____

substitution reaction _____

_____

halogenation _____

_____

Section 1 Alkyl Halides and Aryl Halides (continued)

Main Idea ——— **Details** ————————————————————

Functional Groups

Use with pages 786–787.

Describe *how a functional group can be helpful in determining how a molecule reacts.*

_____

_____

_____

_____

Identify *the meaning of each of the following symbols for functional groups.*

* represents _____

R and R' represent _____

_____

Organize *information about organic compounds and their functional groups by completing the table below.*

Compound Type	General Formula	Functional Group
Halocarbon		Halogen
	R-OH	
		Ether
	R-NH$_2$	
Aldehyde		
		Carbonyl
		Carbonyl
		Ester
		Amide

Section 1 Alkyl Halides and Aryl Halides (continued)

Main Idea	Details

Organic Compounds Containing Halogens

Use with pages 787–788.

Compare and contrast *alkyl halides and aryl halides.*

_____

_____

_____

_____

Naming Halocarbons

Use with page 788.

Describe *how to name halocarbons by completing the following paragraph.*

Organic molecules containing functional groups are given IUPAC names

based on their _____. For the alkyl halides, a

prefix indicates which _____ is present. The prefixes are formed by

_____

Properties and Uses of Halocarbons

Use with page 789.

Examine *Table 2 on page 789. Write three observations you make regarding the compounds listed in the table.*

1. _____

2. _____

3. _____

Substitution Reactions

Use with page 790.

Sequence *the steps needed to add Cl_2 to ethane to create chloroethane. Use the reaction in Table 3 at the top of the page in your text as a reference.*

1. _____

2. _____

3. _____

4. _____

Create *another substitution reaction using Br_2 and methane. Label molecules in each part of the reaction.*

_____

_____

Substituted Hydrocarbons and Their Reactions
Section 2 Alcohols, Ethers, and Amines

Main Idea ———— **Details** ———————————————————————————

Scan *Section 2 of your text. Use the checklist below as a guide.*

- Read all section titles.

- Read all boldfaced words.

- Read all formulas.

- Look at all figures and read the captions.

- Think about what you already know about alcohols, ethers, and amines.

Write *three facts you discovered about alcohols as you scanned the section.*

1. _____

2. _____

3. _____

New Vocabulary   *Use your text to define each term.*

hydroxyl group   _____

alcohol   _____

_____

denatured alcohol   _____

_____

Define *the following terms and write the general formula for each term.*

ether   _____

_____

amine   _____

_____

Academic Vocabulary   *Define the following term.*

bond   _____

Section 2 Alcohols, Ethers, and Amines (continued)

(Main Idea)	(Details)
Alcohols *Use with pages 792–793.*	**Describe** *alcohol by completing the following sentence.* Because they readily form hydrogen bonds, alcohols have _____ boiling points and _____ water solubility than other organic compounds. **Write** *the general formula for alcohol:* _____ **Draw** structures for the following molecules. 1-butanol 2-butanol
Ethers *Use with page 794.*	**Describe** *ethers by completing the following sentence.* Ethers are similar to _____ as they are compounds in which oxygen is bonded to _____. Ethers are different from alcohols because the oxygen atom bonds with _____ carbon atoms. Ethers are much less _____ in water than alcohol because they have no _____ to donate to a hydrogen bond.

Section 2 Alcohols, Ethers, and Amines (continued)

(Main Idea)	(Details)
	Write *the general formula for ethers:*
	_____
	Draw *a structure for the following molecule.*
	ethyl ether
Amines	**Complete** *the following sentence.*
Use with page 795.	Amines contain _____ atoms bonded to carbon atoms in
	_____ chains or _____ rings. Amines are responsible for
	many of the _____ associated with decay.
	Write *the general formula for amines:*
	_____
	Draw *a structure for the following molecule.*
	ethylamine

Substituted Hydrocarbons and Their Reactions
Section 3 Carbonyl Compounds

(Main Idea) ———— (Details) ————————————————

Skim *Section 3 of your text. Write two questions that come to mind from reading the headings and the illustration captions.*

1. _____

2. _____

(New Vocabulary) *Use your text to define each term.*

ketone _____

_____

carboxylic acid _____

carboxyl group _____

ester _____

_____

amide _____

_____

Define *the following terms and write the general formula of each.*

carbonyl group _____

_____

aldehyde _____

_____

_____

condensation reaction _____

_____

_____

Section 3 Carbonyl Compounds (continued)

Main Idea	Details
Organic Compounds Containing the Carbonyl Group *Use with pages 796–800.*	**Identify** *five important classes of organic compounds containing or made from carbonyl compounds:* a. _____ b. _____ c. _____ d. _____ e. _____ **Describe** *the common structure of aldehydes and ketones.* _____ _____ _____
Carboxylic Acids *Use with page 798.*	**Draw** *a molecule of a carboxylic acid.*
Organic Compounds Derived From Carboxylic Acids *Use with pages 799–800.*	**Describe** *organic compounds that are derived from carboxylic acids by completing the following paragraph.* Several classes of organic compound have structures in which the _____ of a carboxylic acid is replaced by _____ or _____. The two most common types are _____.

Section 3 Carbonyl Compounds (continued)

Main Idea ———— **Details** ————————————

Condensation Reactions

Use with page 801.

Sequence *the steps for a condensation reaction.*

_____ A small molecule, such as water, is lost.

_____ Two organic molecules combine.

_____ A more complex molecule is formed.

Complete *the following condensation reaction.*

RCOOH + R′OH → _____

Summarize

Identify *the functional group that corresponds to each of the following:*

a. *-ine* at the end of each halogen name to *–o* _____

b. adding *–amine* as the suffix _____

c. *-ane* of the parent alkane to *–ol* _____

d. replacing *–e* ending with *–amide* _____

e. *–e* at the end of the name to *–al* _____

f. *–ane* of the parent alkane to *–anolic* acid _____

g. *-ic* acid ending replaced by *–ate* _____

h. *–e* end of the alkane replaced by *–one* _____

Substituted Hydrocarbons and Their Reactions
Section 4 Other Reactions of Organic Compounds

Main Idea ——— **Details** ———————————————————————

Scan *Section 4 of your text. Use the checklist below as a guide.*

- Read all section titles.
- Read all boldfaced words.
- Read all formulas.
- Look at all figures and read the captions.

Write *three facts you discovered about organic reactions.*

1. _____

2. _____

3. _____

New Vocabulary *Use your text to define each term.*

elimination reaction _____

_____

_____

dehydrogenation reaction _____

dehydration reaction _____

_____

addition reaction _____

_____

hydration reaction _____

_____

hydrogenation reaction _____

_____

Chemistry: Matter and Change

Science Notebook

Section 4 Other Reactions of Organic Compounds (continued)

(Main Idea)

Classifying Reactions of Organic Substances

Use with pages 802–805.

(Details)

List *what needs to happen for chemical reactions of organic substances to occur. Include when and why a catalyst might be needed.*

1._____

2._____

3._____

_____

_____

Review *the section and give an example formula for each of the following reaction types.*

addition reaction

_____

_____

hydration reaction

_____

_____

dehydrogenation reaction

_____

_____

dehydration reaction

_____

_____

hydrogenation reaction

_____

_____

elimination reaction

_____

_____

Section 4 Other Reactions of Organic Compounds (continued)

⟨**Main Idea**⟩ ———— ⟨**Details**⟩ ————————————————————

Oxidation-Reduction Reactions

Use with pages 806–807.

Describe *oxidation-reduction reactions by completing the following statements.*

Many _____ compounds can be converted to other compounds by _____ and _____ reactions. _____ is the loss of _____. A substance is oxidized when it gains _____ or loses _____. Reduction is the _____ of electrons. A substance is reduced when it loses _____ or gains _____.

Predicting Products of Organic Reactions

Use with pages 807–808.

Write *the generic equation representing an addition reaction between an alkene and an alkyl halide.*

_____

Substitute *the structure for cyclopentene and the formula for hydrogen bromide. From the equation, you can see that:*

A _____ and a _____ add across the _____ to form an _____.

Draw *the formula for the likely product.*

Substituted Hydrocarbons and Their Reactions
Section 5 Polymers

Main Idea ——— **Details** ————————————————————————

Scan *Section 5 of your text. Use the checklist below as a guide.*

- Read all section titles.
- Read all boldfaced words.
- Read all tables and formulas.
- Look at all figures and read the captions.

Write *three facts you discovered about polymers.*

1. _____

2. _____

3. _____

New Vocabulary *Use your text to define each term.*

polymer _____

monomer _____

polymerization reaction _____

addition polymerization _____

_____

condensation polymerization _____

_____

thermoplastic _____

_____

thermosetting _____

_____

Section 5 Polymers (continued)

Main Idea	Details

The Age of Polymers

Use with page 809.

Identify *three common polymers described in the text. Include their uses.*

1. _____

2. _____

3. _____

Reactions Used to Make Polymers

Use with pages 810–811.

Identify *the monomers or polymers.*

Monomer (s)	Polymer (s)
Ethylene	
	Nylon 6, 6
Urethane	

Compare and contrast *condensation polymerization with addition polymerization by placing the terms below into the Venn diagram.*

- all atoms present in final product

- small by-product, usually water

- involves the bonding of monomers

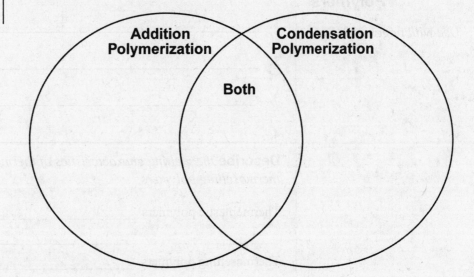

Section 5 Polymers (continued)

Main Idea	Details

Common Polymers

Use with page 812.

Identify *the common polymer. Use Table 14 in your text as a reference.*

Use	Polymers
Foam furniture cushions	
A planter	
Nonstick cookware	
Food wrap	
Windows	
Clothing	
Carpet	
Water pipes	
Beverage containers	

Properties and Recycling of Polymers

Use with pages 813–814.

Identify *four reasons that many different polymers are widely used in manufacturing.*

1. _____

2. _____

3. _____

4. _____

Describe *the melting characteristics of thermoplastic polymers and thermosetting polymers.*

Thermoplastic polymers _____

_____

Thermosetting polymers _____

_____

Section 5 Polymers (continued)

Main Idea ———— **Details** —————————————————————————

Discuss *recycling by completing the following paragraph.*

Americans are not efficient at recycling their plastics. Currently, only

_____ of plastic waste is recycled. This low rate of _____

_____ is due in part to the _____.

Plastics must be _____ according to _____, which

is _____ and _____. The plastic industry has

_____ that indicate the _____ of

each plastic product to make the process easier on individuals.

Describe *what the code of recycling polymers does. Give an example of the code from the textbook.*

_____

_____

_____

REAL-WORLD CONNECTION

Describe some common polymers that you use every day.

_____

_____

_____

_____

_____

_____

Substituted Hydrocarbons and Their Reactions   Chapter Wrap-Up

After reading this chapter, list three things you have learned about substituted hydrocarbons and their reactions.

1. _____

2. _____

3. _____

Review

Use this checklist to help you study.

☐ Study your Science Notebook for this chapter.

☐ Study the definitions of vocabulary words.

☐ Review daily homework assignments.

☐ Reread the chapter and review the tables, graphs, and illustrations.

☐ Answer the Section Review questions at the end of each section.

☐ Look over the Study Guide at the end of the chapter.

REAL-WORLD CONNECTION

Examine a picture of spooled threads. Explain how monomers might be a part of the process that produces these spooled polymer threads.

_____

_____

_____

The Chemistry of Life

Before You Read

Review Vocabulary *Define the following terms.*

hydrogen bond _____

_____

_____

isomers _____

_____

functional group _____

_____

polymers _____

_____

Chapter 12 **Illustrate** *the hydrogen bonding between water molecules.*

Chapter 22 **Illustrate** *the molecules for fluoroethane and 1,2 difluoropropane.*

The Chemistry of Life

Section 1 Proteins

Main Idea ——— **Details** ——————————————————

Skim *Section 1 of your text. Focus on the headings, subheadings, boldfaced words, and the main ideas. Summarize three main ideas of this section.*

_____

_____

_____

New Vocabulary

Use your text to define each term.

protein _____

amino acid _____

_____

peptide bond _____

peptide _____

denaturation _____

enzyme _____

substrate _____

active site _____

Section 1 Proteins (continued)

⟨Main Idea⟩ ——— **⟨Details⟩** ————————————————————

Protein Structure

Use with pages 826–829.

Draw *and label a general amino acid with a variable side chain, an amino group, and a carboxyl group.*

Describe *the structure of a dipeptide and its functional units.*

Rewrite *each of the following statements, making each true.*

To function properly, each protein must be flat.

_____

_____

A dipeptide consists of an amino acid with two side chains.

_____

_____

Complete *the following statements about peptide bonds.*

When a peptide bond is formed, _____ is released in the process.

This type of reaction is known as a _____ reaction.

Section 1 Proteins (continued)

Main Idea ———

Details ————————————————

Identify *the peptide bond between the following amino acids.*

$$
\begin{array}{c}
\text{H} \quad\ \text{R}_1 \qquad \text{H} \ \text{R}_2 \\
\backslash \ \ | \qquad\quad | \ \ | \\
\text{N—C—C—N—C—C—OH} \\
/ \quad\ | \ \ \| \qquad | \ \ \| \\
\text{H} \quad\ \text{H} \ \text{O} \qquad \text{H} \ \text{O}
\end{array}
$$

Explain *why Gly-Phe is a different molecule than the Phe-Gly.*

_____

_____

Describe *three changes in environment that will uncoil or otherwise denature a protein.*

1. _____

2. _____

3. _____

The Many Functions of Proteins

Use with pages 829–831.

Draw *an enzyme/substrate complex with the enzyme and substrates labeled.*

Section 1 Proteins (continued)

⟨**Main Idea**⟩ —— ⟨**Details**⟩ ————————————————————

Describe *how the following functions affect living organisms by giving an example from your text.*

Enzymes: _____

_____

Transport proteins: _____

_____

Structural proteins: _____

_____

Hormones: _____

_____

_____

Review *the statements below and revise to make them correct.*

1. Substrates bind to an enzyme site.

    _____

2. An active site changes shape a great deal to accommodate the substrate.

    _____

    _____

3. An enzyme-substrate complex changes the enzyme, and it becomes part of the new molecule.

    _____

    _____

The Chemistry of Life

Section 2 Carbohydrates

⟨ **Main Idea** ⟩ ———— ⟨ **Details** ⟩ ——————————————

Scan *Section 2 of your text. Use the checklist below as a guide.*

- Read all section titles.
- Read all boldfaced words.
- Look at all figures and read the captions.
- Think about what you already know about carbohydrates.

Write *three facts you discovered about carbohydrates as you scanned the section.*

1. _____

2. _____

3. _____

⟨ **New Vocabulary** ⟩   *Use your text to define each term.*

carbohydrate _____

_____

monosaccharide _____

disaccharide _____

polysaccharide _____

_____

Section 2 Carbohydrates (continued)

(Main Idea) —— (Details) ————————————

Kinds of Carbohydrates

Use with pages 832–834.

Draw *the cyclic and open-chain structures of the monosaccharide glucose.*

Explain *how the monosaccharides glucose and galactose differ. Discuss why they would not react the same way in nature.*

_____

_____

_____

_____

Describe *the structure and composition of the following types of carbohydrates by completing this table.*

Carbohydrate	Example	Structure and composition
starch		
cellulose		
glycogen		
glucose		

The Chemistry of Life

Section 3 Lipids

⊂ **Main Idea** ⊃ ——— ⊂ **Details** ⊃ ————————————————————

Scan *Section 3 of your text. Use the checklist below as a guide.*

• Read all section titles.

• Read all boldfaced words.

• Look at all figures and read the captions.

• Think about what you already know about lipids.

Write *three facts you discovered about lipids as you scanned the section.*

1. _____

2. _____

3. _____

⊂ **New Vocabulary** ⊃   *Use your text to define each term.*

lipid _____

fatty acid _____

triglyceride _____

saponification _____

_____

phospholipid _____

_____

wax _____

steroid _____

Section 3 Lipids (continued)

Main Idea ————— **Details** ————————————————————————————————

What is a lipid?

Use with pages 835–839.

Describe *how a lipid differs from a protein or carbohydrate.*

_____

_____

Compare and contrast *saturated and unsaturated fatty acids. Give an example of each.*

_____

_____

_____

_____

_____

_____

_____

_____

Explain *the reactions that form triglycerides. Give the type of reaction as well as the substrates.*

_____

_____

_____

_____

_____

_____

Section 3 Lipids (continued)

<table>
<tr><td>**Main Idea**</td><td>**Details**</td></tr>
</table>

Describe *how waxes are made and what their specific properties include.*

_____

_____

_____

Describe *a lipid that is not composed of fatty acid chains. Give an example.*

_____

_____

_____

SYNTHESIZE

List the important functions for each of the following types of lipids.

triglycerides _____

phospholipids _____

waxes _____

steroids _____

_____

The Chemistry of Life
Section 4 Nucleic Acids

Main Idea ———— **Details** —————————————————————

Skim *Section 4 of your text. Write three questions that come to mind from reading the headings and the illustration captions.*

1. _____

   _____

2. _____

3. _____

New Vocabulary   *Use your text to define each term.*

nucleic acid _____

   _____

   _____

nucleotide _____

Section 4 Nucleic Acids (continued)

⟨ **Main Idea** ⟩ —————— ⟨ **Details** ⟩ ——————————————————————

Structure of Nucleic Acids

Use with page 840.

Draw *a diagram of a nucleotide. Label all of the parts: sugar, phosphate group, and nitrogen-containing base.*

DNA: The Double Helix

Use with pages 841–842.

Write *a statement that differentiates between nucleotides and nucleic acids.*

_____

Sequence *the events of DNA replication. The first one has been done for you.*

_____ Hydrogen bonds form between new nitrogen bases and the existing strand.

_____ Two nucleotide strands unzip.

_____ Nitrogen bases pair adenine with thymine, cytosine with guanine.

___1___ An enzyme breaks the hydrogen bonds between the nitrogen bases.

_____ The nucleotide strands separate to expose the nitrogen bases.

_____ Free nucleotides are delivered by enzymes from the surrounding environment.

Predict *the complimentary base pairing given the following strand of nucleotides.*

A T C T A T C G G A T A T C T G

_____

Section 4 Nucleic Acids (continued)

Main Idea —————— **Details** ————————————————

RNA

Use with page 843.

Identify *differences in DNA and RNA.*

	DNA	RNA
Sugar		
Nitrogen Bases		
Function		
Form of strand		

State *whether you would find each of the following in DNA, RNA, both, or neither. Explain your answer.*

A-A	
A-T	
C-G	
G-A	
A-U	
U-A	

REAL-WORLD CONNECTION

Suppose you are an assistant to a forensic scientist who has found an unknown sample of DNA at a crime scene. Upon analysis, he finds it contains 22% thymine molecules. A DNA sample that contains 40% guanine is obtained from a suspect who is brought in. You ask for the suspect's release. Explain your reasoning based on the bonding patterns of DNA nucleotides.

_____

_____

_____

_____

_____

_____

The Chemistry of Life
Section 5 Metabolism

Main Idea ——— **Details** ——————————————————————

Skim *Section 5 of your text. Focus on the headings, subheadings, boldfaced words, and the main ideas. List three main ideas of this section.*

1. _____

2. _____

3. _____

New Vocabulary

Use your text to define each term.

metabolism _____

catabolism _____

_____

anabolism _____

_____

ATP _____

_____

photosynthesis _____

_____

cellular respiration _____

_____

fermentation _____

Academic Vocabulary

Define the following term.

conceptualize _____

Section 5 Metabolism (continued)

⟨Main Idea⟩ ——— ⟨Details⟩ ————————————————————

Anabolism and Catabolism

Use with pages 844–845.

Explain *the relationship between metabolism, catabolism, and anabolism.*

_____

_____

_____

Explain *how ATP is able to store and release energy in the cells of organisms.*

_____

_____

_____

_____

Photosynthesis

Use with page 846.

Write *the reaction of photosynthesis. Label the individual molecules.*

Identify *the redox process that occurs during photosynthesis.*

_____

_____

Section 5 Metabolism (continued)

⟨**Main Idea**⟩ —— ⟨**Details**⟩ ————————————————

Cellular Respiration

Use with page 846.

Write *the reaction of cellular respiration. Be sure to label the individual molecules.*

Identify *the redox process that occurs during cellular respiration.*

_____

_____

_____

Summarize *the relationship between photosynthesis and cellular respiration.*

_____

_____

_____

_____

_____

_____

_____

_____

_____

_____

Section 5 Metabolism (continued)

Main Idea ——— **Details** —————————————————————

Fermentation

Use with pages 847–848.

Compare and contrast *alcoholic fermentation and lactic acid fermentation.*

_____

_____

_____

_____

_____

_____

_____

_____

_____

REAL-WORLD CONNECTION

Explain why the redox processes that occur during photosynthesis are vital to life.

_____

_____

_____

_____

_____

_____

The Chemistry of Life Chapter Wrap-up

Now that you have read the chapter, review what you have learned. Write out the major concepts from the chapter.

_____

_____

_____

_____

Review

Use this checklist to help you study.

☐ Study your Science Notebook for this chapter.

☐ Study the definitions of vocabulary words.

☐ Review daily homework assignments.

☐ Reread the chapter and review the tables, graphs, and illustrations.

☐ Answer the Section Review questions at the end of each section.

☐ Look over the Study Guide at the end of the chapter.

REAL-WORLD CONNECTION

Explain why someone with a liver disorder might be advised to avoid overexertion.

_____

_____

_____

_____

Nuclear Chemistry

Before You Read

Review Vocabulary   *Define the following terms.*

Isotopes   _____

_____

_____

nuclear reaction   _____

electron   _____

Chapter 4   *Use your text to review the following concepts which will help you understand this chapter.*

List *the three kinds of subatomic particles.*

1. _____

2. _____

3. _____

Draw and label *a nuclear model of the atom.*

_____

_____

Identify *the primary factor in determining an atom's stability.*

_____

_____

Nuclear Chemistry
Section 1 Nuclear Radiation

Main Idea ——— **Details** ——————————————————

Skim *Section 1 of your text. Write three questions that come to mind from reading the headings and the illustration captions.*

1. _____

2. _____

3. _____

New Vocabulary   *Use your text to define each term.*

radioisotope   _____

X-ray   _____

penetrating power   _____

Section 1 Nuclear Radiation (continued)

Main Idea ———— **Details** ————————————————————————

Comparison of Chemical and Nuclear Reactions

Use with page 860.

Contrast *chemical and nuclear reactions.*

Chemical Reactions	Nuclear Reactions
bonds are and formed	nuclei emit
atoms are , though they may be rearranged	are converted into atoms of another element
reaction rate by pressure, temperature, concentration, and catalyst	reaction rate by pressure, temperature, concentration, or catalyst
involve only valence	may involve protons,
energy changes	energy changes

The Discovery of Radioactivity

Use with pages 860–861.

Summarize *the discovery of radioactivity. Review the dates on the timeline below. Use your text to fill in the important achievements in radioactive research on those dates.*

1895 Roentgen _____

_____

1895 Becquerel _____

_____

1898 The Curies _____

_____

1903 The Curies and Becquerel _____

1911 Marie Curie _____

Section 1 Nuclear Radiation (continued)

Main Idea ———— **Details** ————————————————

Types of Radiation

Use with pages 861–864.

Identify *the common type of radiation signified by each symbol.*

α _____

β _____

γ _____

Differentiate *between each of the subatomic radiation particles mentioned in the chapter.*

Radiation Type	Charge	Mass	Relative Penetrating Power
Alpha			
Beta			
Gamma			

Describe *what happens when a radioactive nucleus emits an alpha particle.*

_____

_____

Describe *beta particles by completing the following statements.*

A beta particle is a very fast-moving _____. To represent its

insignificant mass, beta particles have a superscript of _____. A

subscript of –1 denotes the _____ charge of beta particles. Beta

particles have greater _____ than alpha particles.

Describe *what the subscript and superscript of zero tell you about gamma particles.*

_____

_____

Nuclear Chemistry
Section 2 Radioactive Decay

Main Idea	Details

Scan *Section 2, using the checklist below as a guide.*

- Read all section titles and boldfaced words.
- Study all tables, graphs, and figures.

Write *two facts you discovered about transmutation.*

1. _____

2. _____

New Vocabulary

Use your text to define each term.

transmutation _____

nucleon _____

strong nuclear force _____

_____

band of stability _____

positron emission _____

_____

positron _____

electron capture _____

_____

radioactive decay
series _____

_____

half-life _____

_____

radiochemical dating _____

_____

Section 2 Radioactive Decay (continued)

Main Idea	**Details**

Nuclear Stability

Use with pages 865–866.

Contrast *the properties of isotopes by imagining two eggs as models. One isotope would be created using hard-boiled eggs as building blocks, the other using raw eggs as building blocks. Explain which model would be more stable, and which would be more typical of known isotopes.*

_____

_____

_____

Summarize *how the strong nuclear force helps to keep protons in a nucleus.*

_____

_____

Describe *the neutron-to-proton (n/p) ratio in nuclear stability.*

The number of protons compared to the number of _____ in a

ratio identifies the nuclear ratio. To some degree, the _____ of a

nucleus can be correlated with its _____ ratio. As

atomic number _____, more _____ are needed to

balance the _____ forces. Plotting the number of

neutrons versus the number of _____ for all stable nuclei

illustrates the _____.

Types of Radioactive Decay

Use with page 866–868.

Analyze *the relative stability of radioisotopes. Use Figure 7 as a guide.*

1. a radioisotope with too many neutrons relative to its protons

_____

2. a radioactive isotope _____

3. a nucleus with more than 83 protons _____

4. a nucleus with a high atomic number and a neutron-to-proton

   ratio of 1:5:1. _____

Section 2 Radioactive Decay (continued)

Main Idea	Details

Writing and Balancing Nuclear Equations

Use with page 869.

Compare *positron emission with electron capture.*

Positron emission is _____ that involves the emission

of a _____ (particle with the same mass as an electron but

opposite charge) from a nucleus. During this process, a _____ in

the nucleus is converted into a neutron and a positron, and then the

_____ is emitted.

Electron capture is _____ that decreases the number of

_____ in unstable nuclei lying below the _____. This occurs

when the nucleus of an atom draws in a surrounding _____,

usually from the lowest energy level. The captured electron combines with a

_____ to form a _____.

Contrast *balanced chemical equations with balanced nuclear equations.*

Balanced chemical equations conserve _____

_____.

Balanced nuclear equations conserve _____

_____.

Balancing a Nuclear Equation

Use with Example Problem 1, page 869.

Solve *Read Example Problem 1 in your text.*

You Try It

Problem

Write a balanced nuclear equation for the alpha decay of uranium-238 ($^{238}_{92}U$).

1. **Analyze the Problem**

   Known: _____

   decay type: _____

   Unknown: _____

Section 2 Radioactive Decay (continued)

Main Idea	Details

2. Solve for the Unknown

Using each particle's mass number, make sure the mass number is conserved on each side of the reaction arrow.

Mass number: $238 = X + $ _____ $X = 238 - 4$

Mass number of $X = $ _____

Using each particle's atomic number, make sure the atomic number is conserved on each side of the reaction arrow.

Atomic number: $92 = $ _____ $X = 92 - $ _____

Atomic number of $X = $ _____

Use the periodic table to identify the unknown element.

_____

Write the balanced nuclear equation.

_____

Radioactive Series

Use with page 870.

Describe *a radioactive decay series by completing the following paragraph.*

A radioactive decay series is a series of _____ that begins with a(n) _____ nucleus and ends in the formation of a stable _____. Both alpha decay and _____ are involved in the process.

Section 2 Radioactive Decay (continued)

⌐Main Idea⌐ ——— ⌐Details⌐ ————————————————

Radioactive Decay Rates

Use with pages 870–871.

Describe *how Ernest Rutherford's early experiments in inducing nuclear reactions led to modern particle accelerators.*

Rutherford discovered that particles must move at extremely _____

_____ to overcome electrostatic _____ and affect a target

nucleus. Scientists have built on this to develop methods to accelerate

particles to extreme speed using _____ and _____ fields.

Particle accelerators use conventional and _____ magnets to

force particles to move at high speeds.

Explain *why some naturally occurring radioactive substances still remain on Earth.*

_____

_____

REAL-WORLD CONNECTION

Suppose you want to join an after-school club. Two clubs interest you. In the photography club, there are a lot of members, but only a few who are truly interested (or proactive) about the topic. Most members just seem to have joined to be involved in an activity (or are neutral). The chemistry club, on the other hand, has fewer members, but there seems to be an equal number of truly interested (proactive) students as there are students without a lot of interest (neutrals). If human interactions followed the same laws as radioisotopes, explain which group would be more stable over the school year.

_____

_____

_____

_____

Section 2 Chemistry and Matter (continued)

Main Idea ——— Details ——————————————————

Calculating the Amount of Remaining Isotope

Use with Example Problem 2, page 872.

Solve *Read Example Problem 2 in your text.*

You Try It

Problem

Determine the amount of an original sample of 2.0 grams of thorium-234 after 49 days. The half-life of thorium-234 is 24.5 days.

1. **Analyze the Problem**

   Known: Unknown:

   Initial amount = _____   Amount remaining = ? g

   Elapsed time (t) = _____

   Half-life (T) = _____

2. **Solve for the Unknown**

   Number of half-lives (n) = Elapsed time/Half-life

   n = 49/24.5 = _____

   Amount remaining = _____

   Amount remaining = _____

   Amount remaining = _____

   Amount remaining = _____

3. **Evaluate the Answer**

   After 49 days, _____ half-lives of thorium-234 have

   elapsed. The number of half-lives is equivalent to (1/2)(1/2) or

   _____ . The answer, _____ is equal to

   _____ the original quantity.

Radiochemical Dating

Use with pages 873–874.

Write *the balanced nuclear equation for carbon dating.*

_____

Name _____   Date _____

Nuclear Chemistry
Section 3 Nuclear Reactions

Main Idea ——— **Details** ——————————————————————

Skim *Section 3 of your text. Write three questions that come to mind from reading the headings and the illustration captions.*

1. _____

2. _____

3. _____

New Vocabulary   *Use your text to define each term.*

induced transmutation   _____

transuranium elements   _____

_____

_____

mass defect   _____

nuclear fission   _____

critical mass   _____

breeder reactor   _____

nuclear fusion   _____

thermonuclear reaction   _____

Academic Vocabulary   *Define the following term.*

generate   _____

Section 3 Nuclear Reactions (continued)

(Main Idea) ——   (Details)

Induced Transmutation

Use with pages 875–876.

Sequence the steps in Rutherford's induced transformation of nitrogen-14 into oxygen.

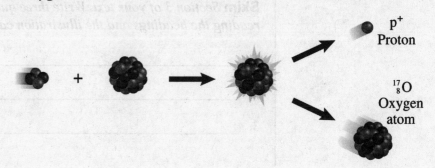

1 $_{2}^{4}$He bombarding alpha particle _____ +

2 _____ →

3 _____ →

4 _____ *and* →

5 p⁺ proton _____

Section 3 Nuclear Reactions (continued)

⟨**Main Idea**⟩ ———— ⟨**Details**⟩ ————————————————————————

Nuclear Reactions and Energy

Use with pages 877–878.

Write *Einstein's equation. Be sure to include the measurement units.*

_____

Identify *the three things you need to know to calculate mass defects.*

a. _____

b. _____

c. _____

Nuclear Fission

Use with pages 878–880.

Organize *the steps in a nuclear fission reaction involving uranium.*

1. A neutron _____

2. The uranium _____

3. The nucleus _____

_____

Explain *why a fissionable material must have sufficient mass before a sustained reaction can take place.*

_____

_____

Explain *why a fissionable material must not have an excess of mass.*

_____

_____

Section 3 Nuclear Reactions (continued)

⟨Main Idea⟩ —— **⟨Details⟩** ————————————

Nuclear Reactors

Use with pages 880–882.

Describe *how a nuclear reactor creates energy. Include how the environment is protected from nuclear waste.*

Nuclear fission produces _____.

A common fuel is _____.

_____. A neutron-emitting source

_____ and control rods absorb virtually all of the

_____ produced in the reaction. Heat from a reaction is used

to power _____ which produce electrical power.

Nuclear Fusion

Use with pages 883–884.

Describe *nuclear fusion by completing the following paragraph.*

Nuclear fusion is the combining of atomic _____. Nuclear fusion reactions

are capable of _____. The most

common fusion reaction is the _____. Because of the energy requirements,

fusion reactions are also known as _____.

Explain *why fusion reaction is not yet a practical source of everyday energy.*

_____

_____

_____

_____

REAL-WORLD CONNECTION

Create a metaphor from everyday life that will show the difference between nuclear fission and nuclear fusion.

Nuclear fission requires   _____

Nuclear fusion requires   _____

Fission is like: _____

Fusion is like: _____

Nuclear Chemistry
Section 4 Applications and Effects of Nuclear Reaction

Main Idea —— **Details** ————————————————

Scan *Section 4, using the checklist below as a guide.*

- Read all section titles.
- Read all boldfaced words.
- Read all tables and graphs.
- Look at all pictures and read the captions.
- Think about what you already know about radioactive decay.

Write *three questions you have about nuclear radiation.*

1. _____

2. _____

3. _____

New Vocabulary *Use your text to define each term.*

ionizing radiation _____

_____

radiotracer _____

_____

Section 4 Application and Effects of Nuclear Reactions (continued)

Main Idea ———— **Details**

Detecting Radioactivity

Use with pages 885–886.

List *and describe three methods of detecting radiation.*

1. _____

   _____

2. _____

   _____

3. _____

   _____

Uses of Radiation

Use with pages 886–888.

Describe *how a radiotracer works.*

A radiotracer is a _____ that emits _____

and is used to signal the presence of _____ or specific

substance. The fact that all of an element's isotopes have the same

_____ makes the use of radioisotopes possible.

Discuss *a common radiotracer that is used in medicine.*

Iodine-131 is commonly used to detect _____ associated with

the _____. A doctor will give the patient a drink containing a

small amount of iodine-131. The iodine-containing _____ is

then used to monitor the function of the thyroid gland.

Section 4 Application and Effects of Nuclear Reactions (continued)

⊂Main Idea⊃ ——— ⊂Details⊃ ————————————————————

Biological Effects of Radiation

Use with pages 888–890.

Identify *three factors that affect the possible damage to the body caused by ionizing radiation discussed in the textbook.*

1. _____

2. _____

3. _____

Discuss *genetic and somatic damage caused by ionizing radiation.*

Somatic damage affects _____

_____

Genetic damage can affect _____

_____

_____

_____

REAL-WORLD CONNECTION

Create a warning label that will identify the dangers of a radioactive material to users.

_____

_____

_____

Nuclear Chemistry Chapter Wrap-Up

After reading this chapter, list three important facts you have learned about nuclear chemistry.

1. _____

2. _____

3. _____

Review

Use this checklist to help you study.

☐ Study your Science Notebook for this chapter.

☐ Study the definitions of vocabulary words.

☐ Review daily homework assignments.

☐ Reread the chapter and review the tables, graphs, and illustrations.

☐ Answer the Section Review questions at the end of each section.

☐ Look over the Study Guide at the end of the chapter.

REAL-WORLD CONNECTION

Imagine you are watching a program on radiation with a friend. Your friend is afraid of all radiation. Explain to your friend some of the common useful applications of radiation.

_____

_____

_____

_____

TRENDS IN ECOLOGICAL RESEARCH FOR THE 1980s

NATO CONFERENCE SERIES

I	Ecology
II	Systems Science
III	Human Factors
IV	Marine Sciences
V	Air–Sea Interactions
VI	Materials Science

I ECOLOGY

TRENDS IN ECOLOGICAL RESEARCH FOR THE 1980s

Edited by

June H. Cooley

and

Frank B. Golley

International Association for Ecology (INTECOL)
Institute of Ecology
University of Georgia
Athens, Georgia

Published in cooperation with NATO Scientific Affairs Division

PLENUM PRESS · NEW YORK AND LONDON

Library of Congress Cataloging in Publication Data

Main entry under title:

Trends in ecological research for the 1980s.

(NATO conference series. I, Ecology; v. 7)
"Proceedings of a NATO ARW and INTECOL Workshop on the Future and Use of
Ecology after the Decade of the Environment, held April 7–9, 1983, at Louvain-la-
Neuve, Belgium"—CIP copyright p.
Bibliography: p.
Includes index.
1. Ecology—Research—Congresses. I. Cooley, June H. II. Golley, Frank B. III.
NATO ARW and INTECOL Workshop on the Future and Use of Ecology after the
Decade of the Environment (1983: Louvain-la-Neuve, Belgium). IV. North Atlantic
Treaty Organization. V. International Association for Ecology. VI. Series.
QH541.2.T74 1985   574.5′072   84-26316
ISBN 0-306-41889-4

Proceedings of a NATO ARW and INTECOL Workshop on the Future and Use of
Ecology after the Decade of the Environment, held April 7–9, 1983,
at Louvain-la-Neuve, Belgium

© 1984 Plenum Press, New York
A Division of Plenum Publishing Corporation
233 Spring Street, New York, N.Y. 10013

Printed in the United States of America

PREFACE

Is ecology at a crossroad? After three decades of rapid,
though somewhat anarchic development, many ecologists now are
beginning to ask this question. They have the feeling of no longer
belonging to a unified and mature scientific discipline. Many of
them claim to be mere empiricists, whereas others are proud to be
considered theoreticians. Each side has its own journals and holds
its own specialists' meetings, tending to disregard the achievements
of the other. The communication gap between the two schools is
quickly widening, to the detriment of both. To make things worse,
the word "ecology" now has a different meaning for the professional
biologists and the general public. Ecology is still considered as
a creditable (though rather "soft") scientific discipline by the
former, whereas it is perceived as a new, non-conformist political
philosophy by the latter.

Empirical ecologists are fundamentally naturalists who enjoy
the immense complexity of the natural world and devote their lifetimes
to the description of the many adaptive characteristics--morpholog-
ical, biological, or behavioral--of the hundreds of thousands of
species sharing the earth with us. They generally are ignorant of,
if not allergic to, the use of any mathematical representation of
living phenomena. They feel that ecological theory is rapidly
becoming a mathematical game that has lost any contact with the
"realities of life."

At the opposite extreme, theoretical ecologists contend that
if mathematics without natural history is indeed sterile, natural
history without mathematics is muddled--to paraphrase Maynard
Smith's famous sentence. They emphasize the importance of the
"hypothetico-deductive method" and advocate the use of inductive
models. Such models are not limited to reproducing the behavior of
a given system under various conditions, but can provide a real
insight into the organization and operation of the system itself.
Obviously, the ideal way to bridge the gap between theory and
practice would be to train new generations of research students to
become equally good field workers and clever mathematicians.
Unfortunately, this strategy has seldom proved feasible up to now.

Furthermore, the eco-activists and other believers in "pop ecology" are more concerned with political action than with intellectual ventures. They cling to some of the very simplistic articles of faith of many of our founding fathers, such as the "balance of nature," and are convinced a *priori* of the negative side effects of all modern human activities.

Such is the situation of ecology at the present time. Therefore, the offer of the Ecosciences Panel of the NATO Science Committee to discuss the future of the science was readily accepted as a unique opportunity by the International Association for Ecology (INTECOL). This is indeed the specific role of INTECOL within the International Council of Scientific Unions' family--to assist ecologists in solving their professional problems and to improve communication, not only among national societies but also among the major schools of thought within the discipline itself.

Previously, INTECOL's action had been limited to the organization of the first two international congresses of ecology (The Hague in 1974 and Jerusalem in 1978) and to the preparation of the one to be held in Syracuse, New York, in 1986. However, the new Executive Committee of our Association feels that INTECOL's role, under present circumstances, should not be limited only to the planning of such large meetings. We now intend to organize symposia and workshops on topical problems between international congresses, during which a limited number of carefully selected specialists can explore together new avenues of research and discuss the best ways and means to tackle them adequately. Short proceedings of these meetings will be published quickly in our *INTECOL Newsletter*, and the most important contributions will appear in full later, either in our *INTECOL Bulletin* or in book form.

The Louvain-la-Neuve workshop (7-9 April 1983) was the first of these small meetings to be held. Its objective was to look ahead and to focus discussions on the major ecological issues of tomorrow. Our group was purposefully made up of people of different generations and linguistic groups, belonging to the major schools of thought of Western Europe and North America. It included both empiricists and theoreticians, but most were convinced that an effort should be made to adopt a hybrid training and research strategy to get the best of both sides. The pleasant setting and accommodations allowed plenty of time to consider the views of all participants, and our lively discussions often continued late into the evenings. I hope that the revised versions of the papers presented at Louvain-la-Neuve, and published in the present volume, will convey to the reader a faithful image of our debates, and convince him or her that the future of scientific ecology is far from being as dark as it looks at first.

Francois Bourliere
President, INTECOL

CONTENTS

INTRODUCTION

Frank B. Golley

Institute of Ecology
University of Georgia
Athens, Georgia   30602 USA

It is tempting to think of science as a linear process of
discovery and development, but this type of model does little
justice to a subject such as ecology. In fact, the very idea of
progress is itself a topic generating reflection and argument.
Nevertheless, human societies expect scientists to contribute to
change in desired directions, and those scientific activities that
appear to be directly and linearly related to desirable social end
products receive special attention and support.

Ecology has just experienced such a popular approbation.
Concern for the human environment (i.e., for all those elements of
the natural world that are not directly controlled by man) has a
long history in this century, but during the past several decades
worldwide public concern reached an intensity and strength that
pushed ecology to a new level of popularity. The decade of the
1970s especially could be termed the "Decade of the Environment."
Following the United Nations Conference on the Human Environment,
Stockholm, 1972, the United Nations Environmental Program, UNESCO's
Man and Biosphere Program, the International Association for Ecology
(INTECOL), the Scientific Committee on Problems of the Environment
(SCOPE), and many other environmental initiatives at regional and
national levels were established.

This widespread public interest, also translated into other
international and national institutions, regulations, and treaties,
resulted in a demand for increased ecological information, which,
in turn, caused an increase in the number of environmental scientists.
University training programs expanded, with the development of
departments of ecology or environmental studies and research centers
and institutes. For example, in the United States, the number of

1

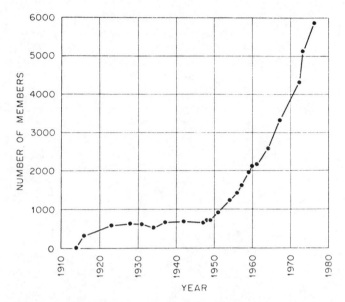

Fig. 1.   Growth in membership in the Ecological Society of America
          from its start in 1914 through 1976, the latest year for
          which data are available (Burgess 1977; reproduced by
          permission of the author).

ecologists increased exceptionally (Fig. 1).   Today, ecologists
occur worldwide in relatively large numbers (Table 1).

Of course, expansion cannot continue indefinitely.   The explosive
growth of the 1970s has given way to the more moderate developments
in the 1980s.   However, it is a mistake to interpret this transition
from the heady experiences of Earth Day to the more sober 1980s as
meaning loss of concern for the environment.   Public interest
remains high and nations are turning to the very difficult task of
solving environmental problems that are deeply embedded in institutions, cultures, and life styles.

To solve these problems requires high quality scientific
information; however, the edifice of ecological research and information worldwide is very complex and burdensome to understand and to
manage.   Even ecologists have difficulty in keeping the entire
field in mind, because research areas are so many and the literature
generated annually is so immense that no one can claim full currency
and expertise.   As an example, in 1982 about 10,000 titles in the
Biosis data base (which produces *Biological Abstracts*) were given a
primary designation as ecology, while about 25,000 more were second-
arily designated as ecology.   Although only part of this output is
cited in later research papers, even an obscure reference to a rare
insect in a local journal may provide the key needed to solve an
important environmental management problem.

Table 1.   Number of ecologists by country, derived from data on
           published papers classed as ecological in *Biological
           Abstracts*, 1981-1982.[a]

Algeria	3	Hungary	51	Poland	286
Argentina	27	Iceland	9	Portugal	12
Australia	699	India	510	Puerto Rico	1
Austria	73	Indonesia	12	Rumania	23
Bangladesh	28	Iran	11	St. Christopher	1
Barbados	4	Iraq	13	Saudi Arabia	14
Belgium	70	Ireland	37	Senegal	13
Bermuda	1	Israel	134	Singapore	1
Botzwana	1	Italy	212	South Africa	253
Brazil	96	Ivory Coast	9	SW Africa	3
Bulgaria	50	Jamaica	14	Spain	137
Burma	2	Japan	930	Sri Lanka	16
Burundi	1	Kenya	14	Sudan	2
Cameroon	1	Korea, South	51	Sweden	253
Canada	1,170	Kuwait	12	Switzerland	91
Chile	50	Madagascar	1	Syria	1
China	86	Malawi	1	Taiwan	18
Colombia	8	Malaysia	33	Tanzania	5
Costa Rica	17	Mauritius	1	Thailand	21
Cuba	6	Mexico	36	Trinidad-Tobago	1
Czechoslovakia	135	Monaco	4	Tunisia	4
Denmark	73	Nauru	1	Turkey	12
Dominica	2	Netherlands, The	255	Uganda	1
Ecuador	4	New Caledonia	2	USSR	1,512
Egypt	32	New Zealand	216	UK	1,145
Ethiopia	2	Niger	2	USA	5,917
Fiji	4	Nigeria	34	Upper Volta	3
Finland	147	Norway	185	Venezuela	30
France	465	Oman	3	Vietnam	1
Fr. Polynesia	5	Pakistan	23	Yemen	4
Germany, East	90	Panama	24	Yugoslavia	52
Germany, West	458	Papua-New Guinea	12	Zaire	1
Ghana	10	Paraguay	2	Zambia	1
Greece	18	Peru	1	Zimbabwe	16
Hong Kong	17	Philippines	17		
				Total	16,579

[a]The author acknowledges the generous contribution of BIOSIS in
providing the data for this table.

There is no simple answer to the problem of striking a proper balance between generality and specificity. The field expands to the limits of the resources available. Ecologists focus on limited topics and create subdivisions of ecology so they can efficiently interact through meetings and the literature. The processes of fission and convergence proceed in a natural way, while research administrations try to follow these processes and guide them toward specific objectives.

Clearly in this environment, the community of ecologists has an important role to play. Ecologists acting through their informal and formal organizations can identify promising areas of study and, through criticism, eliminate the less promising. Professional organizations, such as INTECOL, traditionally provide the setting for informal meetings, where these discussions can take place, and then publish the results for the benefit of the larger scientific community. These professional organizations act as a counterbalance to the pressures and controls of governments and industries, which attempt to channel research to those areas they judge most important.

As a contribution toward ecologists' thoughts on the opportunities for the profession of ecology, INTECOL convened a workshop to focus on the trends in ecological research for the 1980s--after the Decade of the Tropics. This publication is a result of that workshop. The papers printed here represent a wide range of topics, reflecting the different views, backgrounds, and experiences of those present. As arranged, they begin with plant physiological ecology, proceed through microbial, population, and community ecology, and then turn to ecosystems and theoretical ecology. The summary of the workshop deliberations, jointly formulated by the participants and published earlier with the abstracts of the papers in the *INTECOL Bulletin* (Cooley 1983), follows at the end of the volume. This summary is intended to emphasize the main areas of concern for ecological research in this decade as perceived by those present at the workshop.

LITERATURE CITED

Burgess, R.L. 1977. The Ecological Society of America: Historical
     data and some preliminary analyses. Pages 1-22 *in* F.N. Egerton
     and R.P. McIntosh, eds. History of American ecology. Arno
     Press, New York, NY USA.
Cooley, J.H., ed. 1983. Summary. Pages 47-49 *in* INTECOL workshop
     abstracts: Ecology in the 1980s. INTECOL Bull. 9.

PROGRESS AND PROMISE IN PLANT PHYSIOLOGICAL ECOLOGY

H. A. Mooney

Department of Biological Sciences
Stanford University
Stanford, California 94305 USA

ABSTRACT

The study of plant physiological ecology has experienced an explosive development in the past decade. Work has centered primarily on those mechanisms by which plants acquire and utilize resources, particularly carbon and water and, to a much lesser degree, nutrients. Techniques have been developed for assessing costs and benefits of various organic compounds and plant structures; however, problems remain to be solved in this area. Various plant "strategy" groupings have been functionally characterized, although further refinements and extentions to tropical plants are needed. Methods for unraveling past performances of plants have been developed and are being used increasingly. The effects of biological interactions in structuring plant communities recently have received greater attention, particularly plant-herbivore and plant-pollinator interactions, and will be studied more in the coming years. However, further studies of the mechanistic basis of plant-plant interactions are needed.

In the years ahead, plant physiological ecology will be a strong contributor to many applied research areas extending from agroecology to ecosystem management. Further, physiological ecology will continue to serve as an interface between a variety of disciplines, ranging from molecular biology to ecosystem ecology.

INTRODUCTION

During the past two decades, there has been impressive progress in the study of plant physiological ecology. Here I review some of the highlights of this progress and make some observations on

5

likely focal points of research in the decade of the 1980s. I make
no attempt to be all inclusive in this assessment because of limited
space and because of the wide dimensions of the field.

The pace of development in physiological ecology can be measured
in small part by the appearance of textbooks in the area. The
classical monograph of A.F.W. Schimper on plant geography from a
physiological perspective appeared in 1898. It went through several
editions before being succeeded by Walter's similar treatment in
the 1960s (Walter 1964, 1968). R. F. Daubenmire's basic text in
the field, which diverged from the plant geography approach, first
appeared in 1947.

Since the early 1970s, five textbooks have appeared (Larcher
1973, Etherington 1975, Bannister 1976, Osmond et al. 1980, Fitter
and Hay 1981) and a four volume encyclopedia, which summarizes
current knowledge in the field, is now being released (Lange et al.
1981). Clearly, the tempo of progress is accelerating.

Although there is no consensus on the precise delineation of
plant physiological ecology (Tracy and Turner 1982), I define it
here as the study of the structural and physiological mechanisms
involved in the acquisition of resources by plants and the means by
which these resources are used in growth, competition, herbivore
protection, and reproduction. I have examined the progress and
promise in this broad arena in the following sections.

HOW MUCH CAN BE GAINED?

Much of the work in physiological plant ecology in the past
two decades has been devoted to uncovering and evaluating the
efficacy of the mechanisms involved in the acquisition of resources
(light, CO_2, water, and nutrients) by plants. Much of our knowledge
of how plants work has been derived from the study of relatively
few plants, mostly crop species, by plant physiologists. Although
plant physiologists have contributed greatly to this area, it has
been principally the ecologists who have pursued the study of
native plants. However, crop physiologists are now using wild
plants more, searching for genotypes with unusual patterns of
resource use but still with good economic yield (Brown 1979). The
interests of crop physiologists and plant physiological ecologists
have converged in several areas and there certainly will be continued
collaboration in these two areas of research.

Carbon

Much of the work on the mechanisms of resource acquisition by
plants has been on carbon gain and substantial contributions have
been made by physiological ecologists. The impetus for these

studies was the development of the infrared gas analyzer in the
late 1940s (Mooney 1972). This work has been particularly comprehensive, extending from basic studies of the biochemical basis of
differentiation in fixation capacity among different species to
details of the acclimation potential of a given genotype under
different environmental conditions (see, e.g., Berry and Björkman
1980). The studies of the differences in photosynthetic pathways
and the distributional characteristics among plants have been of
particular interest to ecologists this past decade (Ehleringer
1978, Teeri and Stowe 1976). A distinctive feature of much of the
work in photosynthesis by physiological ecologists has been a
vertical view of the problem. That is, individual workers and
groups have pursued these studies from their biochemical and physiological bases up through the significance of a mechanism in whole
plant functioning and distribution. It is because of these integrated
approaches that physiological ecologists have intereacted and
profited from approaches of other disciplines of biology.

Our developing knowledge of the functioning of the photosynthetic
machinery has been synthesized into various models of differing
levels of sophistication (Hesketh and Jones 1980). Because of
common links, these models have been tied with energy exchange
models to provide a tool for predicting photosynthetic gain by a
leaf or by a canopy under specific environmental conditions. The
canopy models, developed from the seminal work of Monsi and Saeki
(1953), have been used particularly by crop physiologists (Duncan
et al. 1967) and also by physiological ecologists (Miller and
Tieszen 1972).

The linking of photosynthesis models, with microenvironmental
models on the one hand and with carbon allocation schemes on the
other, has resulted in the development of a variety of plant growth
models which have been used particularly by crop physiologists
(see, e.g., Fick et al. 1973). Their potential use in ecology is
enormous. However, the main weakness with all of the existing
growth models is that there is no mechanistic basis to account for
allocation patterns. At present, more or less empirical approaches
are used. This is an area which deserves considerable attention in
the coming decade.

Several technological innovations have aided the progress
noted above, particularly methods for making precise measurements
of gas exchange under natural conditions (Field et al. 1982, Schluze
et al. 1982) and the development of highly portable microclimatic
data acquisition systems. We are still seeing substantial new
developments in instrumentation which should promote further work
in this area.

Water

     Our understanding of the differential water use patterns of
species has advanced enormously in the past two decades. We now
have models for linking the movement of water through the soil-plant-
atmospheric continuum (Cowan 1965, Waring and Running 1976). We
have instrumentation for precisely measuring water loss by leaves
(Beardsell et al. 1972) and a theory for calculating water loss
under specified environmental conditions. We can easily measure
the water stress which plants undergo in the field (Scholander et
al. 1964, Ritchie and Hinckley 1975) and can partitión the stress
into various components (Meidner and Sheriff 1976). We now know
that plants can reduce stress by various mechanisms, from biochemical
(osmotic adjustment) and cellular (wall elasticity) to behavioral
(changing leaf orientation or modifying leaf surface area) (Ehleringer and Forseth 1980). Our understanding of how plants modulate
their stomata to control water loss has advanced greatly (Cowan and
Farquhar 1977). Of all of the plant resource phenomena, water
movement through the soil and the plant is perhaps the best understood. We still lack, however, an understanding of how and why
plants budget water through the seasons of their lifetimes. Why
are some plants "extravagant" water users and others conservative,
even when they may co-occur.

Nutrients

     Whereas our knowledge of water as a resource is most complete
that for nutrients is the most deficient. Soil scientists and
plant physiologists are just now converging on similar problems and
approaches (Wild and Breeze 1980). A comprehensive view of the
mineral nutrition of wild plants is fairly recent (Chapin 1980).
Our understanding of the basis for differential availability of
nutrients such as phosphorus and nitrogen and of the associated
implications on the rooting patterns of plants has increased greatly
(Nye and Tinker 1977). Surprisingly, we have a poor understanding
of the mechanisms of uptake of the various nutrients, although
there has been substantial work in this area.

     The nitrogen economy of plants has received considerable
attention because of its great practical application; however, the
ecological aspects have also received attention (Lee and Stewart
1978). The development of the acetylene method for detecting
nitrogen fixation has greatly aided field studies in this area
(Stewart et al. 1967). We have learned how the nitrogen status of
different habitat types varies from nitrate rich to ammonium rich
and how this influences plant adaptive characteristics. The energetic
costs of the different pathways of assimilating nitrogen have been
worked out (Gutshick 1981). However, there have not been adequate
field studies showing why plants may shift from one mode to another.

Though the dependence of nitrogen fixation on photosynthesis has been known for several decades (Wilson et al. 1933), we are just beginning to evaluate the tight links between the nitrogen and carbon economies of plants (Herridge and Pate 1977, Paul and Kucey 1981).

Perhaps the most significant finding of the past decade has been the prevalance of mycorrhizal associations in various plant communities and the role they play in the mineral economy of plants. The dynamics of these associations are just beginning to be understood and there are indications that they may be quite complex (Chiariello et al. 1982).

Our knowledge of the seasonal dynamics of rooting behavior of plants is, in general, quite incomplete. There has been no real breakthrough in technology to make this work other than tedious and imprecise. We clearly need innovations in this area of research.

IS IT POSSIBLE AND HOW MUCH DOES IT COST?

Although most studies in physiological ecology have been on the mechanisms of resource acquisition, there has been substantial progress in other areas. One of these is in the biomechanics of plant structure (e.g., Wilson and Archer 1979). These studies define the structural limitations of various forms and reveal the absolute possibilities in plant architecture. In addition, there has been a real renaissance in the study of functional plant anatomy (Carlquist 1975) and morphology (Hallé et al. 1978) and these studies show promise of developing substantial links with physiological ecology.

One of the more important breakthroughs in physiological plant ecology has been the development of methods for calculating the true carbon costs of constructing (Penning de Vries et al. 1974) and maintaining tissues (Penning de Vries 1975). Thus, we can assess the cost of producing a particular compound or plant structure and relate this to the benefit (carbon gained, seeds produced, etc.) the plant derives from it. This information has been invaluable in growth models as well as in evolutionary assessments. These methods can be used to determine the costs of defense and competitive interactions. For example, one should be able to determine if tropical plants "pay" more for defense from herbivores than do plants from other ecosystem types. A simplified method of making these calculations has been developed recently (McDermitt and Loomis 1981) and undoubtedly the decade ahead will see wide application of these methods in making various cost/benefit assessments. There is still a problem in assessing costs of materials that either turn over rapidly or that are transported. Additionally, an assessment of the ultimate costs of compounds on fitness is needed.

The benefits of only a very few structures can be evaluated fully.
More work obviously is needed in this important area.

IS THERE A BETTER WAY?

Most work in physiological ecology has been directed toward
learning how plants operate in a given environment. Another question
of interest, particularly to evolutionary biologists, is whether
there is a better way of doing things than is found. Simulation
models have been used to probe such questions. For example, using
coupled energy budget and photosynthesis equations, one can ask
what the consequences would be, in carbon and (or) water economy,
if the leaf had characteristics different from those found (Mooney
et al. 1977). Interactions between the environment and such leaf
parameters as shape (Balding and Cunningham 1976), reflectance
(Ehleringer et al. 1976), orientation (Miller 1967), and size
(Smith 1978) also have been examined using quantitative energy
exchange models (Gates 1962). A more theoretical approach to this
problem has been employed using optimization models. In essence,
these models specify what optimal behavior should be and then test
the system to see if it conforms to expectations. Such models have
been employed at levels from leaf behavior to the performance of
whole communities (Cowan and Farquhar 1977, Field 1983, King and
Roughgarden 1982, Schaffer 1977).

HOW MANY WAYS CAN IT BE DONE?

Although it is axiomatic that each species has unique form-functional relationships, it will be difficult to define the resource
utilization patterns of all existing plants because of sheer numbers.
There have been several attempts, since the beginnings of ecology,
to classify plants into groupings of common ecological characteristics. These earlier groupings were principally morphological
(Raunkiaer 1934). In recent times, there has been an effort to
lump plants into form-functional categories. Some of these classifications have very large groupings without much detail (Grime 1979).
As we learn more about the physiological attributes of various
"strategy" types, these classifications could become of great use
in landscape management. The recent system of classifying plants
on the basis of life history traits has already had an important
impact on management strategies in fire prone landscapes (Noble
1981). An extension of such systems to include resource use information could be even more beneficial. For example, recently it was
noted that plants of disturbed habitats share many physiological,
morphological, genetic, and demographic features (Mooney and Godron
1983) that differ profoundly from plants of other habitats. As yet
we do not have enough information to make a complete categorization
of natural groupings. This lack of information is, in part, histori-

cal in that most of the work in agriculture has been devoted to
relatively few plant growth forms, principally annuals, and the
work in forestry has been limited mostly to species of economic
importance. At the same time, the work proceeding in physiological
ecology has focused on plants of a few favored ecosystem types,
notably deserts and arctic and alpine systems. Although one of the
founders of physiological plant ecology, A.W.F. Schimper, studied
tropical plants, remarkably little information has accumulated on
the mechanistic basis for resource utilization by the diverse
growth forms in tropical rain forests since his time. Before we
can fully appreciate the many ways plants acquire and utilize
resources, we need studies on the diverse growth forms of these
neglected ecosystems. Our generalizations of how plants work is,
of course, derived from our current data base. It is quite possible
that these generalizations may have to be altered substantially as
our knowledge expands to encompass a larger variety of plant growth
forms.

HOW WAS IT DONE IN THE PAST?

     Standard tools are still being used to unravel patterns of
species distribution in the past and from these the inferences
about climatic and community relations, including possible plant-
pathogen interactions. In the past decade, an additional, powerful
method of viewing past performance of plants has been developed,
which also has use in interpretations of modern physiological
responses and community interactions. This technique employs
stable carbon isotopes (van der Merwe 1982). Plants having the C_4
photosynthetic pathway use phosphoenol pyruvate (PEP) carboxylase
as their initial carboxylating enzyme, whereas plants with the C_3
pathway use ribulose bisphosphate carboxylase initially. These
plant types discriminate differentially between the isotopes of
carbon during fixation. Thus, from an analysis of the isotopic
composition of plant tissue, one can tell what photosynthetic
pathway the plant possesses. This analysis can be on preserved
carbon of any age. Thus it is possible, as has been done, to work
out past distributions of the different photosynthetic types as
well as food chains of the past and the present. The technique can
be used at a finer level of resolution to determine water relations
of plants, because the isotopic discrimination is based on diffusional
differences of CO_2 into leaves which correlate with rates of water
loss (Farquhar et al. 1982). This method also has been used to
trace the source of CO_2 to plants at different levels within the
forest (Medina and Minchin 1980). Other stable isotopes such as
oxygen and sulfur have been used to similar advantage. These
approaches have wide application in ecology and will be used extensively in the years ahead.

HOW TO BEAT OUT YOUR NEIGHBORS AND SURVIVE YOUR ENEMIES

Past studies in physiological plant ecology have concentrated on the interactions of plants with their physical environment. This was true to such a degree that systems were sought for study, such as deserts, where biotic interactions were thought to be minimal. Competition among plants was thought to act only through alteration of physical factors (Livingston and Shreve 1921). Beginning with the seminal paper of Ehrlich and Raven (1965), a whole field of coevolution of plants with their herbivores and pathogens has been established. The study of chemical interactions among plants also has gained considerable momentum (Rice 1974). It is now widely appreciated that biotic interactions control to varying extents the patterning of plants on the landscape (Harper 1977). Also, it has been established that plants have an array of physical and chemical defenses against herbivores and pathogens and perhaps against their plant neighbors. The study of the chemical interactions between plants and animals (Rosenthal and Janzen 1979) and among different plants is growing rapidly, in part because of the recognition of the problems, but also in part because of the development of new analytical methods for detecting secondary chemicals. There will, no doubt, be much fruitful work on detecting and assessing the effect of various secondary chemicals under natural conditions.

Although the study of plant competition is a classic endeavor in ecology, there has not been a strong link between the knowledge that is derived from physiological ecology with that of other approaches. For example, in plant population ecology, models have been developed to predict the outcome of competition between two species (e.g., Watkinson 1981). A tremendous amount of mechanistic detail is assumed, however, in the various coefficients used in these models. Information on the resource gathering capabilities of two or more species should be coupled with information on the environment, so that competitive outcomes can be predicted on a mechanistic basis.

In recent years, there also has been increasing attention to the study of those mechanisms by which plants share resources (Berendse 1982, Gulmon et al. 1983). Sorting out the complex causes of the maintenance of plant community diversity is a major challenge of the decades ahead.

TOWARD AN APPLICATION OF KNOWLEDGE

Plant physiological ecology has matured enough that practical applications should be expected. Some application is occurring and no doubt will accelerate in the years ahead. Applications will be of two sorts: interfacing with, and hence expanding, the potential of other research disciplines and directly applying knowledge gained in problem solving.

Interfacing will occur in various fields ranging from molecular biology to ecosystem ecology. One thrust in plant molecular biology today is the identification and manipulation of useful genetic traits. The evaluation of the utility of manipulated characters will have to be made on a whole plant basis. Physiological ecology provides the framework to make such evaluations in resource utilization. Such assessments will also bring physiological ecology closer to agroecology. Some research groups in agriculture already are employing the approaches and techniques of physiological ecology in their quest to find potential new crop species as well as in an effort to find cultivars and cultural practices which are efficient at using limited nutrient and water resources. The study of plant demography has the potential to interface with physiological ecology in the study of competition, as noted above, as well as in the study of the physiological independence of various vegetative and reproductive modules of plants. Ecosystem ecology and physiological ecology have many areas of common interest including, to name just a few, the assessments of nutrient balance implications of leaf longevity and internal nutrient cycling and of water and nutrient use efficiencies.

An area which will develop as an interface with other disciplines and which has great practical implications in the managment of ecosystems is the delineation of the suites of physiological, genetic, and reproductive characteristics of various "strategy" groups. This information can be used to predict the potential impact of various management policies which enhance or decrease members of these groups. At present, our knowledge is often biased toward those species of economic interest only, ignoring the ecosystem role of other members of the community.

It is evident that plant physiological ecologists are also beginning to contribute directly to many practical problems in ecology, including fire control management, mine re-vegetation, and atmospheric pollution control. The knowledge of the functioning of wild plants in their native habitats is a necessary ingredient to problem solving in these areas.

During the past few decades, physiological plant ecology has come of age. Undoubtedly, there will be great progress in the development of new concepts in this area in the years ahead. We should expect this field to enlarge the potential of other disciplines and to provide knowledge to solve some major practical problems in the management of our natural resources.

ACKNOWLEDGEMENTS

    I am grateful to J. Berry, N. Chiariello, J. Ehleringer, C.
Field, S. Gulmon, and P. Morrow for comments on this manuscript.
Support from NSF DEB-8102769 contributed to the development of
certain of the areas mentioned. This paper is dedicated to W. D.
Billings who has contributed so much to the development of plant
physiological ecology.

LITERATURE CITED

Balding, F.R., and G.L. Cunningham. 1976. A comparison of heat
    transfer characteristics of simple and pinnate leaf models.
    Bot. Gaz. 137:65-74.
Bannister, P. 1976. Introduction to physiological plant ecology.
    Halsted Press, New York, NY USA.
Beardsell, M.F., P.G. Jarvis, and B. Davidson. 1972. A null-balance
    diffusion porometer suitable for use with leaves of many
    shapes. J. Appl. Ecol. 9:677-690.
Berendse, F. 1982. Competition between plant populations with
    different rooting depths. III. Field experiments. Oecologia
    53:50-55.
Berry, J., and O. Björkman. 1980. Photosynthetic response and
    adaptation to temperature in higher plants. Annu. Rev. Plant
    Physiol. 31:491-543.
Brown, J.C. 1979. Genetic improvement and nutrient uptake in
    plants. Bioscience 29:289-292.
Carlquist, S. 1975. Ecological strategies of xylem evolution.
    Univ. of Calif. Press, Berkeley, CA USA.
Chapin, F.S., III. 1980. The mineral nutrition of wild plants.
    Annu. Rev. Ecol. Syst. 11:233-260.
Chiariello, N., J.C. Hickman, and H.A. Mooney. 1982. Endomycorrhizal
    role for interspecific transfer of phosphorus in a community
    of annual plants. Science 217:941-943.
Cowan, I. 1965. Transport of water in the soil-plant-atmosphere
    system. J. Appl. Ecol. 2:221-239.
Cowan, I., and G. Farquhar. 1977. Stomatal function in relation
    to leaf metabolism and environment. Symp. Soc. Exp. Biol.
    31:471-505.
Daubenmire, R.F. 1947. Plants and environment. Wiley, New York,
    NY USA.
Duncan, W.G., R.S. Loomis, W.A. Williams, and R. Hanan. 1967. A
    model for simulating photosynthesis in plant communities.
    Hilgardia 38:181-205.
Ehleringer, J. 1978. Implications of quantum yield differences on
    the distributions of C_3 and C_4 grasses. Oecologia 31:255-267.
Ehleringer, J., O. Björkman, and H.A. Mooney. 1976. Leaf pubescence:
    Effects on absorptance and photosynthesis in a desert shrub.
    Science 192:376-377.

Ehleringer, J., and I. Foreseth.   1980.   Solar tracking by plants.
     Science 210:1094-1098.
Ehrlich, P., and P. Raven.   1965.   Butterflies and plants: A study
     in coevolution.   Evolution 18:568-608.
Etherington, J.R.   1975.   Environment and plant ecology.   Wiley,
     New York, NY USA.
Farquhar, G., M. Ball, S. von Caemmerer, and Z. Roksandic.   1982.
     Effect of salinity and humidity on $\delta^{13}C$ value of halophytes-
     evidence in diffusional isotope fractionation determined by
     the ratio of intercellular/atmospheric partial pressure of CO_2
     under different environmental conditions.   Oecologia 52:121-124.
Fick, G.W., W.A. Williams, and R.S. Loomis.   1973.   Computer simula-
     tion of dry matter distribution during sugar beet growth.
     Crop Sci. 13:413-417.
Field, C.   1983.   Allocating leaf nitrogen for the maximization of
     carbon gain; leaf age as a control on the allocation program.
     Oecologia 56:341-347.
Field, C., J. Berry, and H.A. Mooney.   1982.   A portable system for
     measuring carbon dioxide and water vapour exchange of leaves:
     Technical report.   Plant, Cell Environ. 5:179-186.
Fitter, A.H., and R.K.M. Hay.   1981.   Environment physiology of
     plants.   Academic Press, London, England.
Gates, D.M.   1962.   Energy exchange in the biosphere.   Harper and
     Row, New York, NY USA.
Grime, P.   1979.   Plant strategies and vegetation processes.
     Wiley, London, England.
Gulmon, S.L., N.R. Chiariello, H.A. Mooney, and C.C. Chu.   1983.
     Phenology and resource use in three co-occurring grassland
     annuals.   Oecologia 58:33-42.
Gutshick, V.P.   1981.   Evolved strategies in nitrogen acquisition
     by plants.   Am. Nat. 118:607-637.
Hallé, F., R.A. Oldeman, and P.B. Tomlinson.   1978.   Tropical trees
     and forests: An architectural analysis.   Springer, Berlin, W.
     Germany.
Harper, J.   1977.   Population biology of plants.   Academic Press,
     London, England.
Herridge, D.F., and J.S. Pate.   1977.   Utilization of net photosyn-
     thate for nitrogen fixation and protein production of an
     annual legume.   Plant Physiol. 60:759-764.
Hesketh, J.D., and J.W. Jones, eds.   1980.   Predicting photosynthesis
     for ecosystem models.   CRC Press, Boca Raton, FL USA.
King, D., and J. Roughgarden.   1982.   Graded allocation between
     vegetative and reproductive growth for annual plants in growing
     seasons of random length.   Theor. Pop. Biol. 22:1-16.
Lange, O., P. Nobel, C.B. Osmond, and H. Ziegler, eds.   1981.
     Encyclopedia of plant physiology, new series.   Vol. 12A,
     Physiological plant ecology.   Springer, Berlin, W. Germany.
Larcher, W.   1973.   Okologie der Pflanzen.   Ulmer, Stuttgart, W.
     Germany.

Lee, J.A., and G.R. Stewart. 1978. Ecological aspects of nitrogen assimilation. Adv. Bot. Res. 6:1-43.

Livingston, B.E., and F. Shreve. 1921. The distribution of vegetation in the United States, as related to climatic conditions. Carnegie Inst. Wash. Publ. 284.

McDermitt, D.K., and R.S. Loomis. 1981. Elemental composition of biomass and its relation to energy content, growth efficiency, and growth yield. Ann. Bot. 48:245-290.

Medina, E., and P. Minchin. 1980. Stratification of $\delta^{13}C$ values of leaves in Amazonian rain forests. Oecologia 45:377-378.

Meidner, H., and D. Sheriff. 1976. Water and plants. Blackie, Glasgow, Scotland.

Miller, P.C. 1967. Leaf orientation and energy exchange in quaking aspen (*Populus tremuloides*) and gambell's oak (*Quercus gambellii*) in central Colorado. Oecol. Plant. 2:241-270.

Miller, P.C., and L. Tieszen. 1972. A preliminary model of processes affecting primary production in the arctic tundra. Arctic Alpine Res. 4:1-18.

Monsi, M., and T. Saiki. 1953. Uber den Lichtfaktor in den Pflanzengesellschaften und seine Bedeutung fur die Stoffproduktion Jap. J. Bot. 14:22-52.

Mooney, H.A. 1972. Carbon dioxide exchange of plants in natural environments. Bot. Rev. 38:455-469.

Mooney, H.A., J. Ehleringer, and O. Björkman. 1977. The energy balance of leaves of the evergreen desert shrub *Atriplex hymenelytra*. Oecologia 29:301-310.

Mooney, H.A., and M. Godron. 1983. Disturbance in ecosystems-components of response. Ecol. Stud. 44. Springer, Heidelberg, W. Germany.

Noble, I. 1981. Predicting successional change. Pages 278-300 *in* H.A. Mooney, T.M. Bonnicksen, N.L. Christensen, I.E. Lotan, and W.A. Reiners, eds. Fire regimes and ecosystem properties. U.S. Dep. of Agric. For. Serv. Gen. Tech. Rep. WO-26.

Nye, P.H., and P.B. Tinker. 1977. Solute movement in the soil-root system. Univ. of Calif. Press, Berkeley, CA USA.

Osmond, B., O. Björkman, and D.J. Anderson. 1980. Physiological processes in plant ecology. Springer, Berlin, W. Germany.

Paul, E.A., and R.M.N. Kucey. 1981. Carbon flow in plant microbial associations. Science 213:473-474.

Penning de Vries, F.W.T. 1975. The cost of maintenance processes in plant cells. Ann. Bot. 39:77-92.

Penning de Vries, F.W.T., A.H.M. Brunsting, and H.H. van Laar. 1974. Products, requirements and efficiency of biosynthesis: A quantitative approach. J. Theor. Biol. 45:339-377.

Raunkiaer, C. 1934. The life forms of plants and statistical plant geography. Clarendon Press, Oxford, England.

Rice, E.L. 1974. Allelopathy. Academic Press, New York, NY USA.

Ritchie, G.A., and T.M. Hinckley. 1975. The pressure chamber as an instrument for ecological research. Adv. Ecol. Res. 9:166-254.

Rosenthal, G., and D. Janzen, eds. 1979. Herbivores: Their interaction with secondary metabolites. Academic Press, New York, NY USA.

Schaffer, W.M. 1977. Some observations on the evolution of reproductive rate and competitive ability in flowering plants. Theor. Pop. Biol. 11:90-104.

Schimper, A.F.W. 1898. Pflanzengeographie auf physiologischer Grundlange. Fischer, Jena, Germany.

Scholander, P.F., H.T. Hammel, E.A. Hemmingsen, and E.D. Bradstreet. 1964. Hydrostatic pressure and osmotic potential in leaves of mangroves and some other plants. Proc. U.S. Nat. Acad. Sci. 52:119-125.

Schulze, E-D., A.E. Hall, O.L. Lange, and H. Walz. 1982. A portable steady-state porometer for measuring the carbon dioxide and water vapour exchanges of leaves under natural conditions. Oecologia 53:141-145.

Smith, W.K. 1978. Temperatures of desert plants: Another perspective on the adaptability of leaf size. Science 201:614-616.

Stewart, W.D.P., G.P. Fitzgerald, and R.H. Burris. 1967. In situ studies on N_2 fixation using the acetylene reduction technique. Proc. U.S. Nat. Acad. Sci. 58:2071-2078.

Teeri, J., and L.G. Stowe. 1976. Climatic patterns of C_4 grasses in North America. Oecologia 23:1-12.

Tracy, C.R., and J.S. Turner. 1982. What is physiological ecology? Bull. Ecol. Soc. 63:340-347.

Van der Merwe, N.J. 1982. Carbon isotopes, photosynthesis, and archeology. Am. Sci. 70:596-606.

Walter, H. 1964. Die Vegetation der Erde in oko-physiologischer Betrachtung. Band I. Die tropischen und subtropischen zonen. Fischer, Jena,

Walter, H. 1968. Die Vegetation der Erde in oko-physiologischer Betrachtung. Band II. Die gemassigten und arkitschen Zonen. Fischer, Jena,

Waring, R.H., and S.W. Running. 1976. Water uptake, storage and transpiration by conifers; a physiological model. Pages 189-202 in O.L. Lange, L. Kappen and E.-D. Schulze, eds. Water and plant life. Springer, Berlin, W. Germany.

Watkinson, A.R. 1981. Interference in pure and mixed populations of Agrostemma githago. J. Appl. Ecol. 18:967-976.

Wild, A., and V.G. Breeze. 1980. Nutrient uptake in relation to growth. Pages 331-344 in C.B. Johnson, ed. Physiological processes limiting plant productivity. Butterworth, London, England.

Wilson, B.F., and R.R. Archer. 1979. Tree design: Some biological solutions to mechanical problems. Bioscience 29:293-298.

Wilson, P.W., E.B. Fred, and M.R. Salmon. 1933. Relations between carbon dioxide and elemental nitrogen assimilation in leguminous plants. Soil Sci. 35:145-165.

INTERACTION AND INTEGRATION--THE ROLE OF MICROBIOLOGY

IN ECOLOGICAL RESEARCH

Thomas Rosswall

Department of Microbiology
Swedish University of Agricultural Sciences
S-750 07 Uppsala, Sweden

INTRODUCTION

In certain ways, science progresses in an erratic fashion.
Key discoveries suddenly advance our knowledge and understanding in
quantum jumps. Many such discoveries are not recognized until much
later. This is illustrated by the history of the Nobel prizes,
which have often honoured discoveries made decades earlier. Other
discoveries are almost immediately perceived as being of paramount
significance, such as the recent use of restriction enzymes in gene
splicing techniques for genetic manipulation.

More common, however, is that scientific research moves along
some major tracks in a fairly predictable fashion. Researchers try
to adapt to the direction of the track and label their research
"ecology," "environment," or "energy," depending on the specific
area that interests society and financing bodies at the time.
Often it is possible to single out specific events that change the
scientific track. Rachel Carson's *Silent Spring*, one such event,
directed our interest to the increasing pollution of our planet.
Similarly, the energy crisis in the early 1970s made renewable
energy sources an item for research. Research funding agencies
established specific programmes for energy research, and large
numbers of scientists tried to climb on the bandwagon by labeling
their projects "energy related." They made subtle changes in the
contents of the programme and drastic changes in its description in
grant proposals.

We will not be able to predict either the quantum jumps or the
sudden changes in direction of ecological research over the next

19

decade; we can base our judgement only on past experience. Our forecast probably will be wrong, as most forecasts are. I do feel, however, that it is important to try to use our scientific knowledge and intuition to point out the roads which might usefully be pursued. Fortunately, not all scientists will follow the tracks, and by identifying some of the other possibilities, which we feel will lead through a beautiful and fascinating scientific country, we hopefully may influence future ecological research.

In this paper, I will confine myself to my own interests and comment, in general, on the role of microbiology in ecological research. In particular, I will briefly discuss four topics: (1) the interactions of microorganisms with plants and animals, (2) the role of soil biota in sustained production in agriculture and forestry, (3) the integration of biogeochemical cycles, and (4) the role of genetic transfer in microbial ecology.

INTERACTION OF MICROORGANISMS WITH PLANTS AND ANIMALS

Interactions of microorganisms with plants and animals are important for many reasons. Proper research on such interactions requires collaboration among microbiologists, botanists, and zoologists; this need frequently has hampered earlier progress. Only by developing multidisciplinary research groups and projects can real progress be made.

Interactions of importance are of various kinds and may be divided into symbiotic relationships, parasitic relationships, and predator-prey relationships.

Symbiotic Relationships

Microorganisms-plants. The two most important symbiotic relationships are those between nitrogen fixing bacteria and plants and between mycorrhizal fungi and plants.

The interactions of *Rhizobium* and *Frankia* species with plants constitute the most important nitrogen fixing symbioses. Our understanding about the factors regulating the survival of individual strains of *Rhizobium* in the rhizosphere, leading to an efficient or inefficient symbiosis, is still scanty . Genetic research probably will produce strains with higher infectivity and higher fixation efficiency. The inoculated strains must, however, be able to survive on the inoculated grain, grow in the soil, and outdo more inefficient strains already existing in the soil. To make this research successful, microbial and plant geneticists need to collaborate to develop suitable pairs of bacteria and host plants which together maximize nitrogen fixation under a given set of environmental constraints.

The recent discovery of methods of cultivating *Frankia* actino-mycetes in pure culture (Callaham et al. 1978, Baker and Torrey 1979) has provided us with the tool needed for further understanding the *Frankia-Alnus* (or *Hippophae*, *Myrica*, etc.) symbiosis, which is important for fixing nitrogen in many northern ecosystems, and the symbiosis with *Casuarina* in tropical climates. The ecology of *Frankia* in soil and the factors regulating the infection process are very poorly known.

The importance of root associations with mycorrhizal fungi, especially for phosphorus uptake, has been known for many years; mycorrhizal infection of roots is the rule under natural conditions (Gerdemann 1975). There are, however, few studies which have quantified the increased nutrient uptake under field conditions. It has been shown that, in tropical forests, mycorrhizal fungi can transfer phosphorus directly from decomposing litter to the roots (Herrera et al. 1978). Such a direct shunt would be especially important for mobile nutrients, such as nitrogen, which could be efficiently adsorbed by the root-fungal symbiosis with a reduced risk for nutrient losses. The competitive advantage offered the plant by an efficient mycorrhizal association is not known. This lack of understanding is further complicated by the findings that mycorrhizal fungi seem able to infect two or more plants of the same or different species and transfer nutrients (32-P) between individual plants (Chiariello et al. 1982).

Total nutrient and energy balances for plants with and without nitrogen fixing and mycorrhizal symbiosis have been investigated (Fig. 1). Information on the quantitative aspects under field conditions probably will facilitate an assessment of the advantages of such symbioses for plants in both natural and managed ecosystems.

Microorganisms-animals. Compared with microorganism-plant interactions, far less is known about the symbiotic relationships between microorganisms and animals, and I will give only two examples.

Many xylophagous insects, especially termites, have in their guts symbiotic microorganisms and protozoa, which are important for satisfying their carbon, nitrogen, and energy requirements. Anaerobic, cellulose degrading protozoa and bacteria are important for wood consuming insects; the activities of these microorganisms result in an unusually high assimilation efficiency of cellulose and hemicelluloses by their hosts (Breznak 1982). Bacteria in the guts of termites also seem to be important for the nitrogen nutrition of these animals, which feed mainly on substrates very low in nitrogen (Breznak 1982). The efficient symbioses between certain insects and microorganisms make it possible for the animals to use food sources otherwise unavailable to them. To further quantify the energy and nutrient metabolism of such animals, close cooperation with microbiologists is needed. This cooperative effort will shed

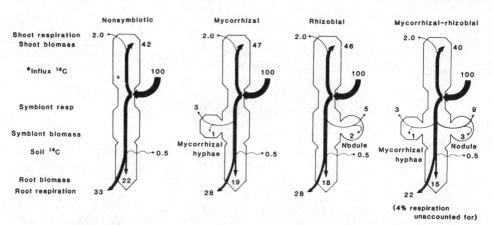

Fig. 1.   The relative 14-C flow to various compartments of symbiotic
          and non-symbiotic faba beans (4 to 5 weeks old) after shoots
          were exposed above ground to 14-CO$_2$ under continuous light.
          The carbon influx has been equalized to 100 units of carbon
          per gram of shoot carbon. (From Paul and Kucey 1981; copy-
          righted 1981 by the AAAS; reprinted by permission.)

light not only on the ecology of these interesting groups of gut
organisms but also on the adaptations of insects to food sources
generally unavailable to them.

The tube worm, *Riftia pachyptila*, which lives abundantly near
the geothermal vents of the Galápagos Rift (Jones 1981), has been
shown to contain enzymes usually connected with chemoautotrophic
growth on reduced sulphur compounds (Felbeck 1981). It has been
suggested that these enzymes are synthesized by chemolithotrophic,
sulphur oxidizing prokaryotes (Cavanaugh et al. 1981), which obtain
energy from the oxidation of hydrogen sulphide emitted from the
vents. The tube worm lacks a mouth and digestive system and seems
to contain the chemolithotrophic bacteria in trophosomal tissue
(Cavanaugh et al. 1981). It is possible that such symbiotic associations occur also in other habitats containing large amounts of
sulphides. This mode of life necessitates protective mechanisms
for sulphide sensitive cytochrome c oxidase. A blood factor is
thought to be important in binding toxic levels of sulphide (Powell
and Somero 1983). This factor seems to be a protein that also
helps in transporting the sulphide to sites of chemolithotrophic
bacteria in the worms (Arp and Childress 1983).

This seemingly obligate animal-microorganism symbiosis has
enabled these animals to live in habitats ruled out for animals
lacking such an association. Obviously, we have not yet discovered
all the various possibilities for life on earth, especially in

Table 1. The effects of plant roots and protozoa on soil organic matter mineralization by soil bacteria and nitrogen uptake by plants (from Clarholm 1984). All values in mg N/pot. Each microcosm contained 3 wheat plants. All microcosms were observed after 42 days.

Treatment	Inorganic N	Bacteria	Protozoa	Plants	Change[a]
Bacteria	0.17	0.36			-0.15
Bacteria + protozoa	0.24	0.28	0.23		-0.07
Bacteria + plants	0.15	0.45		1.61	1.53
Bacteria + protozoa + plants	0.22	0.44	0.16	2.55	2.69

[a]Total observable change in biomass N plus inorganic N.

effect on respiration rates, decomposition, and nitrogen mineralization (van der Drift and Jansen 1977, Hanlon and Anderson 1979, Visser et al. 1981, Ineson et al. 1982). Most studies have used additions of one or two fungal species, but because Parkinson et al. (1977) have shown that soil animals may selectively feed on fungi, extrapolation of the results to field conditions is difficult. Even if "natural" populations of fungi are added to microcosms, it is impossible to obtain a species composition similar to the one existing under field conditions. The results from microcosm experiments also can be erratic in that an effect on respiration can be observed in some microcosm series and not in others (O. Andrén, Department of Ecology and Environmental Research, Swedish University of Agricultural Sciences, Uppsala, and J. Schnürer, Department of Microbiology, Swedish University of Agricultural Sciences, Uppsala, unpublished data). This inconsistency may be because different fungi may be dominant in different series as a result of variation in the inoculants, and the dominant species may have different palatability to the animal grazer used in the experiment. We need experimental designs where results can be related to field conditions to use the possibilities offered by microcosms to the best advantage.

It is possible to calculate carbon and nitrogen budgets for the various soil organism populations and, thus, indirectly determine the effects on element cycling in different ecosystems, as has been done for nitrogen in a pine forest soil (Persson 1983) and in four

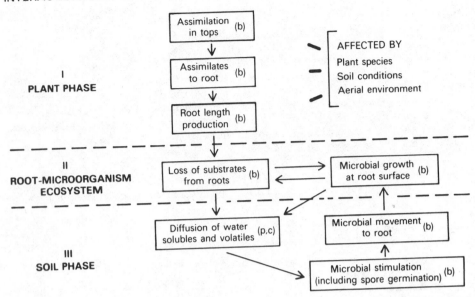

Fig. 2. A conceptual framework for factors affecting microbial growth at the root-soil interface: (b) biological factors, (p,c) physico-chemical factors (from Bowen 1982).

extreme habitats. Such discoveries may make us view other animals in a new light. Symbiotic systems of this type may be more common than previously suspected.

Parasitic Relationships

The ecology of root pathogens was recently summarized by Krupa and Dommergues (1979). Microbial plant diseases are important and it has been estimated that 12-18% of the causes of crop losses; world crop loss is attributable to bacterial diseases (Cramer 1967). Much research is devoted to the selection of resistant plant varieties, the development of new pesticides, and the use of biological control. In the foreseeable future, it will not be possible to control the spread of plant pathogens completely, and it is important to direct research to disease forecasting. Such methods are available (Garrett 1982) but are often too crude because our understanding of the factors regulating the growth and spread of pathogenic microorganisms is inadequate.

The rhizosphere environment plays an important role in determining the success of fungal and bacterial infections (Fig. 2). To devise management practices which optimize the interaction between plants and beneficial microorganisms and minimize the possibilities for infections by pathogens, it is necessary to understand the importance of biotic and abiotic factors in the rhizosphere. Water stress is one such factor (Griffin 1981), and the quantity and quality of root exudates another (Hale et al. 1982). It should be possible to manipulate both the plant and the desired microorganisms genetically to improve the chances of success, but this requires close collaboration between microbiologists, plant physiologists, and plant ecologists. Knowledge of the "rhizosphere ecosystem" is important for understanding processes in natural ecosystems as well as for controlling crop pathogens and increasing food production in managed ecosystems.

The importance of parasitic microorganisms to animals probably is equally important, but I do not feel competent to cover this area in this paper.

Predator-Prey Relationships

The effect of animal predation of microbial populations has received considerable interest in the past 10 years. Many studies have shown how the grazing of animals on bacteria and fungi influences energy flow, organic matter decomposition, and nutrient cycling. I think that most investigations so far have not given us the information necessary to determine the importance of such grazing under field conditions. In a recent review article, Anderson et al. (1981), using microcosm experiments, calculated the amounts of nitrogen (N) that would be mineralized through the grazing of nematodes on soil microorganisms at 19-124 kg N/ha·yr, but extrapolations from microcosm experiments to field conditions are difficult. Soil animals liberate substantial amounts of nitrogen in forests (Persson 1983) and arable soils, especially through grazing of bacteria by protozoa in the rhizosphere (Clarholm 1983, Rosswall and Paustian 1984). Clarholm (1981) has shown in the field that increases in bacterial biomass can be followed by rapid decreases attributed to protozoan grazing (Fig. 3).

Microcosm experiments are important in investigating interactions between animals and microorganisms, and such experiments have received considerable interest in recent years (Giesy 1980). There are, however, several important drawbacks with microcosms which should be realized. In most experiments, they contain only microorganisms and animals. Considering the importance of plant roots, especially for the dynamics of soil bacteria populations, such studies may offer very little resemblance to field conditions. It has been shown that microcosms with plants differ markedly in carbon and nitrogen mineralization rates and in populations of bacteria and protozoa from microcosms without plants (Clarholm 1981, 1984; Table 1).

Using microcosm experiments to study the interaction of fungi and fungal feeders, several investigators have shown a marked

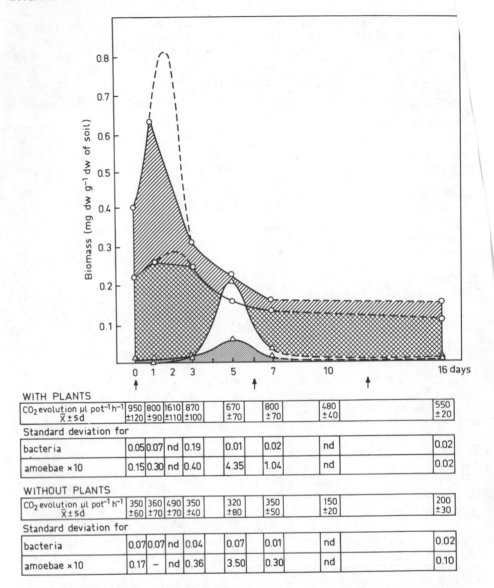

WITH PLANTS

CO_2 evolution µl pot^{-1}h^{-1} $\bar{X} \pm sd$	950 ±120	800 ±90	1610 ±110	870 ±100		670 ±70		800 ±70		480 ±40			550 ±20
Standard deviation for													
bacteria	0.05	0.07	nd	0.19		0.01		0.02		nd			0.02
amoebae ×10	0.15	0.30	nd	0.40		4.35		1.04		nd			0.02

WITHOUT PLANTS

CO_2 evolution µl pot^{-1}h^{-1} $\bar{X} \pm sd$	350 ±60	360 ±70	490 ±70	350 ±40		320 ±80		350 ±50		150 ±20			200 ±30
Standard deviation for													
bacteria	0.07	0.07	nd	0.04		0.07		0.01		nd			0.02
amoebae ×10	0.17	–	nd	0.36		3.50		0.30		nd			0.10

Fig. 3. Development of bacterial (striped, with plants; checked, without plants) and naked amoebic (white, with plants; dotted, without plants) biomass in a pot experiment with an arable soil with and without wheat plants. Watering is indicated by arrows. The dashed lines indicate probable developments not registered because of too infrequent samplings. CO_2 evolution rate the night prior to sampling and standard deviation for the biomass estimates are given in the tables below. Three replicates were sampled at each time. (From Clarholm 1981; reprinted by permission of Springer, New York, NY USA.)

arable cropping systems (Rosswall and Paustian 1984) in Sweden.
Such elemental budgets are important but provide only a static
description and yield no information about the dynamics within the
vegetation period or differences between years. Here, simulation
models offer a powerful tool, but unfortunately most models do not
consider soil organisms (Frissel and van Veen 1981), although there
are models which consider microbial biomass dynamics (Bosatta
et al. 1980, McGill et al. 1981, van Veen and Frissel 1981, Juma
and Paul 1981). Apart from some models on protozoan grazing in
wastewater treatment plants and on continuous cultures (e.g.,
Pavlon and Fredrickson 1983), the role of animals does not seem to
have been considered in simulation models, and this certainly calls
for further model development.

THE ROLE OF SOIL BIOTA FOR SUSTAINED PRODUCTION IN AGRICULTURE AND FORESTRY

Over the past 20 years, ecosystem oriented research has given
us a better understanding of the intricate interactions between the
complex system of biotic and abiotic variables at the ecosystem
level. These studies were initiated largely during the International
Biological Programme (IBP). For various reasons, agroecosystems
were not included in the IBP, and an ecosystem approach has rarely
been used in attempts to study the ecological basis for our food
and fodder production systems. We probably know far more about
energy flow and nutrient cycling in so called natural ecosystems in
spite of the economic importance of agroecosystems.

INTECOL realized this lack of ecological understanding early
and organized a working group for agroecosystems. In spite of this
and other similar attempts, there is still a great need for ecologists
to study agroecosystems from an ecosystem point of view. Probably
one reason for the lack of ecosystem studies in agriculture is the
communication gap between agronomists and ecologists. Agricultural
research has been aimed at maximizing production through the use of
fertilizers, pesticides, and fossil fuel. Much less emphasis has
been placed on attempts to elucidate the fundamental principles
behind the functioning of agroecosystems.

The effects of present management systems on soil microorganisms
and soil fauna is poorly understood. Intensive management of soils
is, for example, thought to have an adverse effect on the occurrence
of earthworms, which are important in aerating and structuring
agricultural soils. Also, they probably are very important for the
decomposition of plant residues, mainly as a consequence of their
soil mixing activity. Recent investigations in Sweden have tried
to elucidate the effects of some cropping systems on microarthropods,
enchytraeids, nematodes (Andrén and Lagerlöf 1983), and earthworms
(Lofs-Holmin 1983), but further investigations are needed in other

climatic regions and soil types before more generally applicable
conclusions can be drawn. This need for further studies is also
true for soil microorganisms, which play a major role in most soil
processes and a crucial role in decomposition and the regulation of
the biogeochemical nitrogen cycle (Rosswall 1982).

Apart from general studies of the role of soil organisms
(including roots) in fertility of agroecosystems, special integrated studies should be initiated to quantify important nutrient
cycles and the factors regulating such biogeochemical cycles in
agroecosystems. In 1976, INTECOL co-sponsored an international
workshop on nutrient cycling in agricultural ecosystems (Frissel
1977). This meeting was very important in trying to bring together
information on nutrient cycling in a wide variety of agricultural
systems. It was, however, evident that few, if any, integrated
studies had been performed. For example, in spite of the great
economic importance of nitrogen as a limiting plant nutrient and
the low efficiency of crops in using nitrogen fertilizer, there
was only scanty quantitative knowledge about nitrogen cycling
processes in agroecosystems. A Swedish project, "Ecology of Arable
Land: The Role of Organisms in Nitrogen Cycling," was started in
1979 (Persson and Rosswall 1983, Rosswall and Paustian 1984, Steen
et al. 1984). I hope that more such studies, focusing on the role
of soil organisms in regulating the biogeochemical cycle, can be
initiated.

One final point regarding agroecosystems should be stressed,
namely that many ecosystem processes can be studied in monocultural, multidisciplinary agroecosystem projects. By selecting such
comparatively simple systems for study, it may be easier to understand some of the basic biological processes.

Considerably more ecosystem oriented research has been carried
out in forest ecosystems (e.g., Persson 1981). There are, however,
many important problems connected with the management of present
day forests which merit attention, such as the ecological effects
of clearcutting and whole tree harvesting. Special attention
should be paid to attempts assessing the effects of a future climate
change on the dynamics of world forest ecosystems. This topic will
be considered in a newly established Scientific Committee on Problems
of the Environment (SCOPE) project.

INTEGRATION OF BIOGEOCHEMICAL CYCLES

Most research on biogeochemical cycles at the ecosystem level
has focused on the fate of single nutrients, although we know that
nutrient cycles and energy flow are interdependent. Present knowledge on the interaction of the biogeochemical cycles was addressed
at recent SCOPE workshops (Freney and Galbally 1982, Bolin and Cook

1983), but most papers inevitabley discussed concepts and hypotheses rather than conclusions because of general lack of reliable scientific results. A few attempts have been made to quantify element inter- actions; such studies could profitably use isotope techniques and--for simplicity and importance--focus on agroecosystems. Till et al. (1982), for example, used clover residues labelled with 14-C, 15-N, 35-S, and 32-P and studied decomposition, mineralization, and plant uptake in an agricultural soil. There is a great need for further studies in this area and, scientifically, they should prove very rewarding.

THE ROLE OF GENETIC TRANSFER IN MICROBIAL ECOLOGY

Many complex organic compounds are decomposed by microorganisms. The genetic coding for these processes is often contained in plas- mids, which have been termed "degradative plasmids" (Chakrabarty 1973). *Pseudomonas* spp. plasmids were originally described to code for degradation of substances such as camphor, octane, naphthalene, toluate, and salicylate (Wheelis 1975), but the degradation of many other compounds, such as aliphatic substances, aromatic and polynu- clear aromatic hydrocarbons, terpenes, alkaloids, and chlorinated hydrocarbons, are also plasmid coded. It has been suggested that such plasmids have evolved as a primary means of detoxification (Farrell and Chakrabarty 1979). Plasmids also have been shown to be important in coding for the degradation of pesticides, such as 2,4-dichlorophenoxyacetic acid (Pemberton et al. 1979) and chloridazon (Ebersprächer and Lingens 1981).

However, xenobiotic compounds are not the only compounds coded for degradation by plasmids; the bacterial metabolism of lignin and lignin derivatives also seems to be plasmid coded (Salkinoja-Salonen et al. 1979). In addition, microbial resistance to heavy metals, including, for example, Hg^{2+} detoxification, is often related to the occurrence of plasmids (Summers and Silver 1978). There are also indications that the determinants for nodulation in *Rhizobium leguminosarum* are found in plasmids (Johnston et al. 1978), as well as the determinants for nitrogen fixing ability and host range specificity (Johnston et al. 1981). The hydrogen uptake by faculta- tive autotropic *Alcaligenes* spp. also has been shown to be regulated by plasmids (Andersen et al. 1981).

The ubiquitous plasmids in bacteria offer interesting opportu- nities for genetic manipulation; this probably will be one important area for biotechnological research. It also may be possible to use bacterial plasmids to transfer genetic information to higher plants, and it already has been shown that part of the Ti plasmid of *Agro- bacterium tumefaciens* is incorporated into plant nuclear DNA and is responsible for the tumor phenotype (Drummond et al. 1977). Plamids are, however, also of great evolutionary and ecological interest.

Probably in the future, microbial ecologists will use many of the
techniques developed in genetics and molecular biology to gain
further information on the role of plasmids in bacteria from different
environments.

CONCLUSIONS

Microorganisms play crucial roles in the ecosystem and should
be considered in that context. The interactions among microorganisms,
animals, and plant roots in soil are of great importance as are
various symbiotic associations of microorganisms with animals and
plants.

The importance of soil microorganisms and soil animals in
agroecosystems should be realized and more ecosystem research in
various types of agroecosystems should be conducted.

There is a need for further interactions between microbial
ecologists and plant and animal ecologists and between microbial
ecologists and microbial geneticists. There is also a need for
integrated studies. For example, further understanding of biogeochemical cycles call for integrated approaches.

REFERENCES CITED

Andersen, K., R.C. Tait, and W.R. King. 1981. Plasmids required
    for utilization of molecular hygrogen by *Alcaligenes eutrophus*.
    Arch. Mikrobiol. 129:384-390.
Anderson, R.V., D.C. Coleman, and C.V. Cole. 1981. Effects of
    saprotrophic grazing on net mineralization. Pages 201-216 *in*
    F.E. Clarke and T. Rosswall, eds. Terrestrial nitrogen cycles:
    Processes, ecosystem strategies and management impacts. Ecol.
    Bull. (Stockholm) 33.
Andrén, O., and J. Lagerlöf. 1983. Soil fauna (microarthropods,
    enchytraeids, nematodes) in Swedish agricultural cropping
    systems. Acta Agric. Scand. 33:33-52.
Arp, A.J., and J.J Childress. 1983. Sulfide binding by the blood
    of the hydrothermal vent tubeworm *Riftia pachyptila*. Science
    219:295-297.
Baker, D., and J.G. Torrey. 1979. The isolation of actinomycetous
    root nodule endophytes. Pages 38-56 *in* J.C. Gordon, C.T.
    Wheeler, and D.A. Perrey, eds. Symbiotic nitrogen fixation in
    management of temperate forests. Ore. State Univ., For. Res.
    Lab., Corvallis, OR USA.
Bolin, B., and R.B Cook, eds. 1983. Interactions of major biogeo-
    chemical cycles. SCOPE Rep. 21. Wiley, Chichester, England.
Bosatta, E., L. Bringmark, and H. Staaf. 1980. Nitrogen transforma-
    tions in a Scots pine forest: Model analysis of mineralization,

uptake by roots and leaching. Pages 565-589 *in* T. Persson,
     ed. Structure and function of northern coniferous forests:
     An ecosystem study. Ecol. Bull. (Stockholm) 32.
Bowen, G.D. 1982. The root-microorganism ecosystem. Pages 3-42
     in Biological and chemical interactions in the rhizosphere.
     Swed. Nat. Sci. Res. Counc., Stockholm.
Breznak, J.A. 1982. Intestinal microbiota of termites and other
     xylophagous insects. Annu. Rev. Microbiol. 36:323-343.
Callaham, D., J.G. Torrey, and P. del Tredici. 1978. Isolation
     and cultivation *in vitro* of the actinomycete causing root
     nodulation in *Comptonia*. Science 199:899-902.
Cavanaugh, C.M., S.L. Gardiner, M.L. Jones, H.W. Jannasch, and
     J.B. Waterbury. 1981. Prakaryotic cells in the hydrothermal
     vent tube worm *Riftia pachyptila* Jones: Possible chemautotrophic
     symbionts. Science 209:340-341.
Chakrabarty, A.M. 1973. Genetic fusion of incompatible plasmids
     in *Pseudomonas*. Proc. U.S. Nat. Acad. Sci. 70:1641-1644.
Chiariello, N., J.C. Hickman, and H.A. Mooney. 1982. Endomycorrhizal
     role for interspecific transfer of phosphorus in a community
     of annual plants. Science 217:941-943.
Clarholm, M. 1981. Protozoan grazing of bacteria in soil--impact
     and importance. Microb. Ecol. 7:343-350.
Clarholm, M. 1983. Dynamics of soil bacteria in relation to
     plants, protozoa and inorganic nitrogen. Ph. D. Dissert.
     Dep. of Microbiology, Swed. Univ. of Agric. Sci. Rep. 17,
     Uppsala.
Clarholm, M. 1984, *in press*. Interactions of soil bacteria, protozoa
     and plants for mineralization of soil nitrogen. Biol. Biochem.
Cramer, H.H. 1967. Plant protection and world crop production.
     Bayer, Leverkusen, W. Germany.
Drummond, M.H., M.P. Gordon, E.W. Wester, and M-D Chilton. 1977.
     Foreign DNA of bacterial plasmid origin is transcribed in
     crown gall tumours. Nature 269:535-536.
Ebersprächer, J., and F. Lingens. 1981. Microbial degradation of
     the herbicide chloridazon. Pages 271-285 *in* T. Leisinger,
     R. Hütter, A.M. Cook, and J. Nüesch, eds. Microbial degradation
     of xenobiotics and recalcitrant compounds. Academic Press,
     London, England.
Farrell, R., and A.M. Chakrabarty. 1979. Degradative plasmids:
     Molecular nature and mode of evolution. Pages 97-109 *in*
     K.N. Timmis and A. Pühler, eds. Plasmids of medical, environ-
     mental and commercial importance. Developments in genetics,
     vol. 1. Elsevier/North Holland Biomedical Press, Amsterdam,
     The Netherlands.
Felbeck, H. 1981. Chemoautotrophic potential of the hydrothermal
     vent tube worm, *Riftia pachyptila* Jones (Vestimentifera).
     Science 209:336-338.
Freney, J.R., and I.E. Galbally, eds. 1982. Cycling of carbon,
     nitrogen, sulphur and phosphorus in terrestrial and aquatic
     ecosystems. Springer, Berlin, W. Germany.

Frissel, M.J., ed. 1977. Cycling of mineral nutrients in agricul-
    tural ecosystems. Agro-ecosystems 4:1-354.
Frissel, M.J., and J.A. van Veen, eds. 1981. Simulation of nitrogen
    behaviour in soil-plant systems. Pudoc, Wageningen, The Nether-
    lands.
Garrett, C.M.E. 1982. Bacterial diseases of food plants--an over-
    view. Pages 115-132 *in* M.E. Rhodes-Roberts and F.A. Skinner,
    eds. Bacteria and plants. Soc. Appl. Bact. Symp. Ser. 10.
Gerdemann, J.W. 1975. Vesicular-arbuscular mycorrhizae. Pages 565-
    591 *in* J.G. Torrey and D.E. Clarkson, eds. The development
    and function of roots. Academic Press, London, England.
Giesy, J.P. 1980. Microcosms in ecological research. U.S. Dep.
    of Energy Symp. Ser. 52, Nat. Tech. Inf. Serv., Springfield,
    VA USA.
Griffin, D.M. 1981. Water and microbial stress. Adv. Microb.
    Ecol. 5:91-136.
Hale, M.G., L.D. Moore, and G.J. Griffin. 1982. Factors affecting
    root exudation and significance for the rhizosphere ecosystem.
    Pages 43-71 *in* Biological and chemical interactions in the
    rhizosphere. Swed. Nat. Sci. Res. Counc., Stockholm.
Hanlon, R.D.G., and J.M. Anderson. 1979. The effects of Collembola
    grazing on microbial activity in decomposing leaf litter.
    Oecologia 38:93-99.
Herrera, R., T. Merida, N. Stark, and C. Jordan. 1978. Direct
    phosphorus transfer from leaf litter to roots. Naturwissen-
    schaften 65:208-209.
Ineson, P., M.A. Leonard, and J.M. Anderson. 1982. Effect of
    collembolan grazing upon nitrogen and cation leaching from
    decomposing leaf litter. Soil Biol. Biochem. 14:601-605.
Johnston, A.W.B., J. L. Beynon, A.V. Buchanan-Wollaston, S.M.
    Setchell, P.R. Hirsch, and J.E. Beringer. 1978. High frequency
    transfer of nodulating ability between strains and species of
    Rhizobium. Nature 276:634-636.
Johnston, A.W.B., G. Hombrecher, and N.J. Brewin. 1981. *Rhizobium*
    plasmids: Their role in the nodulation of legumes. *In* S.B.
    Levy, ed. Molecular biology, pathogenicity and ecology of
    bacterial plasmids. Plenum, New York, NY USA.
Jones, M.L. 1981. *Riftia pachyptila* Jones: Observations on the
    vestimentiferan worm from the Galápagos Rift. Science 209:333-
    336.
Juma, N.G., and E.A. Paul. 1981. Use of tracers and computer
    simulation techniques to assess mineralization and immobiliza-
    tion of soil nitrogen. Pages 145-154 *in* M.J. Frissel and
    J.A. van Veen, eds. Simulation of nitrogen behaviour in
    soil-plant systems. Pudoc, Wageningen, The Netherlands.
Krupa, S.V., and Y. Dommergues, eds. 1979. Ecology of root patho-
    gens. Developments in agriculture and managed-forest ecology,
    vol. 5. Elsevier, Amsterdam, The Netherlands.
Lofs-Holmin, A. 1983. Influence of agricultural practices on
    earthworms. Acta Agric. Scand. 33:225-234.

McGill, W.B., H. W. Hunt, R. G. Woodmansee, and J.O. Reuss. 1981.
    Phoenix, a model of the dynamics of carbon and nitrogen in
    grassland soils. Pages 49-115 in F.E. Clark and T. Rosswall,
    eds. Terrestrial nitrogen cycles: Processes, ecosystem
    strategies and management impacts. Ecol. Bull. (Stockholm)
    33.
Parkinson, D., S. Visser, and J.B. Whittaker. 1977. Effects of
    collembolan grazing on fungal colonization of leaf litter.
    Pages 75-79 in U. Lohm and T. Persson, eds. Soil organisms as
    components of ecosystems. Ecol. Bull. (Stockholm) 25.
Paul, E.A., and R.M.N. Kucey. 1981. Carbon flow in microbial
    associations. Science 213:473-474.
Pavlon, S., and A.G. Fredrickson. 1983. Effects of the inability
    of suspension-feeding protozoa to collect all cell sizes of a
    bacterial population. Biotechnol. Bioeng. 25:1747-1772.
Pemberton, J.M., B. Corney, and R.H. Don. 1979. Evolution of
    spread of pesticides degrading ability among soil micro-
    organisms. Pages 287-299 in K.N. Timmis and A. Pühler, eds.
    Plasmids of medical, environmental and commercial importance.
    Developments in genetics, vol. 1. Elsevier/North Holland
    Biomedical Press, Amsterdam, The Netherlands.
Persson, T., ed. 1981. Structure and function of northern coniferous
    forests: An ecosystem study. Ecol. Bull. (Stockholm) 32:1-611.
Persson, T. 1983. Influence of soil animals on nitrogen minerali-
    zation in a northern Scots pine forest. Pages 117-126 in
    Ph. Lebrun, H.M. André, A. De Medts, C. Grégoire-Wibo, and
    G. Wauthy, eds. New trends in soil biology. Imprimeur Dieu:
    Brichart, Louvain-la-Neuve, Belgium.
Persson, J., and T. Rosswall. 1983, in press. Opportunities for
    research in agricultural ecosystems--Sweden. In R.R. Lowrance,
    R.L. Todd, L.E. Asmussen, and R.A. Leonard, eds. Nutrient
    cycling in agricultural ecosystems. Univ. of Georgia Coll.
    of Agric. Spec. Publ. 23. Athens, GA USA.
Powell, M.A., and G.N. Somero. 1983. Blood components prevent
    sulphide poisoning of respiration of the hydrothermal vent
    tube worm Riftia pachyptila. Science 219:297-299.
Rosswall, T. 1982. Microbiological regulation of the biogeochemical
    nitrogen cycle. Plant Soil 67:15-34.
Rosswall, T., and K. Paustian. 1984. Cycling of nitrogen in
    modern agricultural systems. Plant Soil 76:3-21.
Salkinoja-Salonen, M.S., E. Väisänen, and A. Paterson. 1979.
    Involvement of plasmids in the bacterial decomposition of
    lignin-derived compounds. Pages 301-314 in K.N. Timmis and
    A. Pühler, eds. Plasmids of medical, environmental and commer-
    cial importance. Developments in genetics, vol. 1. Elsevier/
    North Holland Biomedical Press, Amsterdam, The Netherlands.
Steen, E., P.-E. Jansson, and J. Persson. 1984. Site description
    of the experimental field of the 'Ecology of Arable Land'
    project. Acta Agric. Scand. 34(2).

Summers, A.O., and S. Silver. 1978. Microbial transformations of
    metals. Annu. Rev. Microbiol. 32:637-672.
Till, A.R., G.J. Blair, and R.C. Dalal. 1982. Isotopic studies of
    the recycling of carbon, nitrogen, sulfur and phosphorus from
    plant material. Pages 51-59 in J.R. Freney and I.E. Galbally,
    eds. Cycling of carbon, nitrogen, sulphur and phosphorus in
    terrestrial and aquatic ecosystems. Springer, Berlin, W.
    Germany.
van der Drift, J., and E. Jansen. 1977. Grazing of springtails on
    hyphal mats and its influence on fungal growth and respiration.
    Pages 203-209 in U. Lohm and T. Persson, eds. Soil organisms
    as components of ecosystems. Ecol. Bull. (Stockholm) 25.
van Veen, J.A., and M.J. Frissel. 1981. Simulation model of the
    behaviour of N in soil. Pages 126-144 in M.J. Frissel and
    J.A. van Veen, eds. Simulation of nitrogen behaviour in
    soil-plant systems. Pudoc, Wageningen, The Netherlands.
Visser, S., J.B. Whittaker, and D. Parkinson. 1981. Effects of
    collembolan grazing on nutrient release and respiration of a
    leaf litter inhabiting fungus. Soil Biol. Biochem. 13:215-218.
Wheelis, M.L. 1975. The genetics of dissimilatory pathways in
    Pseudomonas. Annu. Rev. Microbiol. 29:505-524.

AQUATIC MICROBIAL ECOLOGY--RESEARCH QUESTIONS AND OPPORTUNITIES

W. J. Wiebe

Department of Microbiology
University of Georgia
Athens, Georgia 30602 USA

In this paper, I will restrict my comments to heterotrophic microorganisms. While the marine environment will be stressed, I will try to show where the questions, concepts, and techniques are applicable to fresh waters, as well as other environments.

In the past two decades, aquatic microbial ecology has advanced greatly. The field entered the 1960s mostly as the esoteric area of research for a few microbiologists who isolated, identified, and examined the physiological and biochemical responses of strains of bacteria from different habitats. (There were, of course, some notable exceptions to this generalization, e.g., Bass-Becking, Wood, etc., but most investigators fell into this category.) It entered the 1980s with hundreds of microbial ecologists working in virtually all environments. Over this period, the concepts of the roles which microorganisms play in nature changed dramatically, the techniques by which their biomass, growth rates, and metabolic activities could be measured expanded many-fold, and the focus shifted from mostly bacterial studies to include fungi, yeast, viruses, and protozoa. Microbiologists, in addition, moved away from pure culture studies toward examination of in situ phenomena.

Before discussing current research questions and opportunities, I would like to review, briefly, what I think have been some of the major advances in aquatic microbiology over the past two decades. Although I am restricting this discussion to aquatic habitats, perhaps the greatest advance has been the development of a common, recognized field of microbial ecology that encompasses studies from all environments. Advancements in one field (e.g., soil, rumen, marine) are now rapidly applied to other environments. Common aims, techniques, and questions concerning the roles of microorgan-

35

isms in nature are now recognized, regardless of the environment
under study.

ADVANCES IN THE 1960s AND 1970s

One great advancement, in my view, has been a conceptual one.
Bacteria and other microheterotrophs now are recognized to perform
a wide variety of activities in nature; no longer are they considered
exclusively as mineralizers. In fact, their role as classical
mineralizers has been challenged (e.g., Johannes 1968). In the
early 1970s, marine biologists thought that the numbers of bacteria
in the coastal and open ocean were few and their functions relatively
unimportant. Today (e.g., Pomeroy 1979) there is a much greater
appreciation of their roles in nutrient cycling, in food web dynamics, and as biogeochemical agents.

One reason for this change of attitude has been the enormous
advances in techniques by which microorganisms can be studied *in
situ*. Biochemical and other approaches have been applied, and in
some cases developed, to deal with specific questions of microbial
distribution and activity. Biomass of microorganisms can be measured
by various methods: chemically by ATP content, muramic acid concentration, lipid spectra, and limulus lysate analysis; and microscopi-
cally using fluorescent stains, scanning electron microscopy, and
particle counters. Rates of bacterial growth, unmeasureable in the
mid-1970s, are assessed routinely now by several methods: thymidine
uptake, incubation with naladixic acid to prevent cell division,
lipid fingerprinting, phospholipid production, frequency of dividing
cells, ribonucleic acid production, and growth of bacteria in
incubated, filtered samples. Growth state, that is the "health" of
the microorganisms, has been measured by the adenylate energy
charge ratio, the electron transport system technique, and the
reduction of a tetrazolium dye followed by direct microscopic
examination.

This revolution was started by two non-microbiologists, Strickland and Parsons (1962), who proposed a technique to measure,
directly, the *in situ* rate of substrate utilization by marine
microheterotrophs. Their procedure entailed adding a radioactive
organic substrate to a water sample and measuring the amount of
label incorporated into the particulate material after incubation.
This technique extended the method of Steeman Nielsen (1952), who
developed the use of ^{14}C-$NaHCO_3$ to measure primary production in
the ocean. The heterotrophic potential technique was subsequently
improved by Wright and Hobbie (1965, 1966). Today it and its
variations are used widely to measure substrate turnover and heterotrophic activity *in situ*.

There also have been advances in instrumentation. Scanning electron microscopes equipped with electron probes make possible the quantification of a variety of metals as inclusions in cells; laser technology and computer controlled particle scanners are revolutionizing the ways in which particles can be counted and assessed. Computer technology is changing the way experiments are done and the types of experiments that can be undertaken. Many of my colleagues feel that we now have more good techniques than good questions. (It is often stated that in all fields of science investigators are limited by techniques, not ideas. This may be less true of aquatic microbiology at the moment.)

Mention should be made of several areas of research that have "come of age" in these past 20 years. Biogeochemistry, a field that few could pronounce in 1960, let alone define, is now a well recognized discipline. Geochemists have embraced the results of studies on microbial processes to explain iron ore and other mineral deposition, oil formation, and organic diagenesis. Concern over pollution of the environment has stimulated advances in microbial metabolism of petroleum products, the response of microorganisms to stress, the degradation of xenobiotics, and the microbial involvement in groundwater chemistry. Study of microbial anaerobes in nature has been initiated; our understanding of sulfate reduction, methane production, denitrification, and nitrogen fixation in nature--all anaerobic processes--has expanded enormously.

Finally, I should point out one area of aquatic microbiology that has not advanced greatly, the study of bacterial taxonomy. Today this topic is almost ignored by microbial ecologists, in part due to their enthusiasm for the advances outlined and in part due to the difficulties inherent in such studies. It is, in my view, a serious omission of effort. In all other areas of ecology, identification of the taxa is a prerequisite to the examination of function. Although we can measure responses of mixed populations and the data provide useful insight into the function of microorganisms, until the taxa are identified, truly reproducible experiments cannot be performed.

With the exception just discussed, the advances of the past two decades have established aquatic microbial ecology as a modern ecological discipline, one that has increasing relevance in the study of ecosystems. Let us turn now to the topic of this conference, the research questions and opportunities for the 1980s.

RESEARCH QUESTIONS AND OPPORTUNITIES

Here I would like to discuss three sets of topics: (1) aspects that apply directly to aquatic microbial ecology, (2) factors that apply to ecological studies in general, and (3) recognition of some "new" fields in ecology.

Aquatic Microbiology

While our understanding of the roles of bacteria and other
microorganisms in aquatic environments has expanded greatly, there
remains a great deal of "unfinished business." In addition, the
advances have opened new lines of research. In this section, I
want to consider some of the possibilities.

Food-web considerations. In 1974 Pomeroy stated that we could
not distinguish whether bacteria are sources of food for animals or
sinks in which substrates are respired without much growth. This
question is still unanswered, for we do not have good in situ
growth efficiency data for bacteria. Nor do we know much about the
assimilation to respiration ratios of various organic compounds,
for the techniques used most often have not been adequate to estimate
true CO_2 evolution.

There is a major division among researchers over whether most
of the bacteria in the ocean are active or dormant. Under the
microscope, many of the bacteria in the open ocean are small and
round (Wiebe and Pomeroy 1972). (Similar forms have also been
described for terrestrial habitats [e.g., Bae et al. 1972].) One
group (e.g., Stevenson 1978, Novitsky and Morita 1976, 1978) states
that these cells are dormant, while another group (e.g., Azam 1984,
Hodson et al. 1981, Sheldon and Sutcliffe 1978) has evidence that
small cells are as active metabolically as larger, more typical
looking bacteria. In organically rich areas, estuaries and coastal
regions, several investigators have shown that bacteria contribute
to the food supply of some animals (e.g., Wetzel 1975, Levinton and
Lopez 1977, Haas and Webb 1979), but most often the significance of
their contribution to animal growth has not been quantified. The
role of bacteria and other microorganisms as energy versus nutritional
sources has not been investigated, but it would be well to distinguish
these two different contributions to animal growth.

Which organisms eat bacteria and other microorganisms? Protozoa
consume bacteria, sometimes exclusively (e.g., Haas and Webb 1979).
While there is evidence that some animals (or their resident gut
microflora) can digest detrital carbon, it is increasingly clear
that much of the food value of detritus is derived from the resident
microflora and microfauna. In this regard, the classical trophic
model of biophages and saprophages (Wiegert and Owen 1971) breaks
down. The simple food chain models that were presented in the
1970s are giving way to more complex, microbially mediated food web
models (Pomeroy 1979).

Fungi, yeast, protozoa, and viruses. The roles of fungi,
yeast, protozoa, and viruses in aquatic environments are slowly
being recognized. Fungi and yeast appear important in localized
areas of estuarine habitats (Newell and Hicks 1982), but their

contributions in the open ocean are uncertain. By all estimates, their biomass in the water is very low. Protozoa have received much attention recently (Sherr et al. 1982, Fenchel 1976), and their study is certain to be expanded greatly in the 1980s. Viruses have received scant attention in aquatic habitats, except as disease agents (e.g., Berg 1977). Viruses have been isolated for aquatic bacteria (Wiebe and Liston 1968, Zachary 1974, Moebus 1980), blue-green and eucaryotic algae (Safferman and Rohr 1979), and animals (Johnson 1978). Their role in population regulation is unknown but of potential importance (see below). For all of these organisms, there are problems in cultivation and identification. Sieburth (1975) has given us a tantalizing visual glimpse of some of these organisms; the next few years should see a much greater understanding develop about their functions.

Plant-microbe relationships. A major question is whether bacteria are competitors or stimulators of plant growth in aquatic habitats. Bacteria generally have better uptake capabilities for inorganic nutrient assimilation than do plants, particularly in low nutrient regimes. This characteristic suggests that they are competitors. However, it has been shown that some algae require the presence of a microflora for growth (Droop 1968); in some, growth in the absence of microorganisms results in aberrant plant morphology (Provosoli and Pinter 1980).

Diseases in aquatic plants occur, but very little information is available about them.

Animal-microbe relationships. I already have mentioned the relationship of microbes in food webs. Another role microorganisms play is in transforming fecal material into living cells and byproducts. Pomeroy and Deibel (1980) observed the rapid colonization of fecal material by bacteria and protozoa, and several workers have shown that most fecal material does not sink to the bottom of the sea. We are only starting to recognize the importance of this intermediate in marine food webs.

Aquatic animals are subject to a variety of diseases. These diseases are better documented than those for plants, probably because of the economic importance of some animals. Bacterial, viral, fungal, and protozoan diseases are well known in fish and oysters, and periodic epidemics have been documented. However, we know little about the role disease plays in controlling populations under natural conditions.

Mutualism between microbes and marine animals is well recognized. Wilkinson (1978) has shown that many sponges harbor heterotrophic and blue-green algae. They are not digested to any extent, but they have been shown to fix nitrogen (letter dated 12 October 1982 from C. Wilkinson, Australian Institute of Marine Science, Townsville,

Queensland, Australia). The guts of many marine animals appear to harbor specific microfloras (e.g., Liston 1957), but there has been little work done on their functional relationships. (There is evidence that a gut microflora is essential for the optimal growth of many animals.) Ruminants, as well as other herbivores, require microbial processing of their food. Although there are many herbivores in marine environments, the specific roles of their microflora are largely unknown.

Rhoads (1974) has demonstrated an indirect way in which microorganisms affect animals. On the subtidal sediment surface, microorganisms often form films, which affect the subsurface redox potential by limiting exchange and bioturbation. These films may be of importance in the development of microhabitats as well as in the reduction of sediment stirring.

Animals also exert control on the density of microorganisms. Habte and Alexander (1978) showed that in broth an introduced bacterial species co-exists with ciliates at a density of 1×10^5 cells/ml, regardless of the number of cells introduced. Berk et al. (1977) demonstrated that ciliates are a good food source for copepods and that the copepod controls the density of the ciliates. I believe that predator-prey studies with microorganisms could be greatly expanded, for they are simple systems with which to develop our understanding of food webs.

Distribution of microorganisms. A prerequisite for analysis of any natural phenomenon is the ability of the investigator to establish order (prediction) for the variables. Temporal and spatial variability of microorganisms in euphotic waters has long plagued investigators. There are several factors that lead to patchiness. Physically these range from surface films, fecal pellets, large organisms, and detritus in the water to day-night difference in metabolism and the structure of water. Platt and Denman (e.g., 1977) have examined the reasons for and the scales of patchiness in phytoplankton. Imberger et al. (1983) examined the distribution of chemical variables (NH_3, silicate, dissolved and particulate carbon) in relation to the different forces that move water. It is becoming clear that often we are not sampling on the right scales. Fine scale structure of water has not been examined until now, because it has been considered too small to matter biologically. I believe much work is needed to resolve the appropriate spatial scales for sampling organisms and chemicals in the sea and that when these scales are resolved we will be saddled with much less variability.

In sediments and deep water, the abundance of organisms appears more predictable (e.g., Christian and Wiebe 1978, Ferguson and Rublee 1976). However, the types of organisms and their functions can differ greatly. We do not know what is responsible for the remarkable homeostasis in density, although grazing by protozoa is being invoked increasingly as an explanation.

Controlling factors. Both physical and biotic factors control the abundance of microorganisms in environments. Factors that act to control the abundance and types of microorganisms are light, pH, type of substrate, water movement patterns, temperature, substrate release, food quality, and predation. The mechanisms for some are well understood, but for others, we have little information on how they exert control.

There are a variety of mechanisms whereby microorganisms can control the distribution and abundance of large organisms. I already have mentioned their role as food and disease agents. In addition, the products of their metabolism substantially affect ecosystem properties. For example, sulfate reducers produce hydrogen sulfide, which is toxic to most plants and animals; denitrifying bacteria convert nitrate, a nutrient, to N_2; acetic acid bacteria substantially alter the pH of the medium. We need to learn much more about these types of control.

Relationship of microbial ecology to classical ecological studies. There has been, and continues to be, resistance on the part of the more classical ecologists to recognizing the necessity for integrating microbial studies in ecosystems studies. And the reverse is true. Microbiologists need to take a much more holistic view of their work. An example of the problem is the structure of the long term ecological research programs of the U.S. National Science Foundation (NSF). Very few microbiological studies are included, yet as I have discussed, microorganisms are an important part of the biota of ecosystems. It is clear that we have a long way to go before microbial ecology is truly integrated in ecological research.

Comparative studies. Too often investigators do not extend their results from one system to others. Comparative studies in biology have proven an essential step in the development of many fields (e.g., comparative anatomy, comparative biochemistry, comparative physiology) and in the development of viable theories. I will return to this subject subsequently.

Field experiments. Ecologists have been said to be the only group of scientists that run controls but no experiments. And it is true that most often we investigate the unperturbed--the more unperturbed the better--ecosystems. Ultimately, this information has limited predictive value. Systems are now being manipulated on several scales from meter plots to hectares. Lakes have been enriched by nutrients either inadvertently (Edmondson and Lehmans 1981) or on purpose (LeBrasseur et al. 1978). In the marine environment, 10 m diameter by 15-30 deep bags have been used to conduct

experiments on effects of pollutants on planktonic populations.
Humans provide us with still other perturbations (see, e.g., Smith
et al. 1981 for an excellent example of how to use a planned sewage
diversion for research purposes). These perturbed areas furnish us
with manipulated systems on realistic scales. Microbial ecologists
may be fortunate, because the small size of their organisms and
their generally fast rates of growth permit field manipulations to
be made for relatively short periods and on small scales. Certainly
in the next decade there will be increasing emphasis on field
experimental studies.

Other topics. In the above sections, I have pointed out some
of the major topics and questions that I believe will be pursued in
the next decade; many are already being pursued. There are other
topics that I would like to mention briefly before closing this
part of the paper.

(1) Prediction of degradation rates of xenobiotics. Consortia of
    organisms together with physical factors such as light and pH
    often transform xenobiotics very differently than do pure
    cultures. The U. S. Environmental Protection Agency is trying
    to develop criteria for predicting the disappearance of xeno-
    biotics; it really is an ecological problem.
(2) Utilizaton of polymeric substances in nature. We have measured
    the concentrations and turnovers of monomers in nature, but
    most organic matter is polymeric, protein, chitin, cellulose,
    lignin, etc., and the metabolism of these compounds needs
    examination.
(3) Microflora of the gut. This topic has received limited attention
    by microbial ecologists, except for work on ruminants. Do
    microorganisms aid (and if so, how) or compete in processing
    food in the guts of animals? Are there specific, resident
    microflora in all animals, and what are their functions?
(4) Marine pharmacology. Many of our modern defenses against
    diseases are derived from compounds isolated from terrestrial
    microorganisms. Much less attention has been paid to marine
    microorganisms. A major facility for this research, the Roche
    Institute of Marine Pharmacology in Australia, has closed but
    research should continue.
(5) Marine biochemistry. Dr. David White has developed numerous
    biochemical assays for different types of organisms, based on
    unique compounds found in different groups (e.g., White et al.
    1979). I believe such studies will go far in the future to
    resolve some of the biomass, growth rate, and identification
    problems cited earlier. These studies provide examples of how
    the application of biochemistry can advance microbial ecology.
(6) Stromatolitic mats. Algal-bacterial mats are being examined
    increasingly from both geochemical and biogeochemical standpoints.
    Stromatolites, 3.5×10^9 million years old, have been identified
    recently in rock (Walters et al. 1980); thus, these present

day communities represent Archean type growth forms. They are
relatively simple, manipulatable systems and their study
should advance several fields, yet few microbiologists are
working on these systems.

(7) Rhizosphere of flooded aquatic plants. While the study of
microbial processes in flooded sediments has increased greatly
over the last decade, little attention has been played to the
rhizosphere specifically, although it represents a major site
of microbial activity. Such studies are badly needed to help
develop an understanding of how plants thrive in anoxic sediments.

In summary, the past two decades have seen a rapid advance in
many areas of microbial ecology. There is now a great opportunity
to build on this foundation and integrate the microbiological
studies into a total ecosystem context. Certainly microbial ecology
should no longer be considered an esoteric or non-essential part of
any ecosystem study.

GENERAL CONSIDERATIONS

In addition to the specific questions and opportunities for
aquatic microbial ecology in the next decade, I believe that there
are some more general aspects of ecological research that deserve
mention, and I want to discuss three of these briefly.

Coastal Zone Research

Although there are many studies of estuaries and continental
shelf and open ocean waters throughout the world, the very nearshore
environments have received limited attention. Mangrove research is
just starting, coral reef research is considerably advanced, but
the processes and organisms that occur in the inner 1-10 km of
rocky and sandy shores have scarcely been investigated. Yet, as
pointed out by Mann (1982):

A large proportion of the world's population is concentrated
along the coastline and along the banks of rivers which drain
into coastal waters. Hence, the effects of pollution are most
marked in nearshore waters. At the same time, man has come to
rely heavily on coastal waters for food and for recreation.
The double impact of adding pollutants while harvesting plants
and animals places a great strain on coastal aquatic ecosystems.
Only by exploring and understanding their modes of operation
shall we be able to preserve their value as a food source and
their recreational potential. (Reprinted by permission of the
University of California Press, Berkely, CA USA.)

Not only is little systematic effort expended on this zone, appropriate platforms from which to conduct research are lacking. Large research vessels most often cannot work in these waters, and small boats often do not permit the use of modern techniques. In addition, the structure of research institutes and the research interests of individual scientists hinder progress. Coastal zone managers recognize the need to examine entire watersheds. In Hawaii, for example, they consider the coastal zone to extend from the tops of the mountains to several kilometers offshore. I know of no single research laboratory that can handle soil, groundwater, and coastal marine problems in a coherent manner. There is an urgent need, for the reasons stated by Mann (1982) as well as others, to focus much more attention on the land-water boundary in the next decade.

Resolving Space and Time Scales

Over the past decade, increasingly sensitive instrumentation coupled with computer assisted analysis have permitted investigators to resolve in real time fine scale physical structure in marine and fresh waters (e.g., Imberger et al. 1983). For the reasons cited previously, it appears that these scales may be the appropriate ones for examining many biological phenomena. There is a need to develop compatible sensors for measuring variables important in biology. Recently, *in situ* pH measurements have been coupled with high resolution conductivity, temperature, depth, and light quanta probes (unpublished data from J. Imberger, Department of Civil Engineering, University of Western Australia, Nedlands, Western Australia). Several specific ion electrodes are now available, including ammonia, nitrate, phosphate, and sulfide. Presently, these are difficult to interface with conductivity/temperature/depth (CTD) measurements because of the longer response times of the electrodes; in addition, they lack sufficient sensitivity for natural waters. Oxygen electrodes have sufficient sensitivity for natural waters, but their response times are longer than the CTD response times. I feel that this area of research offers one of the best possibilities for investigating the relationship between biological and physical processes.

Long Term Research Sites

Perhaps the greatest need for the next decade and beyond is to develop a few sites in different ecosystem types that can be maintained and investigated for long periods, literally hundreds of years. The time scale over which ecological research is now conducted can be compared to a few frames taken from a three hour movie. We simply do not examine ecosystems over time scales relevant to their development. Furthermore, in all other fields of biological research, most of the developmental work, those studies which provide new insights and theories, are confined to a very few organisms. Work on *Escherichia coli* is central to microbiological thought; similarly, work on *Drosophila* in population genetics, rats in mammalian biochemistry, and *Chlorella* in freshwater algal physiology have played pivotal roles in the development of these fields. These organisms are used because of the vast literature on them, not because they are ideal or typical representatives. They serve as "reference organisms" for new research. We have no analogous reference ecosystems, nor can we conveniently package our ecosystems and ship them to colleagues. The U.S. NSF program for the development of long term ecological research sites is a start toward the goal of reference ecosystems, but it suffers several deficiencies as presently conceived.

I propose that a limited number of internationally and broadly supported ecosystem sites be established. Where possible, these sites should be situated in developing nations. A resident staff, responsible for monitoring environmental variables, should be appointed and modern facilities established for use by the resident staff and visiting investigators. Investigators would be supported by outside research grants, as they are presently for use of million volt electron microscopes, research vessels, and cyclotrons.

A meeting is needed to define terms of reference, select sites, and implement studies. Although there are numerous ecological field stations and laboratories, none I know, with the exception of some agricultural institutes (e.g., Rothamsted in Great Britain), are structured to collect coherent, synoptic data for long time periods. I believe that until such sites and data bases are established, ecologists will lack a critical ingredient for the advancement of the field. In addition to providing reference systems for all types of scientists, not just ecologists, the data from these sites should stimulate comparative studies between ecosystems.

NEW FIELDS OF ECOLOGY

Ecologists identify their subdisciplines in a variety of ways, by the organisms (animal, plant, phytoplankton, microbial, etc.), by the level of organization (autoecology, population, community, ecosystem, etc.), by locations of interest (estuarine, soil, terrestrial, coral reef, etc.), and by the approaches they take (evolutionary, genetic, theoretical, energetic, etc.). These titles help identify research interests and provide a focus for investigators.

I believe that there are two fields not specifically represented that deserve recognition, metabolic ecology and comparative ecology.

There has been a rapid increase in studies of metabolic processes and pathways in ecosystems (see Mann 1982 for numerous examples in marine studies), but they are not identified as a legitimate subdiscipline of ecology. The term comparative ecology is necessary, I believe, to focus attention on an important but overlooked field.

ACKNOWLEDGEMENTS

I would like to acknowledge, gratefully, the contributions several colleagues made to the ideas in this paper: Bob Hodson, Dave Kirchman, John Patton, Lex Maccubbin, Rodger Harvey, Bob Murray, Huey-Min Hwang, Emily Peele, Henry Spratt, Jr., and Ron Benner.

The work was supported, in part, by grants from the Sapelo Island Research Foundation, Inc., and Contribution 507 from the University of Georgia Marine Institute, Sapelo Island, Georgia.

REFERENCES CITED

Azam, F.   1984, *in press*.   Growth of bacteria in the ocean.   *In* J.E. Hobbie and P.J.L. Williams, eds.   Heterotrophic metabolism in the sea.   Springer, New York, NY USA.

Bae, H.C., E.H. Cota-Robles, and L.E. Casida, Jr.   1972.   Microflora of soil as viewed by transmission electron microscopy.   Appl. Microbiol. 23:637-648.

Berg, G.   1977.   Microbiology-detection, occurrence and removal of viruses.   J. Water Pollut. Control Fed. 49:1290-1299.

Berk, S.G., D.C. Brownlee, D.R. Heinle, H.J. Kling, and R.R. Colwell.   1977.   Ciliates as a food source for marine planktonic copepods. Microb. Ecol. 4:27-40.

Christian, R.R., and W.J. Wiebe.   1978.   Anaerobic microbial community metabolism in *Spartina alterniflora* soils.   Limnol. Oceanogr. 23:328-336.

Droop, M.R.   1968.   Vitamin B_{12} and marine ecology.   IV. The kinetics of uptake, growth and inhibition in *Monochoysis lutheri*.   J. Mar. Biol. Assoc. U.K. 48:689-733.

Edmonson, W.T., and J.T. Lehman.   1981.   The effect of changes in the nutrient income on the condition of Lake Washington. Limnol. Oceanogr. 26:1-29.

Fenchel, T.   1976.   The significance of bacterivorous Protozoa in the microbial community of detrital particles.   Pages 529-544 *in* J. Cairns, ed.   Aquatic microbial communities.   Garland, VA USA.

Ferguson, R.L., and P. Rublee.   1976.   Contribution of bacteria to standing crop of coastal plankton.   Limnol. Oceanogr. 15:14-20.

Haas, L.W., and K.L. Webb.   1979.   Nutritional mode of several non-pigmented microflagellates from the York River Estuary, Virginia.   J. Exp. Mar. Biol. Ecol. 39:125-134.

Habte, M., and M. Alexander.   1978.   Protozoan density and the coexistence of protozoan predators and bacterial prey.   Ecology 54:140-146.

Hodson, R.E., A.E. Maccubbin, and L.R. Pomeroy. 1981. Dissolved adenosine tri-phosphate utilization by free-living and attached bacterioplankton. Mar. Biol. 64:43-51.

Imberger, J., T. Berman, R.R. Christian, R.B. Hanson, L.R. Pomeroy, E.B. Sherr, D. Whitney, W.J. Wiebe, and R.G. Wiegert. 1983. The influence of water motion on the spatial and temporal variability of chemical and biological substances in a salt marsh estuary. Limnol. Oceanogr. 28:201-214.

Johannes, R.E. 1968. Nutrient regeneration in lakes and oceans. Pages 203-213 in M. Droop and E.F.J. Woods, eds. Advances in marine microbiology. Academic Press, New York, NY USA.

Johnson, P.T. 1978. Viral diseases of the blue crab, *Callinectes sapidus*. Mar. Fish. Rev. 40:13-15.

LeBrasseur, R.J., C.D. McAllister, W.E. Barraclough, O.D. Kennedy, J. Manzer, D. Robinson, and K. Stephens. 1978. Enhancement of sockeye salmon (*Onchorhynchus nerka*) by lake fertilization in Great Central Lake: Summary report. J. Fish. Res. Board Can. 35:1580-1596.

Levinton, J.S., and G.R. Lopez. 1977. A model of renewable resources and limitation of deposit-feeding benthic populations. Oecologia 31:177-190.

Liston, J. 1957. A quantitative and qualitative study of the bacterial flora of skate and lemon sole trawled in the North Sea. Ph.D. Dissert., Univ. of Aberdeen, Aberdeen, Scotland.

Mann, K.H. 1982. Ecology of coastal waters: A systems approach. Stud. Ecol. 8. Univ. of Calif. Press, Berkeley, CA USA.

Moebus, K. 1980. A method for detection of bacteriophages from ocean water. Helgol. Wiss. Meeresunters. 34:1-14.

Newell, S.Y., and R.E. Hicks. 1982. Direct-count estimates of fungal and bacterial biovolume in dead leaves of smooth cordgrass (*Spartina alterniflora* Loisel). Estuaries 5:246-260.

Novitsky, J.A., and R.Y. Morita. 1976. Morphological characterization of small cells resulting from nutrient starvation of a psychrophilic marine *Vibrio*. Appl. Environ. Microbiol. 32:617-622.

Novitsky, J.A., and R.Y. Morita. 1978. Possible strategy for survival of marine bacteria under starvation conditions. Mar. Biol. 48:289-295.

Platt, T., and K.L. Denman. 1977. Organization in the pelagic ecosystem. Helgol. Wiss. Meeresunters. 30:575-581.

Pomeroy, L.R. 1974. The oceans food web, a changing paradigm. Bioscience 24:499-504.

Pomeroy, L.R. 1979. Secondary production mechanisms of continental shelf communities. Pages 163-188 in R.J. Livingston, ed. Ecological processes in coastal and marine systems. Plenum, New York, NY USA.

Pomeroy, L.R., and D. Deibel. 1980. Aggregation of organic matter by pelagic tunicates. Limnol. Oceanogr. 25:643-652.

Provasoli, L., and I.J. Pinter. 1980. Bacteria induced polymorphism in an axenic laboratory strain of *Ulva lactueo* (Chlorophyceae). J. Phycol. 16:196-201.

Rhoads, D.   1974.   Organic-sediment relations on the muddy sea
    floor. Oceanogr. Mar. Biol. Annu. Rev. 12:263-300.
Safferman, R.S., and M.E. Rohr. 1979. The practical directory to
    phycovirus literature. U.S. Environ. Prot. Agency, Cincinnati,
    OH USA.
Sheldon, R.W., and W.H. Sutcliffe, Jr. 1978. Generation times of
    3 h for Sargasso Sea microplankton determined by ATP analysis.
    Limnol. Oceanogr. 23:1051-1055.
Sherr, B.F., E.B. Sherr, and T. Berman. 1982. Decomposition of
    organic detritus: A selective role for microflagellate Protozoa.
    Limnol. Oceanogr. 27:765-769.
Sieburth, J. M. 1975. Microbial seascapes. Univ. Park Press,
    Baltimore, MD USA.
Smith, S.V., W.J. Kommerer, E.A. Laws, R.E. Brock, and T.W. Walsh.
    1981. Kaneoke Bay sewage diversion experiment: Perspective
    on ecosystem responses to nutritional perturbation. Pac. Sci.
    35:279-395.
Steeman Nielsen, E. 1952. The use of radioactive carbon (^{14}C) for
    measuring organic production in the sea. J. Cons. Explor.
    Mer. 18:117-140.
Stevenson, L.H. 1978. A case for bacterial dormancy in aquatic
    systems. Microb. Ecol. 4:127-133.
Strickland, J.D.H., and T.R. Parsons. 1962. On the production of
    particulate organic carbon by heterotrophic processes in sea
    water. Deep Sea Res. 8:211-222.
Walters, M.R., R. Buick, and J.S.R. Dunlop. 1980. Stromatolites
    3,400-3,500 Myr old from the North Pole area, Western Australia.
    Nature 284:443-445.
Wetzel, R.L. 1977. Carbon resources of a benthic salt marsh
    invertebrate *Nassarius absolatus* soy (Mollusca:Nassariidae).
    Estuar. Proc. 2:293-308.
White, D.C., R.J. Bobbie, J.S. Herron, J.S. King, and S.J. Morrison.
    1979. Biochemical measurements of microbial mass and activity
    from environmental samples. Pages 69-81 *in* J.W. Costarton and
    R.R. Colwell, eds. Native aquatic bacteria: Enumeration,
    activity and ecology. Am. Soc. Test. Mater. Spec. Tech. Publ.
    695.
Wiebe, W.J., and J. Liston. 1968. Isolation and characterization
    of a marine bacteriophage. Mar. Biol. 1:244-249.
Wiebe, W.J., and L.R. Pomeroy. 1972. Microorganisms and their
    association with aggregates and detritus in the sea: A micro-
    scopic study. Mem. Ist. Ital. Idrobiol. 29:(Suppl.) 325-352.
Wiegert, R.G., and D.F. Owen. 1971. Trophic structure, available
    resources and population density in terrestrial vs. aquatic
    ecosystems. J. Theor. Biol. 30:69-81.
Wilkinson, C.R. 1978. Microbial associations in sponges. III.
    Ultrastructure of the *in situ* associations in coral reef
    sponges. Mar. Biol. 49:177-185.
Wright, R.T., and J.E. Hobbie. 1965. Uptake of organic solutes by
    bacteria in lake water. Limnol. Oceanogr. 10:22-28.

Wright, R.T., and J.E. Hobbie. 1966. Use of glucose and acetate
    by bacteria and algae in aquatic ecosystems. Ecology 47:447-468.
Zachary, A. 1974. Isolation of bacteriophages of the marine
    bacterium *Beneckea natriegens* from coastal salt marshes.
    Appl. Microbiol. 27:980-982.

SOME GROWTH POINTS IN INVESTIGATIVE PLANT ECOLOGY

Peter J. Grubb

Botany School
Downing Street
Cambridge CB2 3EA, England

ABSTRACT

There are abundant opportunities for progress in the following
fields: (1) maintenance of species richness, particularly in characterizing the factors which chiefly control the population sizes of
co-existing species and in determining the extent to which sparse
species in any given community have an appreciable impact on each
other; (2) the relative importance of shoot and root mediated
interference between plants, and the involvement of various mechanisms
of root mediated interference, in the control of relative abundance
and primary and secondary dominance and in the exclusion of species
from particular communities; and (3) leaf form and function, particularly in understanding the evolution of maximum rates of net assimilation in different kinds of plants and of various structural
characteristics possibly related to intra-leaf hydraulic conductance.
In general, much more long term research is needed in all aspects
of ecology. Plant ecologists still have a great deal to learn from
animal ecologists, and more should be done to make the two groups
aware of each other's ideas.

INTRODUCTION

Investigative plant ecology is a discipline which links the
ecosystem approach and the organism approach, and involves a unification of descriptive and experimental studies. It is based upon
description of the distribution and abundance of species through
space and time, description of the processes in vegetation (fluxes
of individuals or genotypes and energy and nutrients), and experiments designed to elucidate the relationships among various plants

51

and between them and their biotic and physical environment. I shall review very briefly four particular issues: patterns of regeneration in plant communities, maintenance of species richness, the nature of interference between plants in natural and semi-natural communities, and plant form and function.

PATTERNS OF REGENERATION IN PLANT COMMUNITIES

By "regeneration" of a community, I mean the sum total of processes leading to the perpetuation of that community despite the death of individual plants. By "regeneration" of a species, I mean the sum total of processes leading to its perpetuation despite the death of individuals or large parts of individuals. The many different patterns of regeneration now known form a spectrum, from those involving periodic large scale destruction (e.g., by fire), through those involving small scale destruction (e.g., by windthrow of a few trees), to those based on gradual formation of small gaps by the death of single plants from disease or old age. In the last decade, studies of regeneration have merged with studies of demography, but a careful study of the authoritative summary of modern work on the population biology of plants (Harper 1977) shows that there is a long way to go in synthesizing the demographic approach with the experience of other ecologists interested in the dynamics of vegetation.

My chief interest in this area is explaining long term co-existence of many species in a community, but I believe that an understanding of the spatial and temporal pattern of regeneration is essential for any sound study of community or ecosystem function (e.g., production, water use, or nutrient cycling). Equally, an understanding of regeneration is essential for any sound study of the control of relative abundance, that is, the abundance of one species relative to another (which I deal with in the next section under the general topic of interference). The same is true, therefore, for any study of stability or any active practice of management. I believe that this central role of studies on regeneration is now widely appreciated by plant ecologists, but it was not so in the recent past. The gradual appreciation of the point by those working on forest productivity can be traced in the article by Loucks et al. (1981).

Although there is a growing appreciation of the importance of studies on regeneration in all kinds of communities, and we now have an impressive body of knowledge for certain kinds of communities (e.g., forests), we are still woefully ignorant of the patterns and time scales of regeneration in many types of communities (particularly savannas, grasslands, and wetlands) and we have very limited understanding of others (e.g., semi-deserts). In all these little understood communities, much regeneration is by vegetative means and many individuals are long lived. What is needed are new long term studies in the communities concerned. We cannot expect to learn all we want to know in the next decade--it may take at least a century--but the sooner a start is made the better.

The general issue of very great importance here is the need for more funds to be dedicated to long term research in ecology, a theme that is apparent in other contributions to this volume.

MAINTENANCE OF SPECIES RICHNESS

I want to illustrate the value of longer term studies in understanding the maintenance of species richness. In my laboratory, we have been impressed by how much we could learn through monitoring sites in chalk grassland and dune communities for just five years. Our experience has caused me to revise my views on the maintenance of species richness set out in a review six years ago (Grubb 1977a), and I want to show here how I think our understanding is developing. I suggested in 1977 that the most important factor making possible the long term co-existence of species with the same habitat toleran- ces, life forms, and phenologies is differentiation in their require- ments for regeneration--what I called the "regeneration niche." I emphasized the importance of differences between species in their requirements for abundant flowering and seed set, in their degree of dispersal in space and time, and in their requirements for germination, establishment, and "onward growth."

We now have results that illustrate this thesis positively, but also emphasize its limitations. These results are for short lived plants in chalk grassland in southern England: annuals, biennials, and pauciennials. These all regenerate in tiny gaps (diameter 0.5-5 cm) in a matrix of perennials only 2-10 cm tall. At one site, D. Kelly and I have made medium scale observations on numbers of flowering individuals (200 quadrats of 0.25 m^2), small scale observations on seedling fates (8 quadrats of 0.25 m^2), and records of weather and management (Table 1). At another site, Kelly (1982) has made extremely detailed small scale observations on seedling fates and soil temperature and water content, and has carried out certain simple experiments (e.g., with snail poison). As a result, we now have evidence that the species differ appreciably in their sensitivities to water supply, turf height, and predation by molluscs, and in their dormant seed banks and conditions needed to induce germination (Table 2). These factors appear to be the most important in controlling the population sizes of these species. In contrast, certain properties, which are much easier to measure and which attract attention initially as likely useful measures of the regeneration niche, are much less important in the control of population size (e.g., seed size, time of flowering, means of pollination, and means of dispersal in space). The implication is

Table 1.   Numbers of flowering individuals of various short lived
           plants on a transect of 50 x 0.5 m in chalk grassland in
           southern England and information on rainfall and management
           (from Grubb et al. 1982 and unpublished results).

	1978	1979	1980	1981	1982
Annual					
Rhinanthus minor	1,650	3,460	1,620	2,580	210[a]
Biennials					
Blackstonia perfoliata	76	28	76	40	31
Centaurium erythraea	266	622	250[b]	65	4
Gentianella amarella[c]	208	27	939	159[d]	42[d]
Linum catharticum[c]	6,080	1,330	7,940	78[e]	146[f]
Pauciennials					
Carlina vulgaris	8	11	4	8	6
Medicago lupulina	22	19	46	30	10
Picris hieracioides	32	25	96	130	40
Rainfall at nearby site					
April-July (mm)	237	213	292	213	117
April-May (mm)	88	128	46	150	39
Grazing in previous winter	Hard	Hard	Hard	Light	Hard

[a] Low chiefly because many plants died of desiccation.
[b] Heavily infected with a fungus which caused widespread
    malformation of fruits.
[c] Alternation of high values in 1978 and 1980 with low values in
    1979 is possibly a delayed effect of some unfavourable factor
    affecting the flowering individuals of 1977, perhaps the marked
    drought of 1976.
[d] Decidedly concentrated in areas of shortest turf.
[e] Low chiefly because many overwintering plants died in the long turf
    of 1981.
[f] Low chiefly because many seedlings died in the long turf of 1981.

that we still are very far from assessing the importance of differences in regeneration niches in the maintenance of species richness
in many communities we would like to understand (e.g., tropical
rain forests). In the Blue Mountains of Jamaica, for example,
Tanner (1982) has shown that most of the 48 co-existing dicotyledo-
nous tree species in "Mull Ridge forest" and "Wet Slope forest"

(i.e., in temperate grasslands and forests). In recent years, many animal ecologists have questioned the extent of interspecific competition in nature, and that questioning can again be illustrated in the paper of Varley (1949).

THE NATURE OF INTERFERENCE BETWEEN PLANTS

I use the term "interference," following Harper (1961), to refer to infliction of hardship by a plant on its neighbours. I distinguish this phenomenon from competition, which embraces the total relationship between any two species that have the potential to occupy a given place in the landscape, and involves such properties as dispersibility, tolerance during establishment, etc.

I want to consider interference in three contexts: control of relative abundance, dominance, and exclusion of species from particular communities.

Control of Relative Abundance

Although the question of long term co-existence of many species in one community has been widely studied, the interest and importance of explaining long term differences in relative abundance seem hardly perceived by most plant ecologists (for a notable exception, see Rabinowitz 1981), and yet an understanding of this issue is fundamental to any sound practice of management. Indeed, most of the relevant experimental studies that have been published have been by agronomists, but they usually have worked with few species systems (e.g., Haynes 1980). Fortunately, there have been enough studies of relative abundance through time in multispecies, semi-natural systems to document the range of phenomena that we have to explain. In some cases, we have cyclic patterns of change, in others marked fluctuation, and in yet others remarkable constancy (Grubb et al. 1982). Even where there are cyclic effects (e.g., those associated with periodic fire), there may be remarkable constancy in the relative abundance of species that are, respectively, early post-disturbance and late post-disturbance.

My colleagues and I (Grubb et al. 1982) recently wrote

Where one species is persistently less abundant than another, it is because either there are fewer micro-sites in which it can regenerate, or it is less effective in reaching the micro-sites suitable for it. Less abundant species can persist in a community where they find fewer appropriate micro-sites provided they are quicker to invade them, or have a greater potential for interference in the short term.

Table 2. Characteristics most important in governing population size and performance of certain short lived plants in chalk grassland (from Kelly 1982 and unpublished observations).

Plant	Sensitivity of germination & establishment to tall turf	Sensitivity to drought	Sensitivity to molluscan predators	Presence of long lived seed in soil
Rhinanthus minor	+[a]	++++	-[b]	-
Euphrasia officinalis agg.	++	+++	-	-
Linum catharticum	++	++	+	+
Gentianella amarella	+++	+	-	++

[a]+ Characteristic important in governing population size and performance; greatest effect indicated by ++++.
[b]- Characteristic not involved in governing plant population size and performance.

have small, whitish flowers pollinated by generalist insects, bird dispersed fleshy fruits of similar size, periods of flowering extending over several months, and fruit dispersal concentrated into one-half of the year (and thus rather limited phenological separation overall in reproduction); the same plants have a fair degree of shade tolerance as seedlings and saplings (Tanner 1977). What is not known is how their seed production and seedling establishment vary from year to year as a function of differences in the weather or herbivore populations and how much they differ in their shade tolerances at different stages in development.

Future work on the regeneration niche in any community, hot or cold, wet or dry, should concentrate on the responses of plants to physical factors and herbivores, and must center on at least medium term field studies. The importance of an analogous approach to animals was emphasized long ago by Varley (1949). He showed that various co-existing woodland moths had different temporal patterns of reproduction and, therefore (in theory) different susceptibilities

Table 3. Correlation coefficient between number of flowering plants of *Rhinanthus minor* in a quadrat in one year (n) and previous years (n - 1, n - 2, n - 3) in a transect with 100 quadrats (from Kelly 1982).

Year n	n - 1	n - 2	n - 3
1979	0.384[a]		
1980	0.274[b]	-0.047	
1981	0.230[c]	-0.051	-0.142

[a] P < 0.001.
[b] P < 0.01.
[c] P < 0.05.

to weather events at particular times, but questioned whether these differences affected co-existence; the existence of different suites of parasites seemed potentially more important to him. In short, he suggested that observed differences in niche do not necessarily provide the effective basis for co-existence.

The second point to emerge from our studies on short lived plants in chalk grassland is the need to question the extent to which many co-existing species actually interact. I treat species as co-existing when they all have a finite chance of occupying a given microsite within a reasonable time span (e.g., 10 or 100 years). Our studies have shown that the populations of all species studied are markedly patchy and that the patches move about through time like shifting clouds in a skyscape (cf. Table 3 and Fig. 1). Zoologists have long appreciated the importance of this point at all scales from 1 dm^2 to 10^5 ha (cf. Andrewartha and Birch 1954) and interest in it has been rekindled recently by Caswell (1978) and by Taylor and Taylor (1979), but there appear to have been few demonstrations by botanists. The few demonstrations that have been published have been for wandering perennials and not for short lived plants (e.g., Lieth 1960, Bornkamm 1961).

The two points of particular importance in maintenance of species richness are (1) the "clouds" are only rarely dense enough to give appreciable density dependent control of population size, and (2) the "clouds" of different species frequently do not coincide. These findings led us to ask whether interspecific effects among the short lived plants are commonly significant at field densities, and they turned out not to be. When densities were increased by

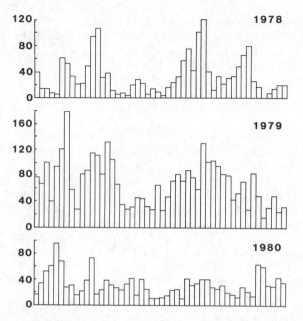

Fig. 1. Number of flowering plants per 0.5 m^2 of the annual *Rhinanthus minor* in 1 m length quadrats on a transect 50 x 0.5 m in three successive years at a chalk grassl site in southern England. The results show marked pat in all years and the movement of "clouds" of abundance year to year.

addition of extra seed, significant interspecific effects could induced. The rarity of interspecific contact must tend to incre as overall rarity increases. Generally important is the fact th in all communities that are extremely species rich most species must be rare. I therefore suggest that although their continued existence is dependent on their having different regeneration niches from the common species (which they constantly meet), the rare species may continue to co-exist for an extremely long time without appreciable differentiation from each other in their rege ation niches because any two of them will encounter each other so infrequently. One important conclusion from this thesis is that, considering survival through a given length of time, rare species do not have to be as different from each other as common species t ensure a high chance of co-existence. The whole issue of the extent to which rare species in one community impinge upon one another is wide open for study, and is critical for understanding species rich communities. It is highly desirable that work aimed at this point should begin in tropical rain forests, but also work is needed in systems nearer to the majority of research scientists

We illustrated these ideas for specific communities, and emphasized the need to disentangle the mechanisms involved in a great many more communities.

I believe now that certain effects of considerable importance were overlooked in the 1982 review, particularly for long lived herbaceous plants. J. Mitchley and I have studied such plants in the special case where there is a strikingly constant hierarchy of relative abundance. We have taken frequently grazed chalk grassland in southern England as our type community (Fig. 2). In contrast to the short lived plants in chalk grassland whose numbers fluctuate greatly in time and space, there is a remarkable constancy from year to year and from site to site in the relative abundance of the perennials (Grubb et al. 1982). As these perennials are long lived, closely intermingled, and show a constant hierarchy on a fine scale (say 0.25 m^2), we have concluded that the interrelations between vegetative adults must hold the key. A field experiment was set up with two abundant species, two intermediate species, and two sparse species in all possible combinations, and a remarkably constant hierarchy of interference effects (as judged by cover) was found after 15 months (Fig. 3). The species with tallest leaves had the greatest potential for interference. A re-examination of thousands of point quadrat data collected without bias showed a striking correlation between relative abundance (or absolute cover) and relative leaf height (Fig. 4). Clearly, it is tempting to suggest that competition for light produces the hierarchy, and to some this might not seem remarkable, but the grassland concerned is characteristically very short of nitrogen and phosphorus and (as mentioned earlier) is only 2-10 cm tall. Other ecologists have suggested that whereas competition for light naturally is overwhelmingly important in grasslands on rich soils, competition for nutrients might be most important on nutrient poor soils (Newman 1973). The importance of competition for light has yet to be established experimentally for chalk grassland. We realize that the correlation between relative leaf height and relative abundance may be secondary and that relative abundance may be determined by belowground competition. Eagles and Williams (1971) found that of two provenances of *Dactylis glomerata* the one with the greater potential to interfere under any particular growing conditions had taller leaves, and they concluded that competition for light was of overwhelming importance. But subsequently, Eagles (1972) separated shoot and root effects experimentally and showed the root effects to be of great, if not greater, importance. It would be wrong, however, to use such results to dismiss the role of competition for light. Pigott (1982), in a study of two woodland herbs, showed that merely by holding aside the taller leaves of *Hyacinthoides non-scripta* he could enable the intermingled shoots of *Anemone nemorosa* to increase their cover substantially (Fig. 5). We badly need experimental studies to disentangle the relative importance of shoot and root mediated interference in a variety of grassland, heathland, and

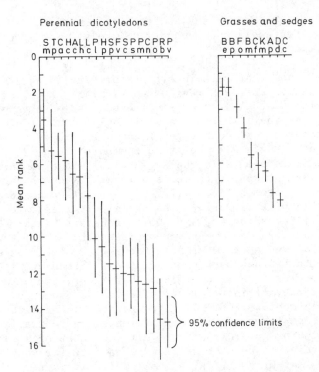

Fig. 2.   Mean rank (± 95% confidence limits) based on cover for 18
          forbs and 9 graminoids of 20 stands of chalk grassland at a
          site in southern England, August 1980:   *Asperula cynanchica*
          (Ac); *Avenula pratensis* (Ap); *Bromus erectus* (Be); *Briza
          media* (Bm); *Brachypodium pinnatum* (Bp); *Cirsium acaule* (Ca);
          Carex caryophyllea (Cc); *Carex flacca* (Cf); *Centaurea nigra*
          (Cn); *Danthonia decumbens* (Dd); *Festuca ovina* + *rubra* (Fo);
          Filipendula vulgaris (Fv); *Hippocrepis comosa* (Hc); *Hieracium
          pilosella* (Hp); *Koeleria macrantha* (Km); *Lotus corniculatus*
          (Lc); *Leontodon hispidus* (Lh); *Plantago lanceolate* (Pl);
          Plantago media (Pm); *Phyteuma orbiculare* (Po); *Pimpinella
          saxifraga* (Ps); *Polygala vulgaris* (Pv); *Ranunculus bulbosus*
          (Rb); *Scabiosa columbaria* (Sc); *Sanguisorba minor* (Sm);
          Succisa pratensis (Sp); *Thymus praecox* (Tp).   (Based on Fig.
          2 of Grubb et al. 1982.)

forest communities.   Besides greater uptake of limiting nutrients
and water, we have to consider other mechanisms in root mediated
interference, such as effects of direct allelopathy (cf. Newman and
Rovira 1975) and effects mediated via associated microorganisms
(Christie et al. 1978).

     The value of prolonged experiments on interference, done under
realistic field conditions on intact soil profiles as in Mitchley's

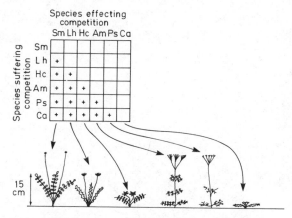

Fig. 3. Results of a field experiment in which six species from
chalk grassland were grown in all possible combinations
on an intact soil profile to test their potential for
interference. Cover was measured 15 months after planting
seedlings, and the "aggressivity" (McGilchrist and Trenbath
1971) was calculated. A positive entry means that the
species at the head of the column suppressed the species
in the line concerned to some degree. Note the regular
hierarchy related to the height of the rosette leaves.
Sanguisorba minor (Sm); *Leontodon hispidus* (Lh); *Hippocrepis
comosa* (Hc); *Achillea millefolium* (Am); *Pimpinella saxifraga*
(Ps); *Cirsium acaule* (Ca). (Results of J. Mitchley,
unpublished.)

study, is now clear. In a recent paper (Grubb 1982), I have reported
the results of another long term outdoor experiment designed to
interpret control of relative abundance in tall roadside grassland
(Arrhenatheretum); the most valuable results did not emerge until
the fifth year. Work in pots in glasshouses is much less likely to
be helpful, although important effects of different conditions
(e.g., mimicking early spring and mid-summer growth periods) can be
demonstrated in such experiments (Fowler 1982). Experiments in
which various species are selectively removed, and the reactions of
others are monitored, can also be revealing (Silander and Antonovics
1982). All experiments in this area should be designed with the
possibility of considerable intraspecific differentiation in mind
(cf. Turkington and Harper 1979).

If leaf height enables a perennial species in chalk grassland
to get ahead, what prevents the species with the lowest leaves from
disappearing? Almost certainly the preferential grazing on the
taller leaves by sheep or cattle is the answer. Where grazing
ceases, the lower growing species are indeed lost. How far the

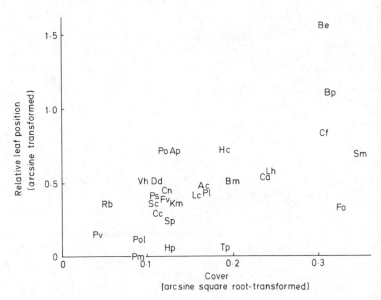

Fig. 4. Results of an analysis of relative leaf position (RLP)
         and cover for 29 species in 20 stands of chalk grassland
         mostly 2-7 cm tall, June 1982 (Kendall's tau 0.431;
         P<0.001). RLP is based on the analysis of about 4,800
         point quadrats; at each point the order in which species
         were hit was recorded. Letters have the same meaning as
         in Figure 2 except Pv = *Prunella vulgaris*, Pol = *Polygala
         vulgaris*, and Vh = *Viola hirta*. (Results of J. Mitchley,
         unpublished.)

maintenance of species richness among perennials in chalk grassland
and similar, grazed grasslands depends on variation in height of
grazing (and on consequent differences in the relative losses by
different species) remains to be seen. It is also unclear whether
the hierarchy among sparse species depends on their potentials for
mutual interference among each other, or mainly on the relationships
they each have with the relatively few most common species. All
these problems need to be answered if a satisfying rational understanding of the control of relative abundance is to be worked out.

     Leaf height plainly does not enable plants to be the most
abundant in other communities (Fig. 6). Where disturbance is not
continual and allover, as it is in frequently grazed grassland, but
patchy and dependent on the trampling or scraping activities of
animals, falling of branches, scouring by water, etc., then we
often find that the less abundant species have taller shoots than
the most abundant species making up the matrix (e.g., various
taller herbs emerging from a carpet of *Mercurialis perennis* in many

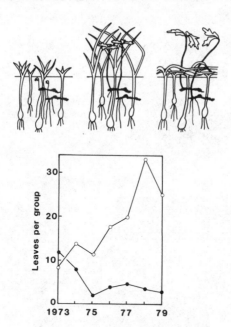

Fig. 5. Profiles of a woodland herb community in late March,
mid-April, and mid-May; the liliaceous *Hyacinthoides
non-scripta* overtops *Anemone nemorosa* in April, when peak
levels of irradiance are experienced at ground level before
canopy closure. Also shown are the results of holding the
leaves of *Hyacinthoides* aside: ● control, o *Hyacinthoides*
leaves held aside. The performance of the intermingled
shoots of *Anemone* improved greatly over a period of five
years. (Based on Figs. 7 and 8 of Pigott 1982.)

northern European woodlands, or from a mat of *Spartina patens* in a
New England salt marsh). In these cases, ability to fill any gap
quickly by sideways spread apparently is more important than leaf
height in determining relative abundance, coupled possibly with an
ability to outlive most tall gap invaders. In neither frequently
grazed chalk grassland nor the *Mercurialis* sward is accumulation of
litter important in determining relative abundance, although in a
derelict, ungrazed chalk grassland it may be very important.

The lesson to be drawn is that different characteristics are
of greatest importance in different communities, depending on the
pattern of disturbance (or lack of it), and thus any unified "dominance index," such as that suggested by Grime (1979), seems to me
to be of limited value.

There are many communities for which we have, as yet, almost
no understanding of the mechanistic control of relative abundance

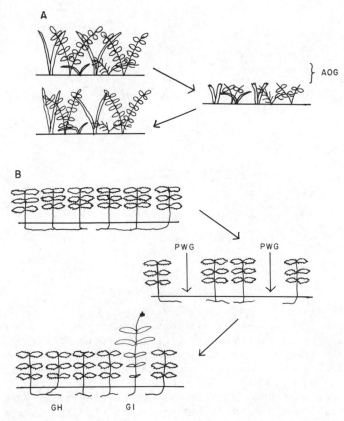

Fig. 6. Comparisons of the effects of two types of "gap" formation.
         In (a), the "allover gap" (AOG) in a pasture created by
         removal of the taller foliage is filled between grazings
         by taller leaved plants. In (b), the "patchwise gaps"
         (PWG) in a woodland herb layer created by falling branches
         or animal foot fall are either healed from the side (GH)
         or invaded by taller growing species (GI). As a result,
         taller species are the more abundant in (a), and the
         lesser abundant in (b).

(e.g., semi-deserts, mediterranean-climate communities, and arctic
and alpine communities).

Dominance

    The dimension I have called relative abundance is called
"dominance" by many. Like Weaver and Clements (1929), I prefer to
reserve dominance for the relationship between different life forms
as seen, for example, between trees and grasses in savannas (Walker

et al. 1981). Much remains to be shown about the relative importance
of interference above and belowground for such systems. When the
grasses are removed and the trees respond by putting on a new flush
of growth in a few weeks, as can happen at certain South African
sites (B.H. Walker, personal communication, May 1980), competition
belowground must be all that is involved (although the nature of
the competition is not clear). But when the trees are cleared and
grass yield is increased, as reported from northern Australia
(Walker et al. 1971), it is not known whether removal of shade or
removal of a sink for water and nutrients is more important.
Similarly, it is not known whether shoot or root competition is
more important where grasses suppress the regeneration of trees and
so maintain the community as savanna rather than letting it succeed
to forest. Equally, we are ignorant about the precise mechanisms
involved when the grasses are overgrazed and the numbers of trees
and shrubs greatly increase (Walter 1973). I hope answers to these
problems will emerge in the remaining 1980s. Such problems clearly
grade into analagous problems that arise in studying the succession
toward forest on abandoned arable land or pasture (old-field succession).

The ability of low stature plants to suppress those of potentially greater height is probably very widespread. I see it as very
important in determining what I call "secondary dominance." For
example, in a forest or woodland, trees are clearly the first
dominants, but the second dominants may be herbaceous (usually the
case on more fertile soils) or woody (usually the case on less
fertile soils). In southern Australia, the *Eucalyptus* woodlands on
soils judged by agriculturalists to be relatively fertile have a
low grassy ground layer, and those on soils considered infertile
have a taller layer of sclerophyll shrubs (Specht 1972). In the
eastern U.S., hardwood forests on fertile soils have a ground layer
of broad leaved herbs and very sparse saplings or shrubs, whereas
those on infertile soils have few herbs and a distinctive abundance
of shrubs and saplings. For an analagous system in beech forests
in southern Germany, Ellenberg (1982) has suggested that the shrubs
and saplings become abundant in "Strauchbuchenwald" on highly
calcareous soil because the tall trees grow poor in comparison and
let through more light. My student, W.J.H. Peace, measured the
light climate critically and found no such effect. More likely the
infertility and (or) droughtiness of the soil prevent the growth of
those broad leaved herbs which would otherwise inhibit the establishment of shrubs or saplings. The model for such a system was provided
long ago in Wardles' (1959) study of the tree *Fraxinus excelsior*
versus the herb *Mercurialis perennis*. In the issue of grass versus
trees, Noble (1980) has produced useful results for the subalpine
zone of eastern Australia, and the review of Wells (1965) provides
an invaluable perspective for the central parts of the U.S. The
whole area of dominance is ripe for experimental study.

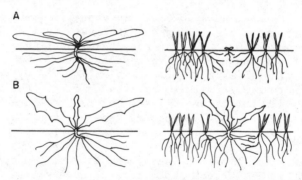

Fig. 7.   Result of an experiment in which contrasted species were
          grown in trays of infertile calcareous soil with and
          without belowground interference from *Festuca rubra*.
          Reseda luteola (a), a colonizer of large gaps under
          natural conditions, proved to be extremely sensitive to
          belowground interference, while *Leontodon hispidus* (b), a
          closed turf species, was much less sensitive. (Results
          of Fenner 1978.)

Exclusion of Species from Particular Communities

      It is common to find that a given species occupies a position
on some environmental gradient such that it is absent from both
"more favourable" and "less favourable" environments (e.g., as
measured by the productivity of the whole community). Many ecologists
think that interference is important in limiting a plant's distribution on the more favourable side of its tolerance range, but that
direct effects of the environment are responsible for its exclusion
on the less favourable side. There must be instances where this is
largely the case (e.g., for some seaweeds on rocky shores [Schonbeck
and Norton 1978, Lubchenco 1980]), but I doubt the generality.

      We have been interested in the exclusion of certain potentially
tall growing, short lived plants from chalk grassland. These
plants (e.g., *Galium aparine*, *Reseda luteola*, and *Verbascum nigrum*)
are normally associated with disturbed sites on more fertile soils,
that is, more favourable habitats ("less stressed" in the sense of
Grime 1979). When seeds are sown into closely cut but intact chalk
grassland with no gaps more than 3-5 cm diameter, they germinate
but the seedlings grow very little and eventually die (Gay et al.
1982). It was shown by Fenner (1978) that such plants are suppressed
by root competition, and can grow well on relatively infertile soil
in the absence of interference from established plants (Fig. 7).
We have confirmed these results, and Taylor (1982) has shown that
the potential for established dicotyledons to suppress seedlings of
Verbascum nigrum is enormously greater when they are infected by
mycorrhizae than when they are uninfected.

These two phenomena--exclusion from communities on infertile soils by interference and the involvement of mycorrhizae--need much further study. They might be important, for example, in the well lit field layer of chestnut-oak forests in the eastern U.S., or in newly burnt pine barrens in New Jersey, which seem not to be invaded by common roadside weeds despite the temporary enrichment of the soil with nutrients from the ash.

LEAF FORM AND FUNCTION

Investigation of the relative importance of root and shoot mediated interference leads naturally to a study of the distribution of activity in the canopy, as has been practised by Kira (1975), Miller and Stoner (1979), and Galoux et al. (1981) for forests, and by Cernusca (1977) for shrub and herb communities, and thence to studies on the form and function of individual leaves. In the latter, I want to draw attention to certain issues that to me seem particularly promising for investigation.

The first is understanding in evolutionary terms the control of the maximum net assimilation rate (MNAR) of CO_2 in leaves. Orians and Solbrig (1977) made an important step forward when they demonstrated the association between high MNAR and short leaf life in sun leaves, but their physiological explanation was unconvincing. Mooney and Gulmon (1982) have summarized more recent work, and have suggested that longer lived leaves allocate more carbon to defense than shorter lived leaves, and therefore use less carbon to build photosynthetic machinery. However, Mooney et al. (1981) and others have shown a close relationship between MNAR and leaf nitrogen concentration. My contention is that the nitrogen economy of the leaf is more likely than the carbon economy to limit the amount of photosynthetic machinery synthesized. Quite simply, sun leaves with the lowest MNAR on an area or weight basis, or on a per unit nitrogen basis, are the ones with the highest C/N ratios (e.g., those of evergreen conifers and mediterranean climate sclerophylls).

The relationship between MNAR and leaf nitrogen concentration in several kinds of plants is shown in Figure 8. The simplest interpretation of the general trend in that figure is that the more nitrogen there is available to the leaf, the more it can allocate to photosynthetic machinery as opposed to other functions, and in particular the more it can allocate to the enzyme ribulose bisphos- phate carboxylase (RuBPCase). This enzyme has very low catalytic activity per unit of nitrogen, and makes up about 20% of the total leaf nitrogen in a fast growing plant. I think that this interpreta- tion is inadequate. First, where MNAR varies with nitrogen concentra- tion in the leaf in any one species (e.g., in a fertilizer trial), the MNAR rises to an asymptote at moderate nitrogen concentration and rises no further. Secondly, and more importantly, where function-

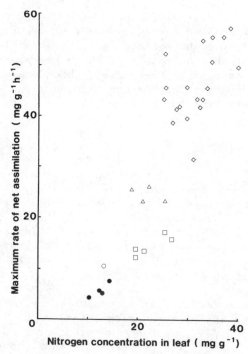

Fig. 8.   Relation between the maximum rate of net assimilation per
          unit dry weight of leaf and the nitrogen concentration in
          the leaf per unit dry weight, shown for contrasting types
          of plants: evergreen conifers (●); deciduous conifers
          (o); non-bog deciduous trees (□); non-bog deciduous
          shrubs (Δ); old-field herbs (◇).   (Data on old-field
          herbs from Mooney et al. 1981, the rest from Small 1972.)

al groups have overlapping leaf nitrogen concentrations, those with
shorter lived leaves have a higher ratio of MNAR to nitrogen (e.g.,
herbs versus shrubs, deciduous shrubs versus deciduous trees, and
deciduous conifers versus evergreen conifers) as shown in Figure 8.
My suggestion is that these observations result from more nitrogen
being allocated to defense against animals and microorganisms and
against environmental hazards, such as drought and frost, in the
leaves which last longer (Fig. 9). Much new work is needed, first
to establish whether the catalytic activity (and amount) of photosynthetic enzymes (particularly RuBPCase) per unit of total leaf
nitrogen is indeed lower as I have suggested, and second (if the
first proposition is true) to identify the enzymes in which so much
"defense nitrogen" is present. Osmond (1983) has suggested that a
similar internal competition for nitrogen may exist between the
light and dark reactions of photosynthesis such that the MNAR per
unit total nitrogen in shade leaves is inevitably low relative to

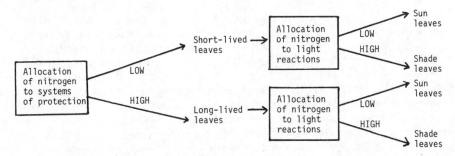

Fig. 9.   Tentative unifying hypothesis which relates the maximum
          rate of net assimilation of a leaf to the allocation of
          nitrogen within it according to expected leaf life and
          degree of shade.

that in equally long lived sun leaves (Fig. 9).   In fact, this
ratio can be as low as 0.2 mg CO_2/mg N·hr in strongly shade tolerant
herbs, and as high as 2.0 mg CO_2/mg N·hr in light demanding herbs.
This issue is of great importance not only for understanding natural
and semi-natural systems, but also for providing a sound perspective
for breeding new crops.

The second issue is more academic, but also far reaching--the
interpretation of stomatal density.   Although there is no tight or
simple relationship between stomatal density and stomatal resistance
to diffusion of CO_2 or water, it is true generally that leaves with
low MNAR and low stomatal conductance (Körner et al. 1979) have low
stomatal density (e.g., shade tolerant herbs and succulents).   My
thesis is that two opposing selective forces effectively determine
stomatal density.   The selective force towards a high density
relates to the advantage of keeping stomatal resistance to CO_2 low
relative to mesophyll resistance (Cowan and Farquhar 1977).   The
selective force towards a low density results from the cost of
providing--both within the leaf lamina and in the petiole, stem,
and root--sufficient conductance to prevent the leaf cells from
suffering excessive desiccation when the driving force for water
loss is unusually high.   That leaves have a limited internal conductance and can be thought of as at risk in this sense is shown by an
experiment of T.A. Mansfield (Kershaw 1982).   An upturned, conical
flask was placed over the two upper, fully expanded leaves of a
healthy plant of *Commelina communis* in a pot of well watered soil,
and the mouth closed with cotton wool.   After an hour, the flask
was removed and the stomata, having been exposed to a very low CO_2
concentration, were gaping widely.   A dry air stream was then
passed over the leaves; in a few minutes the leaves wilted badly
and in another hour the tissues farthest from the midrib were
irreparably damaged.   A control plant kept in light, but not enclosed
in a flask before being subjected to the same dry air stream,

showed no wilting or other damage. Clearly, leaves have limited
conductance, and if transpiration is excessive they are damaged.
In nature plants do not suffer enclosure in upturned, conical
flasks, but they have evolved low stomatal densities in species
which appear superficially to have low potential for intra-leaf
conduction of water (e.g., low density of veins, thin epidermal
walls, and highly lacunose mesophyll). Sophisticated experiments
on the hydraulic conductance of different leaf tissues have been
made (Sherriff and Meidner 1974, 1975) and comparable values are
now needed for a wide variety of species. Inevitably any relationship
between internal conductance and stomatal density will be complicated
by the extent to which the plant has evolved a tolerance of leaf
cell desiccation, a tolerance which must carry its own costs. The
potential importance of this point is brought out by the frequent
sight of marked wilting in some kinds of large leaved woodland
herbs in bright sunshine (cf. Fig. 97 in Ellenberg 1963); at least
some such plants may combine a relatively high stomatal density
with a relatively low hydraulic intra-leaf conductance.

Intra-leaf hydraulic conductance has definitely been understudied. I suggested earlier (1977b) that low intra-leaf conductance
might be responsible, in an evolutionary sense, for the lobing of
leaves of deciduous trees and shrubs. Givnish (1979) independently
reached a similar conclusion, but there has yet to be an experimental
test of the idea.

I have said nothing about ecological studies on structure and
function of stems and roots. In fact, these organs commonly have
been neglected by ecologists studying the "adaptations" of various
types of plants. A new phase of quantitative comparative anatomical
studies on stems and roots probably would reveal many fascinating
issues for physiological and evolutionary interpretation. One
particular issue that has been understudied is the spatial disposition
of tissues that defend the stem or root against animals and (or)
microorganisms. The very idea has been neglected in most traditional
teaching about the roles of various tissues in support, transport,
and storage.

ACKNOWLEDGEMENTS

I thank particularly Jonathan Mitchley for allowing me to
quote from his unpublished results, David Kelly and Pamela Taylor
(née Gay) for their active support in studying aspects of chalk
grassland ecology, and Hal Mooney for helpful discussion of my
ideas on the evolution of maximum rates of net assimilation.

REFERENCES CITED

Andrewartha, H.C., and L.C. Birch. 1954. The distribution and
    abundance of animals. Univ. of Chicago Press, Chicago, IL USA.
Bornkamm, R. 1961. Zur Konkurrenzkraft von *Bromus erectus*. Bot.
    Jahrb. 80:466-479.
Caswell, E. 1978. Predator mediated coexistence: a non-equilibrium
    model. Am. Nat. 112:127-154.
Cernusca, A. 1977. Bestandesstruktur, Mikroklima und Energiehaushalt
    von Pflanzenbeständen des alpinen Grasheidegürtels in den
    Hohen Tauern. Erste Ergebnisse der Projekstudie 1976. Pages
    25-45 *in* A. Cernusca, ed. Veröffentlichungen des Österreichischen
    MaB Hochgebirgsprogrammes Hohe Tauern, vol. I. Alpine Grasheide
    Hohes Tauern. Wagner, Innsbruck, Austria.
Christie, P., E.I. Newman, and R. Campbell. 1978. The influence
    of neighbouring grassland plants on each others' endomycorrhizas
    and root-surface microorganisms. Soil Biol. Biochem. 10:521-527.
Cowan, I.R., and G.D. Farquhar. 1977. Stomatal function in relation
    to leaf metabolism and environment. Pages 471-505 *in* D.H.
    Jennings, ed. Integration of activity in the higher plant.
    Soc. Exp. Biol. Symp. 31. Cambridge Univ. Press, Cambridge,
    England.
Eagles, C.F. 1972. Competition for light and nutrients between
    natural populations of *Dactylis glomerata*. J. Appl. Ecol.
    9:141-151.
Eagles, C.F., and D.H. Williams. 1971. Competition between natural
    populations of *Dactylis glomerata*. J. Agric. Sci. 77:187-193.
Ellenberg, H. 1963. Vegetation Mitteleuropas mit den Alpen.
    Ulmer, Stuttgart, W. Germany.
Ellenberg, H. 1982. Vegetation Mitteleuropas mit den Alpen, 3rd
    ed. Ulmer, Stuttgart, W. Germany.
Fenner, M. 1978. A comparison of the abilities of colonizers and
    closed-turf species to establish from seed in artificial
    swards. J. Ecol. 66:953-963.
Fowler, N. 1982. Competition and coexistence in a North Carolina
    grassland. III. Mixtures of component species. J. Ecol.
    70:77-92.
Galoux, A., P. Benecke, G. Gietl, H. Hager, C. Kayser, O. Kiese,
    K.R. Knoerr, C.E. Murphy, G. Schnock, and T.R. Sinclair. 1981.
    Radiation, heat, water and carbon dioxide balances. Pages
    87-204 *in* D.E. Reichle, ed. Dynamic properties of forest
    ecosystems. Cambridge Univ. Press, Cambridge, England.
Gay, P.E., P.J. Grubb, and H.J. Hudson. 1982. Seasonal changes in
    the concentrations of nitrogen, phosphorus and potassium, and
    in the density of mycorrhiza, in biennial and matrix-forming
    perennial species of closed chalkland turf. J. Ecol. 70:571-593.
Givnish, T.J. 1979. On the adaptive significance of leaf form.
    Pages 375-407 *in* O.T. Solbrig, S. Jain, G.B. Johnson, and P.H.
    Raven, eds. Topics in plant population biology. Macmillan,
    London, England.

Grime, J.P.   1979.   Plant strategies and vegetation processes.
    Wiley, Chichester, England.
Grubb, P.J.   1977a.   The maintenance of species-richness in plant
    communities: the importance of the regeneration niche.   Biol.
    Rev. 52:107-145.
Grubb, P.J.   1977b.   Leaf structure and function.   Pages 317-329 in
    R. Duncan and M. Weston-Smith, eds.   An encyclopedia of igno-
    rance.   Vol. 2, Biological sciences.   Pergamon, Oxford, England.
Grubb, P.J.   1982.   Control of relative abundance in roadside
    Arrhenatheretum: results of a long-term garden experiment.   J.
    Ecol. 70:845-861.
Grubb, P.J., D. Kelly, and J. Mitchley.   1982.   The control of
    relative abundance in communities of herbaceous plants.   Pages
    79-97 in E.I. Newman, ed.   The plant community as a working
    mechanism.   Spec. Publ. Ser. Br. Ecol. Soc. 1.   Blackwell,
    Oxford, England.
Harper, J.L.   1961.   Approaches to the study of plant competition.
    Pages 1-39 in F.L. Milthorpe, ed.   Mechanisms in biological
    competition.   Soc. Exp. Biol. Symp. 15.   Cambridge Univ.
    Press, Cambridge, England.
Harper, J.L.   1977.   Population biology of plants.   Academic Press,
    London, England.
Haynes, R.J.   1980.   Competitive aspects of the grass-legume associa-
    tion.   Adv. Agron. 33:227-261.
Kelly, D.   1982.   Demography, population control and stability in
    short-lived plants of chalk grassland.   Ph.D. Dissert., Univ.
    of Cambridge, England.
Kershaw, T.J. (compiler).   1982.   Some suggestions for practical
    work.   Pages 147-167 in W.J. Davies and P.G. Ayres, eds.
    Biology in the 80s: Plant physiology.   Univ. of Lancaster,
    Lancaster, England.
Kira, T.   1975.   Primary production of forests. Pages 5-40 in J.P.
    Cooper, ed.   Photosynthesis and primary productivity in different
    environments.   Cambridge Univ. Press, Cambridge, England.
Körner, C., J.A. Scheel, and H. Bauer.   1979.   Maximum leaf diffusive
    conductance in vascular plants.   Photosynthetica 13:45-82.
Lieth, H.   1960.   Patterns of change within grassland communities.
    Pages 27-39 in J.L. Harper, ed.   The biology of weeds.   Br.
    Ecol. Soc. Symp. 1.   Blackwell, Oxford, England.
Loucks, O.L., A.R. Ek, W.C. Johnson, and R.A. Monserud.   1981.
    Growth, aging and succession. Pages 37-85 in D.E. Reichle, ed.
    Dynamic properties of forest ecosystems.   Cambridge Univ.
    Press, Cambridge, England.
Lubchenco, J.   1980.   Algal zonation in the New England rocky
    intertidal community: an experimental analysis.   Ecology
    61:333-344.
McGilchrist, C.A., and B.R. Trenbath.   1971.   A revised analysis of
    competition experiments. Biometrics 27:659-671.
Miller, P.C., and W.A. Stoner.   1979.   Canopy structure and environ-
    mental interactions. Pages 428-448 in O.T. Solbrig, S. Jain,

G.B. Johnson, and P.H. Raven, eds. Topics in plant population biology. Macmillan, London, England.

Mooney, H.A., C. Field, S.L. Gulmon, and F.A. Bazzaz. 1981. Photosynthetic capacity in relation to leaf position in desert versus old-field annuals. Oecologia 50:109-112.

Mooney, H.A., and S.L. Gulmon. 1982. Constraints on leaf structure and function in reference to herbivory. Bioscience 32:198-206.

Newman, E.I. 1973. Competition and diversity in herbaceous vegetation. Nature 244:310.

Newman, E.I., and A.D. Rovira. 1975. Allelopathy among some British grassland plants. J. Ecol. 63:727-737.

Noble, I.R. 1980. Interactions between tussock grass (*Poa* spp.) and *Eucalyptus pauciflora* seedlings near the treeline in south-eastern Australia. Oecologia 45:350-353.

Orians, G.H., and O.T. Solbrig. 1977. A cost-income model of leaves and roots with special reference to arid and semi-arid areas. Am. Nat. 111:677-690.

Osmond, C.B. 1983. Interactions between irradiance, nitrogen nutrition, and water stress in the sun-shade responses of *Solanum dulcamara*. Oecologia 57:316-321.

Pigott, C.D. 1982. The experimental study of vegetation. New Phytol. 90:389-404.

Rabinowitz, D. 1981. Seven forms of rarity. Pages 205-217 *in* H. Synge, ed. The biology of rare plant conservation. Wiley, Chichester, England.

Schonbeck, M., and T.A. Norton. 1978. Factors controlling the upper limits of fucoid algae on the shore. J. Exp. Mar. Biol. Ecol. 31:303-313.

Sherriff, D.W., and H. Meidner. 1974. Water pathways in leaves of *Hedera helix* L. and *Tradescantia virginiana* L. J. Exp. Bot. 25:1147-1156.

Sherriff, D.W., and H. Meidner. 1975. Water movement into and through *Tradescantia virginiana* (L.) leaves. I. Uptake during conditions of dynamic equilibrium. J. Exp. Bot. 26:897-902.

Silander, J., and J. Antonovics. 1982. Analysis of interspecific interactions in a coastal plain community--a perturbation approach. Nature 298:557-560.

Small, E. 1972. Photosynthetic rates in relation to nitrogen cycling as an adaptation to nutrient deficiency in peat bog plants. Can. J. Bot. 50:2227-2233.

Specht, R.L. 1972. The vegetation of South Australia, 2nd ed. Gov. Print., Adelaide, South Australia.

Tanner, E.V.J. 1977. Four montane rain forests of Jamaica: a quantitative characterization of the floristics, the soils and the foliar mineral levels, and a discussion of the interrelations. J. Ecol. 65:883-918.

Tanner, E.V.J. 1982. Species diversity and reproductive mechanisms in Jamaican trees. Biol. J. Linn. Soc. 18:263-278.

Taylor, P.E. 1982. The development and ecological significance of mycorrhiza in contrasted groups of chalk grassland plants. Ph.D. Dissert., Univ. of Cambridge, Cambridge, England.

Taylor, R.A.J., and L.R. Taylor. 1979. A behavioural model for
    the evolution of spatial dynamics. Pages 1-27 in R.M. Anderson,
    B.D. Turner, and L.R. Taylor, eds. Population dynamics. Br.
    Ecol. Soc. Symp. 20. Blackwell, Oxford, England.

Turkington, R., and J.L. Harper. 1979. The growth; distribution
    and neighbour relationships of *Trifolium repens* in a permanent
    pasture. IV. Fine-scale biotic differentiation. J. Ecol.
    67:245-254.

Varley, G.C. 1949. Population changes in German forest pests. J.
    Anim. Ecol. 18:117-122.

Walker, B.H., D. Ludwig, C.S. Holling, and R.M. Peterman. 1981.
    Stability in semi-arid savanna grazing systems. J. Ecol.
    69:473-498.

Walker, J., R.M. Moore, and J.A. Robertson. 1971. Herbage response
    to tree and shrub thinning in *Eucalyptus populnea* shrub wood-
    lands. Aust. J. Agric. Res. 23:405-410.

Walter, H. 1973. Ecology of tropical and subtropical vegetation.
    Oliver and Boyd, Edinburgh, Scotland.

Wardle, P. 1959. The regeneration of *Fraxinus excelsior* in woods
    with a field layer of *Mercurialis perennis*. J. Ecol. 47:483-497.

Weaver, J. E., and F.E. Clements. 1929. Plant ecology. McGraw-Hill,
    New York, NY USA.

Wells, P.V. 1965. Scarp woodlands, transported grassland soils,
    and concept of grassland climate in the Great Plains region.
    Science 148:246-249.

RESEARCH QUESTIONS IN ECOLOGY RELATING TO COMMUNITY ECOLOGY, PLANT-

HERBIVORE INTERACTIONS, AND INSECT ECOLOGY IN GENERAL

Peter W. Price

Department of Biological Sciences
Box 5640
Northern Arizona University
Flagstaff, Arizona   86011 USA

ABSTRACT

     Major advances in community ecology could be made in the
remainder of this decade by developing a strongly comparative
approach.   Some suggestions for emphasis include the following five
questions: (1) What are the relative importances of the following
types of interactions between organisms in natural systems: competition, predation, parasitism, mutualism, amensalism, commensalism,
and neutralism?   Does relative importance differ in different kinds
of organisms like bacteria, protozoa, annual and perennial plants,
invertebrates, vertebrates, generalists, and specialists?   (2) What
are the organizing factors in communities?   How do they differ
between communities of different kinds of organisms?   (3) How do
members of the same food web interact directly and indirectly,
particularly between trophic levels?   (4) What roles do microorganisms play in natural systems, and what are their relative importances?
(5) To what extent does the understanding of natural phenotypic and
genotypic variation within and between populations contribute to
our understanding of ecological systems?

     The further development of ecology requires a more general
application of a rigorous scientific method involving the clear
definition of questions or hypotheses, testing among alternatives
simultaneously, more extensive use of experiments, and the reporting
of negative as well as positive results.   Funding to encourage more
teamwork involving several disciplines relevant to a particular
question would be desirable.

INTRODUCTION

The subdisciplines of ecology represented here will be community ecology, with particular emphasis on specialists such as parasites and many insects, and plant-herbivore relationships, with particular emphasis on insect herbivores.

In the last 10-15 years, we have observed an increased emphasis on insect ecology, such that now several of the major figures in ecology have developed their reputations studying these organisms. Part of the attraction of insects lies in their great diversity. But another important aspect is the ease with which experiments can be designed and executed and the commonly large population sizes, such that large sample sizes can be obtained with moderate ease. As Colwell (1983) said, theoretical ecology is no longer the ecology of birds (see also Colwell 1984).

Thus, an increasing emphasis on insect studies has coincided with a change of emphasis in ecology from a very descriptive science to a science seeking to understand mechanisms (see Simberloff 1980). This is no accident. Insects and other small arthropods frequently are ideal as experimental animals, and experimental techniques for understanding mechanisms in ecology are also essential. As the science of ecology has matured, we have seen increasingly sophisticated application of the scientific method, often involving a necessary experimental approach. Connell (1975) and Colwell and Fuentes (1975) published complementary reviews on experimental studies of species interactions, citing a total of about 300 references. But today such a task would be enormous, and would necessarily appear as several reviews (Colwell 1984).

My view of basic ecological research for the remainder of this decade and into the 1990s is one of optimism. As a realization grows among ecologists that a rigorous scientific method must prevail, more emphasis will be placed on organisms that are experimentally highly tractable. Insects and many other arthropods fulfill this requirement admirably, and will be used increasingly in the development of ecological concepts and theory.

Another virtue of the increasing emphasis on insects is that theoretical and conceptual development in ecology can encompass a broader base--including plants, vertebrates, and invertebrates--for a strongly comparative science, and a more realistic perception of diversity of mechanisms can develop. It is, perhaps, my excitement over the development of a strongly comparative ecology that motivates the following questions. These are all questions that are being actively researched so that detailed answers may be available by the end of this decade. A major challenge in the coming century will be the development of a fully comparative theoretical ecology embracing all five kingdoms of organisms and the interactions between them.

However, any effort to facilitate research development almost inevitably requires that some concern be devoted to the methods to be used and the administrative policies to be adopted. These two issues will be discussed briefly after the questions have been addressed.

THE QUESTIONS

(1) *What is the relative importance of each of the following types*
    of interactions between organisms in natural systems: competi-
    tion, predation, parasitism, mutualism, amensalism, commensalism,
    and neutralism? Does relative importance differ in different
    kinds of organisms, e.g., bacteria, protozoa, annual plants,
    perennial plants, invertebrates, vertebrates, generalists,
    specialists?

Ecological theoreticians have been beguiled by the force and simplicity of competition and predation. The tractability of such interactions has colored our perception of nature to such an extent that young ecologists learn from current general texts that these are the major forces resulting in predictable relationships between organisms and patterns in nature. The fact that a subject is theoretically convenient has led to the establishment of paradigms in ecology quite out of proportion to their importance in nature relative to other types of interactions.

This view is changing. The Tallahassee School of Null Hypotheses has established a new level of rigor in the scientific method as applied to ecology, and certainly has undermined the credibility of competition as a universally potent organizing force in communities (e.g. Strong 1980, Strong et al. 1979, Simberloff and Boecklen 1981). Others, such as Connell (1975, 1978, 1980), have contributed importantly in questioning the competition paradigm.

Other types of interaction are also receiving increasing attention. Lawton and Hassell (1981) emphasized the importance of amensalism, stating that amensalism, or near amensalism, is more common than competition for insects in nature by a ratio of at least 2 to 1. Parasitism is much more common than predation among insects and all taxa combined (Price 1980). Mutualism is also gaining in recognition among ecologists and a strong theoretical basis is developing (e.g., Boucher et al. 1982, Addicott 1984, Price 1984a).

However, even for the most simple ecosystems we do not know the relative importance of each interaction type. We cannot even discount the possibility that the distribution of most organisms is individualistic, idiosynchratic, and essentially neutralistic towards other organisms on the same trophic level. Theory cannot

help us in this empirical assessment of interaction in nature, but
theory colors our perception of nature so vividly. This discrepancy
has resulted from a rampant theoretical development in ecology
without the systematic assessment of theory by field studies. The
relationship between these two essential features of ecology is
like the relationship between mutation and natural selection.
Should an organism change by mutation through time without the
influence of natural selection, it will evolve to be totally irrele-
vant to its environment.

The challenge, then, is to assess in nature the relative
importance of each type of interaction, so that theoretical studies
can address more realistic systems, and to evaluate theory with
field studies unbiased by theoretical predictions. (The need for
testing alternative hypotheses is addressed in the section on
research methods.) Theory escaped from the clutches of the real
world is too costly in time and effort. Had competition been
objectively evaluated early in the development of ecology, we would
now know much more about the other important interactions between
organisms.

Another aspect of this question is whether the same types of
interactions are likely to be equally important in all forms of
organisms. Lawton and Strong (1981) made the important point that
competition in vertebrate communities usually has been invoked as a
potent organizing force, but, although community patterns in foliv-
orous insects appear to be similar, the evidence for competition is
largely lacking and is based on much more rigorous approaches. We
are left with an open question: Are vertebrate and invertebrate
communities organized in fundamentally different ways, or has the
role of competition been overemphasized by vertebrate ecologists?

These questions seem to be fundamental to the conceptual
development of ecology, and relate directly to other major questions
in the field, as discussed in the next question.

(2) *What are the organizing factors, if any, in communities? How
     do they differ between communities of different kinds of
     organisms?*

Much of community ecology has involved studies and hypotheses
about factors that influence the distributions and relative abundances
of species in the same community and on the same trophic level
(e.g., plant communities, herbivorous insect communities, parasite
communities). Of course, these hypotheses necessarily involve
members higher and lower in the trophic system of each community.
Many viable hypotheses exist on the organization of such communities
and rapid progress can be made by testing several alternative
hypotheses simultaneously, particularly using experiments.

These hypotheses relate to the number of species present, their relative abundances, their distributions, and, to a certain extent, the kinds of species present. They include (1) the individualistic response hypothesis, (2) the resource heterogeneity hypothesis, (3) the island or patch size hypothesis, (4) the time hypothesis, and (5) the enemy impact hypothesis. These and their supporting evidence have been reviewed by Price (1983, 1984b).

Within these major hypotheses lie alternative explanations. At least three alternatives exist for the island or patch size hypothesis. The empirical observation is that as areas of patches or islands increase, the number of species present increases. One hypothesis has been called the "area *per se* hypothesis." This hypothesis states that area alone has a direct and positive effect on the number of species in a community because a larger area has a higher probability of receiving colonists, and they will persist longer because larger populations on larger islands are less likely to go extinct through the influences of interspecific competition, impact of natural enemies, or stochastic events (MacArthur and Wilson 1967). An alternative hypothesis is that as area increases so does resource heterogeneity, so the "resource heterogeneity hypothesis" is applicable. A third possibility is the "passive sampling hypothesis" (Connor and McCoy 1979), which recognizes that larger areas sample larger numbers of colonizing species, and thus accumulate more species per unit time.

The time hypothesis also has alternatives within it, invoking the importance of evolutionary time without competition as an organizing force (e.g., Southwood 1961), evolutionary time with competition as an organizing force (e.g., Wilson 1969), and ecological time (see Price 1984b).

Systematic testing of several hypotheses and subhypotheses simultaneously, and on several groups of organisms, would yield major new understanding on the organization of communities--the factors that influence the distribution and relative abundance of species.

Particularly interesting are the questions: Are vertebrate communities of, say, birds organized similarly to those of invertebrates such as insects (see Question 1)? Do communities composed mainly of specialists have organizing influences similar to those with mostly generalists? How do communities of microorganisms or plants compare with these other types?

The time for typological thinking in ecology is over (see also Question 5). This means that we have tended to regard a species as invariable in its ecological role or in its environmental interactions, or that an interaction or mechanism is invariable wherever it occurs. It is naive to believe that one factor will play a

major role in the organization of communities of all kinds of
organisms and of communities based on all types of resources. Some
types of resources pre-dispose community members on the same trophic
level to severe competition, whereas other types almost certainly
ensure that competition will be unimportant (Price 1984b). The
challenge is to find pattern in the diverse array of natural systems,
not to find uniformity. Although the stated goal of ecologists is
to seek pattern in nature, too frequently they have claimed to have
discovered factors of universal importance, which are effective in
all locations most of the time. This view simply is unrealistic
and should be rapidly replaced in the realm of community ecology.

(3) *How do members of the same food web interact directly and
     indirectly, particularly between trophic levels?*

, In ecology there has been an overemphasis on obvious and
simple two species antagonistic interactions. These are tractable
theoretically and reward the field ecologist by yielding positive
data and support for theory. This approach does not necessarily
reflect reality. Many subtle influences have profound impact on
these simple relationships to such an extent that the force of
antagonistic relationships may be influenced largely by a third
member in the trophic system.

This concern for influence on interactions by a third member
is particularly valid where the resources are living organisms.
Some of the best examples come from the plant-insect herbivore
literature, where the plant is the living host and the quality of
resources it provides may be influenced by extrinsic factors, or by
herbivores. An example of the latter concerns the feeding by
aphids on Fraser fir (Fedde 1973). Heavy feeding on branches by
aphids causes reduced cone size, making seeds more accessible to
the seed eating chalcid wasps that must oviposit into the seed from
the external surface of the cone. Thus, understanding variation in
susceptibility to the chalcid wasps necessarily involves understanding
variation in aphid populations. One herbivore influences profoundly
the resources for another herbivore and has a positive impact,
although one (e.g., Janzen 1973) could argue that they may well be
in competition for the common pool of energy and nutrients in a
plant. This facilitation of one species by another may be exceedingly
common in nature and akin to commensalism in tightly knit interactive systems.

Plants may also mediate important interactions between herbivores
and their enemies. Such three trophic level interactions are much
more common than is generally realized, and are reviewed by Price
et al. (1980). For example, large galls protect the internal
parasitic herbivore from parasitic wasps that must reach the galler
with an ovipositor pierced through the gall wall. Large galls

provide protection because walls are too thick for the wasps to
pierce, whereas small galls trap the herbivore in a vulnerable
position. Gall size is, in turn, influenced by plant quality and
this, in turn, by physical factors such as precipitation. If
precipitation is plentiful, the plants grow rapidly and the galls
grow large. The result of the herbivore-enemy interaction is that,
on vigorous plants with large galls, enemies are ineffective except
in dry years when galls become significantly smaller. Adjacent
plants with slower growth produce smaller galls whose inmates are
attacked heavily by enemies unless galls become so rare that they
are hard to find (see Price 1984c).

Such subtlety of interaction depends on the interplay between
individual plants and the trophic system based on them. It cannot
be appreciated when plants are used as units in large sampling
programs such as those that were commonly undertaken in the study
of population dynamics of forest insects (e.g., Morris 1955, 1963,
Baltensweiler 1968, Klomp 1968). Although Morris (1955) recognized
that high between tree variation in insect abundance was almost a
rule, the common response was to increase sample sizes until the
resulting variances became acceptably small. The important and
biologically significant variation was submerged in large sample
sizes (see also Question 5). This approach predominanted in the
late 1950s to early 1970s and provided little insight into subtle
interactions that are probably highly significant in the population
dynamics of many herbivores.

Progress could be made by paying much more attention to the
details of resource quantity and quality and how these vary between
living resources and through time (see also Question 5). Only then
will the important, but subtle interactions between members of the
same trophic system become understood.

(4) *What roles do microorganisms play in natural systems, and what
     are their relative importances?*

Microorganisms are ubiquitous in nature and ecologists are
becoming more aware of the important roles they play. They act as
food for a large variety of organisms, and frequently modify interactions between species of plants, between plants and herbivores, and
between animals. These complex interactions frequently are subtle,
but they have profound impact on the population dynamics and community
organization of species. Their study also necessitates some expertise
in microbiology coupled with an ecologist's view of nature. The
challenge is to bring microbiology fully into the ambit of conceptual
ecology and to train ecologists to be fully aware of the ubiquitous
effects of viruses, bacteria, fungi, and protozoa in natural systems.

Some examples may help to illustrate the important effects
microorganisms have in addition to (1) the large role they play in

providing food for a very diverse array of metazoans; (2) the
direct effects they have as mutualists with each other, with plants,
and with animals; and (3) being parasites of each other and of
members of the higher kingdoms.

Mycorrhizal associations are very common, if not universal, in
non-parasitic plants, but these associations have effects beyond
mutualism. Chiariello et al. (1982) recently showed that nutrients
from one plant species pass to about 20% of its close neighbours
and that mycorrhizal fungi connecting the root systems of different
plant species probably mediate this transfer. These authors concluded
that the factors controlling transfer of nutrients and the results
on community structure appear to be complex: a warning that demographic and community studies should be aware of the role that
mycorrhizal fungi play. Some root infecting fungi are serious
pathogens on some plant species, but mycorrhizal on others (e.g.,
Rhizoctonia solani; Harley 1969), again indicating the potential
for complex community interactions. Another between plant interaction
mediated by microbes concerns allelopathy in the chaparral plant,
Adenostoma fasciculatum, where the phytotoxins appear to be of
microbial origin (Kaminsky 1981).

In plant-herbivore relationships, mutualism between the herbivore
and microorganisms is very important. But the associations may be
obligate or facultative, they may relate to nutrition or detoxification of food. Jones (1984) reviews this very diverse and complex
field, making the point forcefully that microbes usually cannot be
ignored in plant-herbivore interactions.

Among interactions in animals microbes also play an important
but, as yet, poorly understood role. Price (1980) and Holmes
(1982) have reviewed some of the effects that parasites have on
species interactions and distributions, and McNeill (1976) emphasized
the global impact of disease on distribution, population dynamics,
and intraspecific competition in human populations. There is
growing awareness that parasites may transfer genetic information
between organisms, enabling adaptive leaps onto new resources
(Anderson 1970, Price 1984a, Rosenthal 1983).

These examples illustrate that if mechanisms of interaction in
ecology are to be understood, then we must be concerned heavily
with the role of microorganisms. A growing awareness of this point
should yield significant new understanding in the next decade.

(5) *To what extent does the understanding of natural phenotypic
      and genotypic variation within and between populations contribute
      to our understanding of ecological systems?*

I have mentioned typological thinking in ecology before (Question 2), and our discipline is not completely rid of this viewpoint.

Much of the population dynamics literature does not acknowledge the importance of natural variation on which natural selection can act; evolutionary change is a significant component in population dynamics. As recently as 1979, Lewontin bemoaned the fact that ecology and evolutionary genetics "remain essentially separate disciplines, travelling separate paths while politely nodding to each other as they pass" (p. 3). As a discipline within evolutionary biology, variation is fundamentally important and should be studied in much more detail.

There are indications that variation is "catching on" in ecology, and at least three areas have developed major themes on this topic. A book was published in June 1983, with the title *Variable Plants and Herbivores in Natural and Managed Systems* (Denno and McClure 1983), which illustrates how rapidly the study of plant variation has contributed to understanding of herbivore population dynamics, competition, and community structure. Important stimulus for this emphasis came from the widely influential papers by Edmunds and Alstad (1978) and Whitham (1978). The irony is that, in the discipline of plant breeding, variation almost always has been considered important, but ecologists, including those studying the population dynamics of forest insects, have ignored this literature until recently (see Question 3). With the publication of Denno and McClure (1983) and Whitham and colleague's (1984) chapter, "The variation principle: Individual plants as temporal and spatial mosaics of resistance to rapidly evolving pests," this field will receive a lot of attention and much progress will be made in the next decade. Another area that comes closest to uniting genetic variation and ecology is the study of life histories. It is now well established that polymorphism in life history traits and variation between populations are important aspects of adaptation to patchy and variable environments (e.g., Dingle and Hegmann 1982, Dingle 1984, Istock 1984). A closely allied area concerns behavioral ecology where alternative strategies of behavior are receiving considerable attention (e.g., Cade 1979, Alcock 1979). For example, a male cricket may sing a love song at night, but the potential mate thus attracted may be waylaid by a silent satellite male. The selective advantage for silent males is provided by a parasitic fly orienting to hosts using phonotaxis (Cade 1979).

Ecology has emerged from a long phase of typological thinking, although pockets remain. Emphasis on variation, alternative strategies, and the genetic basis for these will lead to rapid developments in ecology and its full integration into evolutionary biology.

RESEARCH METHODS

Much of conceptual ecology has developed from work using a non-rigorous scientific method. Empirical observations have been made, and interpretations offered without critical tests. Alternative explanations have not been explored and refuted. Hypotheses and alternatives that are falsifiable have not been erected. Studies have been undertaken that show a high probability of supporting dogma or theory. If they result in negative data, they are not published.

As a result, much of ecological theory and concept is not empirically verified, such that theory and concept have become divorced from reality. This is particularly apparent in the area of competition theory (see Questions 1 and 2).

The science will advance much more rapidly when the scientific method is applied universally. Erection of clearly stated falsifiable hypotheses and alternatives, testing in an unambiguous fashion with the frequent use of experiments, and reporting of the results, be they positive or negative, supportive of or undermining theory, will yield the highest cost/benefit ratio for scarce research funds. This emphasis is expanded upon in Price et al. (1984).

ADMINISTRATIVE POLICIES

Policy at the administration level influences the kind of research undertaken and published. Some major aspects of policy which need consideration are funding decisions, development of teamwork, editorial and peer review, and education of ecologists. These will be treated briefly.

At a time of scarce research funds, one tendency has been to spread the money widely, but thinly. This is justified and sound policy for several reasons, but alternatives should be seriously considered. As ecology as a science matures, the complexity of natural systems will be treated in more detail. There is a growing need for more teamwork that unites investigators from substantially different disciplines. For example, for studies on plant-herbivore relationships, a good team may involve a microbiologist, a phyto-chemist, a plant physiologist, and a herbivore ecologist. Additional members may be very useful, such as a plant demographer, a life history botanist, plant and herbivore parasitologists, and so on. My contention is that the discipline would advance more rapidly if a smaller number of studies were funded at a level that would foster such teamwork. The gamble comes in finding the studies that will really pay off in their abilities to develop effective teams that are highly innovative and productive.

Editorial policy and peer review of research proposals and publications have tended to foster the publication of positive results and have selected against the publication of negative data. This approach permits a warped view of reality to develop and an overemphasis in some areas of ecology while others are almost ignored. The ecologist's emphasis on pattern and conformity makes him very uncomfortable when they cannot be discovered. If a rigorous scientific method is developed in ecology, then the publication of negative data becomes of fundamental importance and just as significant and acceptable as positive data. The science will progress more rapidly when these policies are widely accepted (for more detail, see Price et al. 1984).

The education of ecologists should emphasize the need for an appreciation of theory and concept as well as an understanding that all five kingdoms of organisms play important and interactive roles in almost all natural ecosystems. Perhaps it is impossible to train ecologists in the techniques necessary for studying members of each kingdom, but they should be sensitized to their importance and the subtle roles they play in mediating interactions between other species. This training may pre-dispose later generations of ecologists to policy that fosters more interaction and teamwork involving disciplines not necessarily directly related to ecological concepts and theories. In addition, ecology has developed largely in ignorance of other disciplines that could have made important contributions, such as the fields of plant breeding, parsitology, and crop physiology. Many applied fields of biological research can contribute importantly to the understanding of natural systems. Such areas should be embraced more fully in the education of ecologists.

ACKNOWLEDGMENTS

I am grateful to speakers at a conference, "A new ecology: Novel approaches to interactive systems," held at Northern Arizona University, August 11-13, 1982, for discussing some rapidly developing areas of ecology, and especially to W.S. Gaud and C.N. Slobodchikoff for developing with me some of the ideas expressed here. I also thank R. Fritz for reviewing a draft of this paper and the National Science Foundation for financial support through grant DEB 80-21754.

LITERATURE CITED

Addicott, J.F. 1984. Mutualistic interactions in population and community processes. Pages 437-455 in P.W. Price, C.N. Slobodchikoff, and W.S. Gaud, eds. A new ecology: Novel approaches to interactive systems. Wiley, New York, NY USA.
Alcock, J. 1979. The evolution or intraspecific diversity in male reproductive strategies in some bees and wasps. Pages 381-402 in M.S. Blum and N.A. Blum, eds. Sexual selection and reproductive competition in insects. Academic, New York, NY USA.
Anderson, N.G. 1970. Evolutionary significance of virus infection. Nature 227:1346-1347.

Baltensweiler, W.   1968.   The cyclic population dynamics of the grey larch tortrix, *Zeiraphera griseana* Hübner (= *Semasia diniana* (Guenée) (Lepidoptera: Tortricidae).   Pages 88-97 *in* T.R.E. Southwood, ed.   Insect abundance.   Symp. R. Entomol. Soc. London 4.

Boucher, D.H., S. James, and K.H. Keeler.   1982.   The ecology of mutualism.   Annu. Rev. Ecol. Syst. 13:315-347.

Cade, W.   1979.   The evolution of alternative male reproductive strategies in field crickets.   Pages 343-379 *in* M.S. Blum and N.A. Blum, eds.   Sexual selection and reproductive competition in insects.   Academic, New York, NY USA.

Chiariello, N., J.C. Hickman, and H.A. Mooney.   1982.   Endomy-corrhizal role for interspecific transfer of phosphorus in a community of annual plants.   Science 217:941-943.

Colwell, R.K.   1983.   Review of theoretical ecology: Principles and applications, 2nd ed.   Auk 100:261-262.

Colwell, R.K.   1984.   What's new?   Community ecology discovers biology.   Pages 387-396 *in* P.W. Price, C.N. Slobodchikoff, and W.S. Gaud, eds.   A new ecology: Novel approaches to interactive systems.   Wiley, New York, NY USA.

Colwell, R.K., and E.R. Fuentes.   1975.   Experimental studies of the niche.   Annu. Rev. Ecol. Syst. 6:281-310.

Connell, J.H.   1975.   Some mechanisms producing structure in natural communities: A model and evidence from field experiments.   Pages 460-490 *in* M.L. Cody and J.M. Diamond, eds.   Ecology and evolution of communities.   Belknap, Cambridge, MA USA.

Connell, J.H.   1978.   Diversity in tropical rain forests and coral reefs.   Science 199:1302-1310.

Connell, J.H.   1980.   Diversity and the coevolution of competitors, or the ghost of competition past.   Oikos 35:131-138.

Connor, E.F., and E.D. McCoy.   1979.   The statistics and biology of the species-area relationship.   Am. Nat. 113:791-833.

Denno, R.F., and M.S. McClure, eds.   1983.   Variable plants and herbivores in natural and managed systems.   Academic, New York, NY USA.

Dingle, H.   1984.   Behavior, genes, and life histories: Complex adaptations in uncertain environments.   Pages 169-194 *in* P.W. Price, C.N. Slobodchikoff, and W.S. Gaud, eds.   A new ecology: Novel approaches to interactive systems.   Wiley, New York, NY USA.

Dingle, H., and J.P. Hegmann, eds.   1982.   Evolution and genetics of life histories.   Springer, New York, NY USA.

Edmunds, G.F., and D.N. Alstad.   1978.   Coevolution in insect herbivores and conifers.   Science 199:941-945.

Fedde, G.F.   1973.   Impact of the balsam woody aphid (Homoptera: Phylloxeridae) on cones and seed produced by infested Fraser fir.   Can. Entomol. 105:673-680.

Harley, J.L.   1969.   The biology of mycorrhiza.   Hill, London, England.

Holmes, J.C. 1982. Impact of infectious disease agents on the
     population growth and geographical distribution of animals.
     Pages 37-51 in R.M. Anderson and R.M. May, eds. Population
     biology of infectious diseases. Springer, New York, NY USA.
Istock, C.A. 1984. Boundaries to life history variation and
     evolution. Pages 143-168 in P.W. Price, C.N Slobodchikoff,
     and W.S. Gaud, eds. A new ecology: Novel approaches to
     interactive systems. Wiley, New York, NY USA.
Janzen, D.H. 1973. Host plants as islands. II. Competition in
     evolutionary and contemporary time. Am. Nat. 107:786-790.
Jones, C.G. 1984. Microorganisms as mediators of plant resource
     exploitation by insect herbivores. Pages 53-99 in P.W. Price,
     C.N. Slobodchikoff, and W.S. Gaud, eds. A new ecology: Novel
     approaches to interactive systems. Wiley, New York, NY USA.
Kaminsky, R. 1981. The microbial origin of the allelopathic
     potential of Adenostoma fasciculatum H & A. Ecol. Monogr.
     51:365-382.
Klomp, H. 1968. A seventeen year study of the abundance of the
     pine looper, Bupalus piniarius L. (Lepidoptera: Geometridae).
     Pages 98-108 in T.R.E. Southwood, ed. Insect abundance.
     Symp. R. Entomol. Soc. London 4.
Lawton, J.H., and M.P. Hassell. 1981. Asymmetrical competition in
     insects. Nature 289:793-795.
Lawton, J.H., and D.R. Strong. 1981. Community patterns and
     competition in folivorous insects. Am. Nat. 118:317-338.
Lewontin, R.C. 1979. Fitness, survival and optimality. Pages
     3-21 in D.J. Horn, G.R. Stairs, and R.D. Mitchell, eds.
     Analysis of ecological systems. Ohio State Univ. Press,
     Columbus, OH USA.
MacArthur, R.H., and E.O. Wilson. 1967. The theory of island
     biogeography. Princeton Univ. Press, Princeton, NJ USA.
McNeill, W.H. 1976. Plagues and peoples. Anchor, New York, NY
     USA.
Morris, R.F. 1955. The development of sampling techniques for
     forest insect defoliators, with particular reference to the
     spruce budworm. Can. J. Zool. 33:225-294.
Morris, R.F., ed. 1963. The dynamics of epidemic spruce budworm
     populations. Mem. Entomol. Soc. Can. 31:1-332.
Price, P.W. 1980. Evolutionary biology of parasites. Princeton
     Univ. Press, Princeton, NJ USA.
Price, P.W. 1983. Hypotheses on organization and evolution in
     herbivorous insect communities. Pages 559-596 in R.F. Denno
     and M.S. McClure, eds. Variable plants and herbivores in
     natural and managed systems. Academic, New York, NY USA.
Price, P.W. 1984a, in press. Insect ecology, 2nd ed. Wiley, New
     York, NY USA.
Price, P.W. 1984b. Alternative paradigms in community ecology.
     Pages 353-383 in P.W. Price, C.N. Slobodchikoff, and W.S.
     Gaud, eds. A new ecology: Novel approaches to interactive
     systems. Wiley, New York, NY USA.

Price, P.W. 1984c, *in press*. The gall-inducing Tenthredinidae.
    In J.D. Shorthouse and O. Rohfritsch, eds. Biology of insect
    and Acarina induced galls. Praeger, New York, NY USA.
Price, P.W., C.E. Bouton, P. Gross, B.A. McPheron, J.N. Thompson,
    and A.E. Weis. 1980. Interactions among three trophic levels:
    Influence of plants on interactions between insect herbivores
    and natural enemies. Annu. Rev. Ecol. Syst. 11:41-65.
Price, P.W., W.S. Gaud, and C.N. Slobodchikoff. 1984. Intro-
    duction: Is there a new ecology? Pages 1-11 *in* P.W. Price,
    C.N. Slobodchikoff, and W.S. Gaud, eds. A new ecology: Novel
    approaches to interactive systems. Wiley, New York, NY USA.
Rosenthal, G.A. 1983. Biochemical adaptations of the bruchid
    beetle, *Caryedes brasiliensis* to L-canavanine, a higher plant
    allelochemical. J. Chem. Ecol. 9:803-815.
Simberloff, D. 1980. A succession of paradigms in ecology:
    Essentialism to materialism and probabilism. Synthese 43:3-39.
Simberloff, D.S., and W. Boecklen. 1981. Santa Rosalia recon-
    sidered: Size ratios and competition. Evolution 35:1206-1228.
Southwood, T.R.E. 1961. The number of species of insect associated
    with various trees. J. Anim. Ecol. 30:1-8.
Strong, D.R. 1980. Null hypotheses in ecology. Synthese 43:271-285.
Strong, D.R., L.A. Szyska, and D.S. Simberloff. 1979. Tests of
    community-wide character displacement against null hypotheses.
    Evolution 33:897-913.
Whitham, T.G. 1978. Habitat selection by *Pemphigus* aphids in
    response to resource limitation and competition. Ecology
    59:1164-1176.
Whitham, T.G., A.G. Williams, and A.M. Robinson. 1984. The variation
    principle: Individual plants as temporal and spatial mosaics
    of resistance to rapidly evolving pests. Pages 15-51 *in* P.W.
    Price, C.N. Slobodchikoff, and W.S. Gaud, eds. A new ecology:
    Novel approaches to interactive systems. Wiley, New York, NY
    USA.
Wilson, E.O. 1969. The species equilibrium. Pages 38-47 *in* G.M.
    Woodwell and H.H. Smith, eds. Diversity and stability in
    ecological systems. Brookhaven Symp. Biol. 22.

VEGETATION SCIENCE IN THE 1980s

Eddy van der Maarel

Institute of Ecological Botany
Box 559, S-751 22
Uppsala, Sweden

INTRODUCTION

Vegetation science started as descriptive, geographically orientated phytosociology, with an emphasis on the floristic composition of plant communities and the correlation of community patterns with variations in climate and soil. It is now rapidly broadening to include structural and functional characteristics of plant communities integrated with operational factors from the dynamic environment. Parallel to this development, there has been a shift in related disciplines: from floristic plant geography, soil science, climatology, and nature conservation to population ecology, ecophysiology, paleoecology, mathematics, natural resource ecology, and landscape ecology.

My involvement in vegetation science has changed correspondingly: from scientific roots in classical phytosociology and past research activities in quantification of pattern and process in vegetation to present interests in population dynamics and management ecology. Moreover, I have had the privilege of being in close contact with new streams and trends in vegetation science since the early 1970s as editor of the journal *Vegetatio*.

SURVEY OF DEVELOPMENTS AND TRENDS

This paper will discuss five major developments in vegetation science related to eight trends in some neighbouring disciplines. For each development, I will sketch a short history over the 1960s and 1970s. The developments are as follows:

(1)   the maturation of phytosociology through the adoption of
      numerical techniques and the refinement of structural analysis,
      as well as its application far outside its European origin;
(2)   the development of gradient theory as a general ecological
      theory on the distribution of plant populations and communities
      in relation to environmental variation;
(3)   the development of analytical and statistical methods to
      describe spatial patterns of plant species and species groups
      in plant communities;
(4)   the development of the concept of diversity and the emergence
      of an ecological and historic-geographical theory of diversity;
(5)   the advancement of our understanding of the processes underlying
      vegetation succession and the development of succession modeling.

Along with these developments, at least eight trends in related
fields have emerged:

(1)   the maturation of ecosystem ecology, that is, the study of
      biomass and energy flows;
(2)   the study of plant-plant interactions and the emergence of a
      more general theory of competition and niche differentiation;
(3)   the study of plant-animal interactions and the control mechanisms
      operating at various levels in the food chain;
(4)   the study of species' life cycles;
(5)   the development of ecophysiology;
(6)   the emergence of a community oriented biosystematics and
      genecology of plants;
(7)   the maturation of plant population ecology, especially its
      demographic aspects;
(8)   the emergence of the theory of island biogeography and the
      gradual inclusion of plants and plant communities in that
      theory.

THE MATURATION OF PHYTOSOCIOLOGY

Most approaches in phytosociology arose in Europe around the
turn of this century (see Shimwell 1971, Mueller-Dombois and Ellenberg
1974, Whittaker 1978a, McIntosh 1978). Undoubtedly the Braun-Blanquet
(1965) approach, with its consistent vegetation description, has
proven the most successful (Westhoff and van der Maarel 1978). By
1970 a fairly stable system for describing plant communities was
reached and the higher units described for Europe have become
widely accepted and ecologically well known (see Ellenberg 1978 for
an outstanding survey).

In 1976 an elaborate code of phytosociological nomenclature
appeared (Barkman et al. 1976). This school of thought, usually
linked to Zürich-Montpellier, has not been widely accepted among
ecologists in the Anglo-American world or in Europe (e.g., Egler

1954, Poore 1955-1956). Phytosociology was identified too easily
with only one of its branches, syntaxonomy (i.e., a complicated and
formalized hierarchy of plant communities), and sampling procedures
were not considered as fully objective. However, an entirely
objective sampling system does not exist yet in vegetation science
and if it ever develops it will be prohibitively ineffective.

Progress in classical phytosociology has been mainly in the
refinement of local typologics and in its application in nature
management. It can not be followed easily by those not mastering
German, the major language from the beginning of the International
Society of Vegetation Science. The symposium volumes organized and
edited largely by the late R. Tüxen, who influenced the society for
more than 30 years, is a benchmark series. The volume on community
complexes (Tüxen 1978), for example, introduces landscape ecology.

Within the society, the Working Group for Succession Research
and the Working Group for Data Processing in Phytosociology gathered
new ideas and stimulated new initiatives. The latter group sought
and found cross fertilization with Anglo-American centers of quantitative ecology. It developed a data base and a coding system for
storing and processing the data (the so called relevés), and it
began storing salt marsh and woodland data. It also developed
guidelines for numerical syntaxonomy (see the series of papers by
members of the group in *Vegetatio* 1976 and 1980, collected as a
special publication edited by van der Maarel et al. 1980).

At least two recent trends are connected with these activities.
One trend is the re-orientation of syntaxonomy among young European
phytosociologists. They describe their units on a much more objective
basis. They use all the field data available and not just that
part suiting the already accepted ideas on the classification
system. They apply clustering techniques to arrive at numerically
based hierarchies and table structuring programs to present community
types and their characterizing species blocks. They also apply
ordination techniques--mainly principal component analysis and
related techniques--to determine floristic variation over a range
of stands or community types.

One example of this effort is the numerical treatment of
Spartina dominated salt marsh vegetation (Kortekaas et al. 1976).
Over 300 relevés from all over Europe were subjected to hierarchical
clustering and the resulting clusters (Fig. 1) were interpreted
with the syntaxonomical scheme of Beeftink and Géhu (1973). The
interpretation showed that the two classification systems were
largely the same, but the numerical system showed a new unit which
was interpreted as a subassociation of *Spartinetum townsendii* with
Aster tripolium. Another example is the treatment of ruderal plant
communities by Mucina (1982) (Fig. 2A). The main line of variation
from Malvion via Sisymbrietalia towards Onopordetalia is clearly

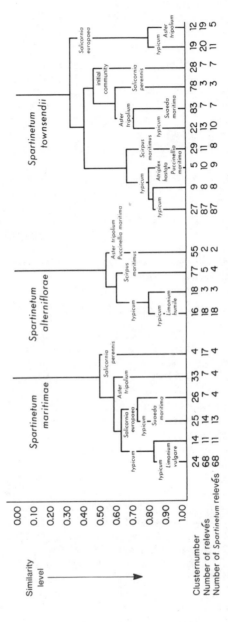

Fig. 1. Agglomerative clustering of European *Spartina* communities derived from agglomerative cluster-ing with relocation of a larger salt marsh data set (from Kortekaas et al. 1976; reprinted by permission of Junk, The Hague, The Netherlands). Similarities between clusters expressed by the similarity ratio (SR), a Jaccard derivative. Community names reflect the corresponding syntaxonomical unit from the classical treatment by Beeftink and Géhu (1973).

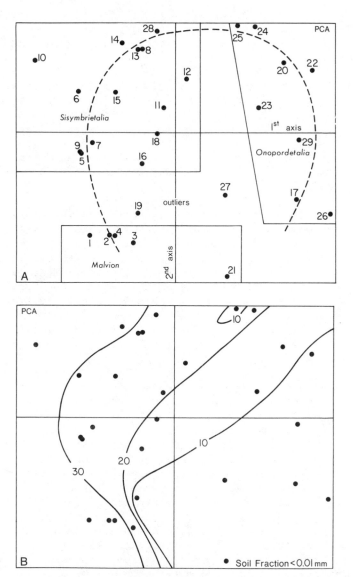

Fig. 2.   (A) Principal component analysis (PCA) of 29 ruderal plant
community types in W. Slovakia, with axes 1 and 2 represented. Results of a hierarchic clustering on the three
cluster levels (with outliers) are superimposed. Floristic-
sociological variation (coenocline) runs along the dashed
line.   (B) Average values for soil fraction of diameter <0.01
mm in the community types, superimposed on the same diagram.
(From Mucina 1982; reprinted by permission of Junk, The
Hague, The Netherlands.)

related to the variation in soil texture (Fig. 2B; values for soil
skeleton percent increasing and values for percent of fine soil
fraction decreasing from left to right, following the same pattern
as the community variation).

The second trend is the initiation of numerical vegetation
surveys, with the same detail and with similar plant community
concepts, in countries that until recently disregarded the Braun-
Blanquet approach. The best example is the National Vegetation
Classification project in Great Britain which soon will be published
as a thoroughly documented account of British plant communities.
Some other countries with similar developments are Norway, Mexico,
and Australia.

By using these numerical techniques, it is much easier to
relate plant community variation to environmental variation, which
has always been a major issue in phytosociology. A wealth of
computer programs, theoretical papers and reviews, and textbooks
are now available on the topic (e.g., Whittaker 1978a, 1978b,
Orlóci 1978, van der Maarel 1979, Gauch 1982), all covering the
techniques used most frequently in phytosociology.

I foresee a further extension of the original ideas of Braun-
Blanquet, which I summarized a few years ago (van der Maarel 1975,
Westhoff and van der Maarel 1978) as follows:

(1) Plant communities are conceived as types of vegetation recognized
     by their floristic composition. The full species composition
     of communities express their relationships to one another and
     to their environments better than any other characteristic.
(2) Some species in the floristic composition of a community are
     more sensitive to certain relationships than others, and thus
     are more effective bio-indicators. These diagnostic species
     comprise character species, differential species, and constant
     companions and together form the characteristic species combina-
     tion.
(3) Diagnostic species are used to organize communities into a
     hierarchical classification where the association is the basic
     unit. The hierarchy is invaluable for understanding and
     communicating community relationships. The integrated classifi-
     cation and ecology of community types will be elaborated
     further in relation to recent theoretical developments (Pignatti
     1980), especially those on the use of grammars (Dale 1980,
     1981).

In conclusion, I predict phytosociology a bright future if it
continues to broaden its scope and its organization (see McIntosh
1978). I hope the International Society for Vegetation Science,
which recently adopted a new statute and has *Vegetatio* as its
official journal for international publications, will serve as a
platform for the future expansion of phytosociology.

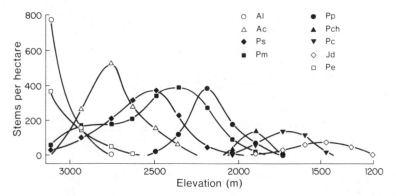

Fig. 3. Ecological response curves for coniferous tree species along
an elevation gradient on north facing slopes in the Santa
Catalina and Pinaleno Mountains, Arizona: *Abies lasiocarpa*
(Al); *Abies concolor* (Ac), *Pinus strobiformis* (Ps), *Pseudot-
suga menziesii* (Pm), *Pinus ponderosa* (Pp), *Pinus chihauhuana*
(Pch), *Pinus cembroides* (Pc), *Juniperus deppeana* (Jd), *Picea
engelmanni* (Pe) (from Whittaker 1978c, as modified by Austin
1980; reprinted by permission of Junk, The Hague, The Netherlands).

THE DEVELOPMENT OF GRADIENT THEORY

     Gradient theory, based on the so called individualistic behaviour
of plant species (see, e.g., MacIntosh 1967, Whittaker 1967),
originated independently in North America (with H.A. Gleason) and
Russia (with L.G. Ramensky). "Individualistic" refers to the
observation that no two species show the same distribution pattern
along an environmental gradient. This idea comes close to Gause's
principle of competitive exclusion (see McIntosh 1980 for a historical
review).

     It was primarily Whittaker (1967, 1978c; see also Westman and
Peet 1982) who originated the all encompassing scheme of ecological
interpretation of plant community variation. Often environmental
gradients (e.g., altitude, moisture, fertility) can explain or at
least make plausible the variation in vegetation. Such gradients
can be detected directly or indirectly: directly through measurement
or estimation of an environmental factor and then alignment of
community samples along the factor and indirectly through ordination
of samples and subsequent correlation of components with environmental
factors. A gradient may be elucidated by plotting species performance
on the y-axis and the environmental factor on the x-axis (Fig. 3).
The species distribution should approximate a Gaussian curve.

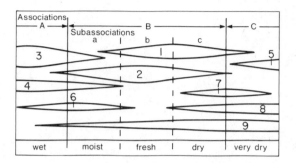

Fig. 4.   Schematic representation of ecological response curves of
          species along an environmental gradient which also can be
          used to characterize plant community types based on combina-
          tion versus exclusion of relative optima of 17 species (after
          Westhoff and van der Maarel 1978; reprinted by permission
          of Junk, The Hague, The Netherlands).

     According to Whittaker, the view that variation in the floristic
composition of communities is continuous does not exclude the
concept of the community as a relatively discontinuous unit based
on similarities in species distribution patterns. And indeed,
gradient theory and phytosociology are now being integrated (Fig. 4).

     Two major and interrelated problems remain to be solved in the
further expansion of gradient theory: the refinement of the environmental measurements and the development of more realistic parameters
expressing species performance. One problem with environmental
measures is that gradients as traditionally considered are very
broadly defined topographically. The performance of a species
along such a conditional gradient depends on various operational
factors, including resources (notably nutrients and directly available
moisture) and stresses (Austin 1980). Related to this problem is
the question of how to plot the environmental values. Is the
variation in a resource factor linear or geometrical?

     Is a relative measure of species performance more realistic
than an absolute measure (e.g., species biomass relative to the
total biomass of the stand)? (See Austin 1980.) A relative measure,
which is useful in competition studies, may be of interest for
phytosociology also where species quantities are usually absolute
cover or cover/abundance estimates. As Austin has shown, different
types of performance curves can arise using relative measures,
including a wedge type. This type was presumed for a gradient of
increasing environmental dynamics, or stress. Figure 5 shows wedge
type relative performances of three different grass species grown
in pots with logarithmically increasing nutrient concentrations
(Austin and Austin 1980).

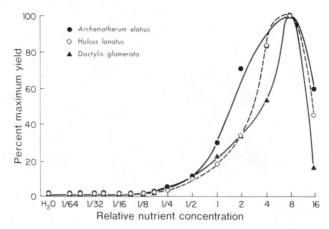

Fig. 5. Physiological (mono-species culture) response curves of three
         grass species expressed as relative yield in relation to a
         nutrient gradient (from Austin 1980, after Austin and Austin
         1980; reprinted by permission of Junk, The Hague, The Nether-
         lands).

     More recently Austin (1982) suggested the ratio between shoot
dry weight yield of a species (in a mixed growth experiment) and
the yield of the most productive species (at the particular level
of the environmental factor) as a new normalized ecological performance value. Ecological maxima along the gradient according to
this parameter do not correspond with the species physiological
optimum (i.e., if grown separately) when the optimum is measured
using absolute yield. Clearly, the performance of species in
natural communities has to be reconsidered.

     The connection between performance along gradients and environ-
mental stress was elaborated by Grime (1973, 1979). Figure 6
presents a model in which both relative dominance of species and
species diversity (i.e., number of species per unit area) are a
function of stress, which Grime defines as conditions limiting the
photosynthesis of plants. High dominance may occur under conditions
of both low and high stress and diversity is optimal under moderate
stress. This latter aspect of the model agrees with the general
observation that species diversity reaches a local maximum in the
middle of an environmental gradient (e.g., van der Maarel and
Leertouwer 1967).

THE DEVELOPMENT OF PATTERN ECOLOGY

     Although not traditionally a priority item in phytosociology,
the analysis of pattern (i.e., deviation from a random configuration)

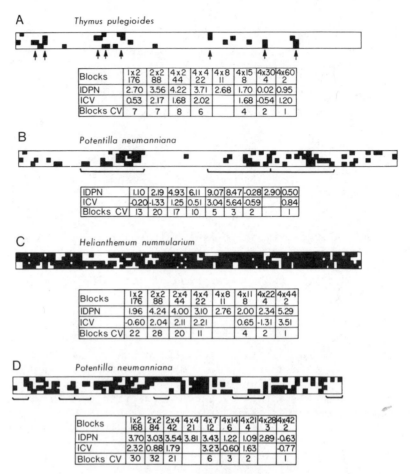

Fig. 6.   Examples of distribution patterns (based on presence or
          absence in 1/16 m²) of species in Belgian limestone grass-
          lands (from Bouxin and Gautier 1982; reprinted by permission
          of Junk, The Hague, The Netherlands). (A) *Thymus pulegioides*
          --agregate pattern; (B) *Potentilla neumanniana*--clump pat-
          tern; (C) *Helianthemum nummularium*--small gap pattern;
          (D) *Potentilla neumanniana* (other transect)--large gap
          pattern. The smallest block combines two quadrats of
          1/16 m² each.   IDPN is the non-parametric dispersion index,
          ICV is the contagion index used, Blocks CV is the number of
          pairs of blocks used for the calculation of ICV.

is an intrinsic and important aspect of vegetation science. The
usual approach is to lay out a transect or grid of very small
quadrats in a relatively homogeneous stand of vegetation and note
the presence or absence of each species in each quadrat, or use

some quantitative measure. Numerous methods detect patterns. Most
of them calculate some index (e.g., statistical variance, dispersion
index) for blocks of quadrats with increasing block sizes. Well
known general (but now outdated) introductions to pattern analysis
are found in Greig-Smith (1964) and Kershaw (1973) and in a more
recent review presented by Greig-Smith (1979). The application of
information statistics for detecting patterns has been promoted by
Montpellier ecologists (Godron 1966, also Godron, this volume).
Results have been combined with environmental data to arrive at so
called ecological profiles of species (Guillerm 1971, Bottliková et
al. 1976). For newer methods, recent issues of the *Journal of
Ecology* and *Vegetatio* should be consulted.

Coherence of patterns of species distribution with environmental
variation is interpreted as environmental pattern. As Kershaw
(1973) made clear, there are three main types of pattern: morphological, ecological, and sociological. Morphological pattern
arises directly from the growth of the plant individual, ecological
pattern is a direct response of the plant population to environmental
variation, sociological pattern arises from interaction between
plant species, and between animals and plants. According to Greig-
Smith's (1979) review, many species patterns can be interpreted as
ecological.

There are two manifestations of pattern: the occurrence of
coherent fields containing a species, pattern in the proper sense,
and the occurrence of such fields without the species, called gaps.
Figure 6 shows some examples taken from Bouxin and Gautier (1982).
Two indices were used in this study, a non-parametric dispersion
index based on the variability of the frequency of a species within
blocks of quadrats, and a contagion index based on the heterogeneity
of two adjacent blocks of quadrats. Most patterns in this study
were contagious patterns, either small scale aggregrates or larger
scale clumps.

The occurrence of random configurations is relatively rare,
even in apparently homogeneous vegetation (see Greig-Smith 1979).
Of the 289 cases Bouxin and Gautier (1982) included, only 32 were
random. This has a direct bearing on phytosociology, on the concept
of character species, and on diversity and minimal area problems.
Bouxin and Gautier (1982) suggested that most of the good character
species in their Mesobromion limestone grassland showed clear,
clumped patterns. This is just one example of the future significance
pattern analysis will have for vegetation analysis.

THE DEVELOPMENT OF THE CONCEPT OF DIVERSITY

Since the mid 1960s, the notion of species diversity has
become apparent in vegetation studies, mainly through papers by

Whittaker (1965, 1972, 1977). Because plant individuals in average
stands of vegetation are difficult to distinguish and because
vegetation is approached on an area basis rather than on a sample
size basis, it is no wonder that diversity in vegetation is usually
related to area. Indeed species-area relationships are a common
element in vegetation diversity studies. This relationship again
means a link to classical phytosociology where the concept of
minimal area was developed (see Mueller-Dombois and Ellenberg
1974).

According to the classical view of plant communities, the
number of species in any stand will reach a saturation level; the
area on which this occurs is called the minimal area. Swedish
authors (Arrhenius, Romell, Kylin) demonstrated in the 1920s that
at least two other models are realistic (see van der Maarel 1970).
The first is the semi-logarithmic relation where species number
never reaches an upper limit but continues to increase with the
geometrical increase of area. The second, the log-log model, goes
even farther and assumes a geometrical increase in species number
with a geometrical increase in area.

We recognize two of the recent species-individual distribution
functions in these old models: the log series distribution where
the parameter is linked directly to the species-log area relation
(Williams 1964) and the canonical log normal distribution where the
parameter 2 lies behind the log-log relation (Preston 1962). We
might even interpret the saturation curve, on which phytosociology
relies, as an extreme case of a third well known distribution
function, the geometrical one.

Vegetation scientists have not yet contributed to the theory
of species-abundance distributions. If any, they mainly use one of
the two simple distribution free measures of Shannon and Simpson
(which are correlated in most plant communities), and base them on
cover estimates (see Pielou 1966, 1975, Hill 1973, Peet 1974).
There is a growing preference for Shannon's H', which really cannot
be defended (Hurlbert 1971), but which is generally justified by
the relative preponderance of its use.

A more direct approach concerns species richness on standard
areas. Whittaker's standard analysis of a 0.1 ha plot has become
popular for comparing Mediterranean scrub types and temperate
woodlands (Glenn-Lewin 1975, 1977, Peet 1978, Whittaker et al.
1979, Naveh and Whittaker 1980). Some general patterns that emerge
are

(1) An optimum richness occurs along the fertility gradient.
(2) Internal environmental heterogeneity (including heterogeneity
     of regeneration sites, which I will discuss later) is important,
     but still not measured adequately.

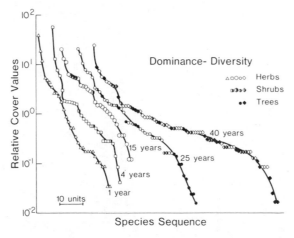

Fig. 7.   Dominance-diversity relations based on relative cover values
          for plant species in different stages of an old-field
          succession.   The pattern changes from geometrical to log
          normal.   (From Bazzaz 1975; copyright © 1966 by Ecological
          Society of America; reprinted by permission.)

(3)   History of the area is important.   This history includes both
      the evolutionary development of a rich flora comprising various
      phytogeographical elements and a continuity in general environ-
      mental setting.   The latter may include disturbance (e.g.,
      grazing) if it is a constant factor.
(4)   Richness in some (not all) tropical rain forests still surpasses
      our imagination and our analytical power.

      Another approach which has become common is the dominance-diver-
sity relation (Whittaker 1965):   the species of a stand are ranked
according to their relative proportions (e.g., to total biomass).
The curve connecting the respective points can be interpreted as a
broken stick model reflecting random niche boundaries, or as a
geometrical series reflecting the niche pre-emption hypothesis, or
as a log normal distribution for intermediate cases (Whittaker
1965, May 1975, 1976).   Figure 7 shows a roughly geometrical distribution in an early stage of succession on an abandoned field and an
approximate distribution for older stages of succession (Bazzaz
1975).

      Both the richness and the dominance aspect of diversity can be
used to characterize a stand of vegetation (Fig. 8).   The minimal
area, although bound to species saturation, can be approached more
pragmatically as the area where the more commonly distributed
species are represented (qualitative minimal area).   The similarity
between samples of the minimal area size (calculated on the basis

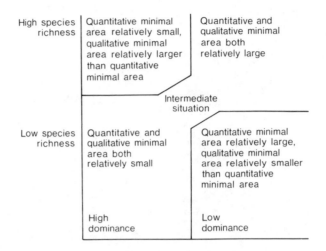

Fig. 8. Model for relating the qualitative and quantitative minimal
area to species richness and dominance (from Dietvorst et al.
1982).

of species quantities: quantitative minimal area) reaches values
considered to indicate coherence on the lower community type (association) level.

In conclusion, vegetation scientists should begin discussing
sound ecological models for various distributions, specifically
whether the log normal distribution is the universal function for
equilibrium situations as postulated by Gray (1981, Ugland and Gray
1982).

THE ADVANCEMENT OF SUCCESSION THEORY

Interest in succession is almost as old as vegetation science
itself and some far reaching elements of successional theory,
notably from Clements, are still subject to vivid discussions (see
Miles 1979 for a recent review). A few major lines of development
can be discussed here.

One particularly European phytosociological development is the
study of permanent plots, where floristic compositions are described
yearly (or at least every few years). Most plots are situated in
semi-natural vegetation subject to changes brought about by variation
in management. After 8-10 years, such studies become extremely
useful, because by then some extreme weather conditions have occurred
and the effect and extent of fluctuations become clearer. The
proceedings of three symposia, published in *Vegetatio* and issued
simultaneously as separate volumes (Beeftink 1980, van der Maarel

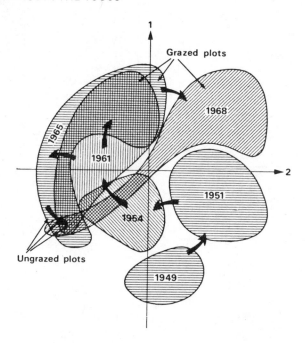

Fig. 9.   Principal component analysis of Australian grasslands under
          different grazing regimes surveyed between 1949 and 1968
          (from Austin et al. 1981; reprinted by permission of Junk,
          The Hague, The Netherlands).

1980, Poissonet et al. 1981), have brought together many case
studies and interpretations. One general finding of these European
studies is the far reaching effect of such fluctuations, which
causes vegetational change to be an irregularly cyclic succession.

    Often permanent plots are described in spatial series (transects)
and thus spatio-temporal developments also can be followed. If
numerical methods are applied to such spatio-temporal series,
interpretable and fascinating patterns of change emerge. Transition
probability studies are made and Markovian chains applied as an
explanation (the latter with moderate success because of the everchanging transition rates in time).

    One example of a numerical analysis concerns the dynamics of
an Australian grassland under sheep grazing (Austin et al. 1981).
Permanent plots were followed from 1949 to 1968 in a *Danthonia
semiannularis* grassland where different grazing intensities were
maintained. Six years of quantitative floristic data from 325
quadrats were subjected to divisive cluster analysis (Table 1).
Communities A-D characterize early and (or) dry years, E-G wet
years, and H-I late dry years. Figure 9 shows axes 1 and 2 of a

Table 1.   Relationship between numerical classification communities
           and years in an Australian grassland.   Frequency of each
           community in a year is shown as a percentage of total
           occurrences of the community.   (From Austin et al. 1981.)

	Community	Year					
		1949	1951	1954	1961	1965	1968
A	-*Plantago varia*, -*Medicago polymorpha*	80.1[a]	2.2	9.6	1.5	6.6	
E	+*P. varia*, -*Hypochaeris radicata*, -*Hedyponis cretica*	12.4[b]	72.7	4.7	0.6	1.8	8.3
B	+*P. varia*, +*M. polymorpha*, -*Avena fatua*, -*Euphorbia drummondii*	23.7	10.3	36.1	11.3	17.0	1.5
C	-*P. varia*, *M. polymorpha*, -*A. fatua*, +*E. drummondii*	8.8		89.6	1.6		
I	+*P. varia*, *H. radicata*, +*Isoetopsis graminifolia*		6.0	59.1	24.2	10.7	
H	+*P. varia*, +*H. radicata*, -*I. graminifolia*			3.2	42.4	49.8	4.6
D	-*P. varia*, +*M. polymorpha*, +*A. fatua*		0.7	7.6	20.8	52.1	18.8
F	+*P. varia*, -*H. radicata*, +*H. cretica*, -*I. graminifolia*	23.6	9.7	2.1	26.4	38.2	
G	+*P. varia*, -*H. radicata*, +*H. cretica*, +*I. graminifolia*	7.9	1.9	0.5	4.7	85.0	
	Rainfall for May-July (mm)	264	715	166	276	293	449

[a] Year of primary occurrence underlined.
[b] Year of secondary occurrence italicized.

principal components analysis of the data. The first component, accounting for 20% of the total variance, can be interpreted as a general trend throughout the years, which both grazed and ungrazed quadrats undergo. Whether this is real succession of "prolonged fluctuation" is unclear. The second component, accounting for 17% of the total variance, clearly reflects seasonality, that is, the variation in winter rainfall.

Some successional studies are now combined with population demographic studies of the major species developing during primary succession (e.g., Pickett 1982). This will certainly become a major line of research in vegetation dynamics (see Harper 1977, 1981).

A different approach has been developed through modeling, mainly of the establishment, growth, and death of dominant forest species (for surveys, see Shugart and West 1980, West et al. 1981). A new development is the modeling of regeneration after disturbance (Connell and Slatyer 1977, Nobel and Slatyer 1980). Both from classical permanent plot studies in semi-natural grassland and heathland and from tree regeneration studies in natural temperate and tropical forests, the overriding impact of disturbances becomes evident. Three types of regeneration pathways may be distinguished: the facilitation pathway in which the presence of early occupants facilitates the entry of successive species, the tolerance pathway in which later species become established in the presence of others, and the inhibition pathway in which later species cannot grow to maturity in the presence of early ones.

Through disturbance, we may link succession and diversity studies. From the work of Grime (1977, 1979), Grubb (1979), Hubbell (1979), White (1979), and others we begin to understand how species diversity is very much an outcome of the continuous creation of niches ("regeneration niches") through local disturbances. Again we may conclude that this will be an important line of research and ecological theory!

CONCLUSION

The most important or at least the most exciting development in vegetation science, I think, is that of vegetation dynamics. Through the combined effort of permanent plot description and repeated vegetation mapping on one hand, and population dynamics and paleoecology on the other hand, we will understand much more about the present composition of vegetation and its future develop-ment. We will learn much more about the role of natural disturbances through intensified research on succession. This effort will give us more clues to the proper management of our semi-natural and post-cultural vegetation and to the driving forces, abiotic and biotic, of vegetation succession.

In this context, it will be both inevitable and fascinating to study the pattern and process of individual species populations. The general idea is that there is a continuous formation of gaps through dying of individuals or vegetative parts of plants, usually in combination with some sort of disturbance. Certain species can enter the gap if they are on the spot and find their exact establishment conditions. This study will include seed bank, growth, strategy, and interaction analyses of species. This approach is a long way from the original phytosociology! But it still has the community as a whole as the common prime interest.

And now my plea. In view of the rapid advancement in the theory and techniques of classification and ordination and the rapid expansion of floristic-structural-functional classification, it would be very appropriate to develop a world survey of vegetation types. If this survey could be combined with our present views on bioclimatology and historical plant geography, we could, perhaps, detect very broad lines of world vegetation development. To mention just one of the perspectives, we could obtain the distribution pattern of the dominant species of the world. The survey also would contribute much to biosphere reserve programs. Vegetation is still a black box, or rather a green box. The behavior of this green box can be studied on different levels of abstraction and it can be systematically described. Both vegetation typology and dynamics will be major elements in the vegetation science of the 1980s.

LITERATURE CITED

Austin, M.P. 1980. Searching for a model for use in vegetation analysis. Vegetatio 42:11-21.
Austin, M.P. 1982. Use of a relative performance value in the prediction of performance in multispecies mixtures from monoculture performance. J. Ecol. 70:559-570.
Austin, M.P., and B.O. Austin. 1980. Behaviour of experimental plant communities along a nutrient gradient. J. Ecol. 68:891-918.
Austin, M.P., O.B. Williams, and L. Belbin. 1981. Grassland dynamics under sheep grazing in an Australian Mediterranean type climate. Vegetatio 46/47:201-211.
Barkman, J.J., J. Moravec, and S. Rauschert. 1976. Code of phytosociological nomenclature. Vegetatio 32:131-185.
Bazzaz, F.A. 1975. Plant species diversity in old-field successional ecosystems in southern Illinois. Ecology 56:485-488.
Beeftink, W.G., ed. 1980. Vegetation dynamics. Proc. 2nd Symp. Working Group on Succession Research on Permanent Plots, Delta Inst. Hydrobiol. Res., Oct. 1-3, 1975. Junk, The Hague, The Netherlands.

Beeftink, W.G., and J.-M. Géhu. 1973. *Spartinetea maritimae* Prodromes des groupements végétaux d'Europe, vol. 1. Cramer, Lehre, W. Germany.

Bottlikova, A., Ph. Daget, J. Dross, J.-L. Guillerm, F. Romane, and M. Ruzickova. 1976. Quelques resultats obtenus par l'analyse factorielle et les profiles écologiques dans la vallée de Liptov. Vegetatio 31:79-96.

Bouxin, G., and N. Gautier. 1982. Pattern analysis in Belgian limestone grasslands. Vegetatio 49:65-83.

Braun-Blanquet, J. 1965. Plant sociology: The study of plant communities. (Transl., rev. and ed. by C.D. Fuller and H.S. Conard.) Hafner, London, England.

Connell, J.H., and R.O. Slatyer. 1977. Mechanisms of succession in natural communities and their role in community stability and organization. Am. Nat. 111:1119-1144.

Dale, M.B. 1980. A syntactic basis of classification. Vegetatio 42:93-98.

Dale, M.B. 1981. A grammatical approach to vegetation classification. Pages 141-149 *in* A.N. Gillison and D.J. Anderson, eds. Vegetation classification in Australia. CSIRO, Canberra, Australia.

Dietvorst, P., E. van der Maarel, and H. van der Putten. 1982. A new approach to the minimal area of a plant community. Vegetatio 50:77-91.

Egler, F.W. 1954. Philosophical and practical considerations of the Braun-Blanquet system of plant sociology. Castanea 19:45-60.

Ellenberg, H. 1978. Vegetation Mitteleuropas mit den Alpen, 2 Aufl. Ulmer, Stuttgart, W. Germany.

Gauch, H.G., Jr. 1982. Multivariate analysis in community ecology. Cambridge Univ. Press, Cambridge, England.

Glenn-Lewin, D.C. 1975. Plant species diversity in ravines of southern Finger Lakes region, New York. Can. J. Bot. 53:1465-1472.

Glenn-Lewin, D.C. 1977. Species diversity in North American temperate forests. Vegetatio 33:153-162.

Godron, M. 1966. Application de la théorie de l'information a l'étude de l'homogénéité et de la structure de la végétation. Oecol. Plant. 1:187-197.

Gray, J.S. 1981. Detecting pollution induced changes in communities using the log-normal distribution of individuals among species. Mar. Pollut. Bull. 12:173-176.

Greig-Smith, P. 1964. Quantitative plant ecology, 2nd ed. Butterworths, London, England.

Greig-Smith, P. 1979. Pattern in vegetation. J. Ecol. 67:755-779.

Grime, J.P. 1973. Control of species density in herbaceous vegetation. J. Environ. Manage. 1:151-167.

Grime, J.P. 1977. Evidence for the existence of three primary strategies in plants and its relevance to ecological and evolutionary theory. Am. Nat. 111:1169-1194.

Grime, J.P.   1979.   Plant strategies and vegetation processes.
    Wiley, Chichester, England.
Grubb, P.J.   1977.   The maintenance of species-richness in plant
    communities: The importance of the regeneration niche.   Biol.
    Rev. 52:107-145.
Guillerm, J.L.   1971.   Calcul de l'information fournie par un
    profil écologique et valeur indicatrice des especès.   Oecol.
    Plant. 6:209-226.
Harper, J.L.   1977. Population biology of plants.   Academic Press,
    London, England.
Harper, J.L.   1981.   After description.   Pages 11-25 *in* E.I. Newman,
    ed.   The plant community as a working mechanism.   Blackwell,
    Oxford, England.
Hill, M.O.   1973.   Diversity and evenness: A unifying notation and
    its consequences.   Ecology 54:427-432.
Hubbell, S.P.   1979.   Tree dispersion, abundance, and diversity in
    a tropical dry forest.   Science 203:1299-1309.
Hurlbert, S.H.   1971.   The non-concept of species diversity: A
    critique and alternative parameters.   Ecology 52:577-586.
Kershaw, K.A.   1973.   Quanitative and dynamic ecology, 2nd ed.
    Elsevier, New York, NY USA.
Kortekaas, W.M., E. van der Maarel, and W.G. Beeftink.   1976.   A
    numerical classification of European spartina communities.
    Vegetatio 33:51-60.
MacIntosh, R.P.   1967.   The continuum concept of vegetation.   Bot.
    Rev. 33(2):131-187.
MacIntosh, R.P.   1978.   Phytosociology.   Benchmark papers in ecology
    16.   Dowden, Hutchinson, and Ross, Stroudsburg, PA USA.
MacIntosh, R.P.   1980.   The background and some current problems of
    theoretical ecology.   Synthese 43:195-255.
May, R.M.   1975.   Patterns of species abundance and diversity.
    Pages 81-120 *in* M.L. Cody and J.M. Diamond, eds.   Ecology and
    evolution of communities.   Belknap, Cambridge, MA USA.
May, R.M.   1976.   Patterns in multi-species communities.   Pages
    142-162 *in* R.M. May, ed.   Theoretical ecology.   Blackwell,
    Oxford, England.
Miles, J.   1979.   Vegetation dynamics.   Chapman and Hall, London,
    England.
Mucina, L.   1982.   Numerical classification and ordination of
    ruderal plant communities (Sisymbrietalia, Onopordietalia) in
    the Western part of Slovakia.   Vegetatio 48:267-275.
Mueller-Dombois, D., and H. Ellenberg.   1974.   Aims and methods of
    vegetation ecology.   Wiley, New York, NY USA.
Naveh, Z., and R.H. Whittaker.   1980.   Structural and floristic
    diversity of shrublands and woodlands in Northern Israel and
    other Mediterranean areas.   Vegetatio 41:171-190.
Noble, I.R., and R.O. Slatyer.   1980.   The use of vital attributes
    to predict successional changes in plant communities subject
    to recurrent disturbances.   Vegetatio 43:5-21.

Orlóci, L. 1978. Multivariate analysis in vegetation research,
     2nd ed. Junk, The Hague, The Netherlands.
Peet, R.K. 1974. The measurement of species diversity. Annu.
     Rev. Ecol. Syst. 5:285-307.
Peet, R.K. 1978. Forest vegetation of the Colorado front range:
     Patterns of species diversity. Vegetatio 37:65-78.
Pickett, S.T.A. 1982. Population processes through twenty years
     of old field succession. Vegetatio 49:45-59.
Pielou, E.C. 1966. The measurement of diversity in different
     types of biological collections. J. Theoret. Biol. 13:131-144.
Pielou, E.C. 1975. Ecological diversity. Wiley Interscience,
     New York, NY USA.
Pignatti, S. 1980. Reflections on the phytosociological approach
     and the epistemological basis of vegetation science. Classifi-
     cation and ordination. Vegetatio 42:181-185.
Poissonet, P., F. Romane, M.P. Austin, E. van der Maarel, and
     W. Schmidt, eds. 1981. Vegetation dynamics in grasslands,
     heathlands and mediterranean lignous formations. Vegetatio
     46/47 and Adv. Veg. Sci. 4. Junk, The Hague, The Netherlands.
Poore, M.E. 1955-1956. The use of phytosociological methods in
     ecological investigations, parts 1-4. J. Ecol. 43:226-244,
     245-269, 606-651, and 44:28-50.
Preston, F.W. 1962. The canonical distribution of commonness and
     rarity. Ecology 43:185-215, 410-432.
Shimwell, D.W. 1971. The description and classification of vegeta-
     tion. Sidgwick and Jackson, London, England.
Shugart, H.H., Jr, and D.C. West. 1980. Forest succession models.
     Bioscience 30:308-313.
Tüxen, R., ed. 1978. Assoziationskomplexe (Sigmeten) und ihre
     praktische Anwendung. Ber. Int. Symp. Rinteln 1977. Cramer,
     Vaduz, Liechtenstein.
Ugland, K.I., and J.S. Gray. 1982. Lognormal distributions and
     the concept of community equilibrium. Oikos 39:171-178.
van der Maarel, E. 1970. Vegetationsstruktur und Minimum-Areal in
     einem Dünentrockenrasen. Pages 218-239 in R. Tüxen, ed.
     Gesellschaftsmorphologie. Junk, The Hague, The Netherlands.
van der Maarel, E. 1975. The Braun-Blanquet approach in perspective.
     Vegetatio 30:213-219.
van der Maarel, E. 1979. Multivariate methods in phytosociology,
     with reference to the Netherlands. Pages 161-225 in M.J.A.
     Werger, ed. The study of vegetation. Junk, The Hague, The
     Netherlands.
van der Maarel, E., ed. 1980. Succession. Vegetatio 43(1/2) and
     Adv. Veg. Sci. 3. Junk, The Hague, The Netherlands.
van der Maarel, E., and J. Leertouwer. 1967. Variation in vegeta-
     tion and species diversity along a local environmental gradient.
     Acta Bot. Neerl. 16:211-221.
van der Maarel, E., L. Orlóci, and S. Pignatti, eds. 1980. Data-
     processing in phytosociology. Junk, The Hague, The Netherlands.

West, D.C., H.H. Shugart, and D.B. Botkin, eds. 1981. Forest
    succession: Concepts and application. Springer, New York, NY
    USA.
Westhoff, V., and E. van der Maarel. 1978. The Braun-Blanquet
    approach, 2nd ed. Pages 287-399 *in* R.H. Whittaker, ed.
    Classification of plant communities. Junk, The Hague, The
    Netherlands.
Westman, W.E., and R.K. Peet. 1982. Robert H. Whittaker (1920-1980):
    The man and his work. Vegetatio 48:97-122.
White, P.S. 1979. Pattern, process, and natural disturbance in
    vegetation. Bot. Rev. 45:229-299.
Whittaker, R.H. 1965. Dominance and diversity in land plant
    communities. Science 147:250-260.
Whittaker, R.H. 1967. Gradient analysis of vegetation. Biol.
    Rev. (London) 42:207-264.
Whittaker, R.H. 1972. Evolution and measurement of species diver-
    sity. Taxon 21:213-251.
Whittaker, R.H. 1977. Evolution of species diversity in land
    communities. Pages 1-67 *in* M.K. Hecht, W.C. Steere, and B.
    Wallace, eds. Evolutionary biology, vol. 10. Plenum, New York,
    NY USA.
Whittaker, R.H., ed. 1978a. Classification of plant communities.
    Junk, The Hague, The Netherlands.
Whittaker, R.H., ed. 1978b. Ordination of plant communities.
    Junk, The Hague, The Netherlands.
Whittaker, R.H. 1978c. Direct gradient analysis. Pages 7-15 *in*
    R.H. Whittaker, ed. Ordination of plant communities. Junk,
    The Hague, The Netherlands.
Whittaker, R.H., W.A. Niering, and M.D. Crisp. 1979. Structure,
    pattern and diversity of a mallee community in New South
    Wales. Vegetatio 39:65-76.
Williams, C.B. 1964. Patterns in the balance of nature and related
    problems in quantitative ecology. Academic Press, New York,
    NY USA.

ECOLOGICAL RESEARCH ON ARTHROPODS IN CENTRAL AMAZONIAN FOREST

ECOSYSTEMS WITH RECOMMENDATIONS FOR STUDY PROCEDURES

Joachim Adis

Max Planck Institute for Limnology
Tropical Ecology Working Group
D-2320 Plön, West Germany, in cooperation with
Instituto Nacional de Pesquisas da Amazônia (INPA)
Caixa Postal 478, 69 000 Manaus/AM, Brazil
(Convênio INPA/Max Planck)

Herbert O.R. Schubart

Instituto Nacional de Pesquisas da Amazônia (INPA)
Caixa Postal 478, 69 000 Manaus/AM, Brazil

ABSTRACT

In this work, we conducted three comparative studies on arthropods in Central Amazonian forest ecosystems.

(1) The soil faunas from seven forest types were compared using the soil extraction method of Kempson. Anthropod abundance varied between 3,500 and 9,000 ind/m^2. Abundance generally was higher in inundation forests when compared to dryland forests. Acari and Collembola represented 60-80% of the soil fauna collected. Dominance of frequent taxa varied significantly among forest types. For preliminary studies of ecosystems, particularly orientation projects such as these, we recommend modern standing crop methods.

(2) Abundance of arthropods and dominance of "key taxa" in different strata (soil/litter, trunk, and canopy) of riverine inundation forests, one in the Black Water Region and one in the White Water Region, were compared with a primary forest in the Dryland Region during the rainy season. Sampling methods were Kempson soil extraction, ground and arboreal photo-eclectors, and canopy fogging. Relative abundance of arthropods (Acari

and Collembola disregarded) ranged between 1,000 and 1,500 ind/m^2 in the soil and from 1,000 to 7,000 ind/m^2 in ground photo-eclectors. Between 1,000 and 8,000 arthropods occurred per tree trunk and 30 to 60 ind/m^2 were collected from the canopy. Relative abundance of arthropods was somewhat higher in the black water inundation forest, intermediate in the primary dryland forest, and lower in the white water inundation forest. Dominance of taxa varied significantly among strata within the forests and among forest types as well. Zoophagous groups were predominant. On the forest floor, primary decompos- ers other than Diptera (in particular, larvae) were almost absent. We recommend a combination of permanent collecting methods, complemented by standing crop samplings, for future comparative studies on arthropod populations in neotropical forest ecosystems.

(3) Population fluctuations of arthropods may correspond somewhat with altering abiotic conditions. The activity density of arthropods on trunks was studied during and between seasons both in a black and a white water inundation forest, as well as in a primary and a secondary dryland forest. More arthropods were caught during the rainy season, except in the primary dryland forest. Formicidae represented the predominant group with 36-81% of the total catch. Dominance of taxa in samples varied considerably within and between seasons. Pronounced trunk ascents of ants were observed during the dry season in all forest types. A dry, warm climate is believed to stimulate tree inhabiting, meso-xerophilous ant species to become highly active and to migrate between canopy, trunk region, and forest floor. With the beginning of the rainy season and a changing microclimate, their activity density decreased steadily. In the black water inundation forest, numerous non-flying (terri- colous) arthropods migrated into the trunk-canopy region with the beginning of the rainy season, where they remained throughout forest inundation (5-6 months duration). Trunk ascents of hygrophilous species apparently are influenced by rising wetness on the forest floor and increasing relative humidity in the lower trunk region. Thus, for the Central Amazonian forests we studied, we could not confirm the hypothesis that tropical ecosystems with sufficiently high numbers of species lack population fluctuations.

We conclude this paper by discussing theories about the diver- sity of Amazonian arthropods and by posing timely research questions for tropical studies. Integrated long term studies are required to achieve indepth insights into the structure and function of neotropi- cal forest ecosystems. Information obtained by basic research is the foundation for ecological criteria to be used in biological management plans for Amazonia.

1. PRESENT SITUATION

Tropical moist forests are threatened worldwide and they likely will be mostly destroyed within 50 years (National Research Council 1980, Council on Environmental Quality 1980, Meyers 1980). In the Brazilian Amazon, deforestation is proceeding at an unprecedented rate. However, cutting intensity is very uneven and forecast of total conversion is difficult (Fearnside 1982a, 1982b). There may be as many as 30 million tropical arthropod species in the world (Erwin 1982) for which there are few professionals trained to collect and to describe them (National Research Council 1980). Most species are still unknown. It is believed that many will become extinct before being documented. The function of distinct arthropod populations in neotropical ecosystems remains to be studied. Insights into dominance, phenology, and activity dynamics of terrestrial arthropods have to be obtained for different types of ecosystems. Basic values to estimate energy flows within arthropod-cenoses (depending on data from, e.g., emergence traps [Funke 1971, 1973]) are totally missing. Even the applicability and efficiency of modern collecting methods, developed in monocultural (forest) ecosystems of temperate zones during the International Biological Program (IBP) (e.g., Funke 1977, Phillipson 1971), still have to be tested in neotropical areas rich in plant species.

This paper presents the first information on modern collecting methods used during long term studies in inundation and dryland forests of Central Amazonia (Adis 1981, Penny and Arias 1982), the bias on data interpretation, and, most important, recommendations for study procedures and study durations for future projects.

2. FOREST TYPES VERSUS STUDY DURATION

About 3.7 million km^2 of the Brazilian Amazon are covered by vegetation: 89% by dryland forests (Terra Firme), 2% by inundation forests in the white and black water areas (Várzea and Igapó, respectively) (Sioli 1956, Prance 1979, 1980), and 9% by other vegetation types (Braga 1979). Floristically, each major forest type is composed of various subdivisions (Braga 1979) or associations (cf. Walter 1973) such as, for example, campina forest, campinarana forest, bamboo forest, and dense forest on Terra Firme areas (Braga 1979, Klinge 1983) and lower and upper Igapó forests in black water areas (Adis 1984a). Considering that geochemical differences (e.g., Fittkau 1970, 1971) occur within associations, that climatic conditions are not uniform over the Amazon Basin (Salati et al. 1978), and that associations themselves are composed of different, but characteristic, plant species (subassociations) (Anderson and Benson 1980, Beck 1971, Irmler 1977, Klinge et al. 1983, Prance et al. 1976, Revilla 1981, Worbes 1983), diversity and endemism of arthropods in neotropical forest ecosystems are expected to be high.

Ecosystem analysis--a basic requirement for understanding the function of neotropical forest types--can be realized by short term visits to distinct areas or by permanent presence (i.e., long term studies) in selected areas. In the Brazilian Amazon, investigations depend mostly on logistical problems. In recent years, research activities by international ecologists collaborating with the Instituto Nacional de Pesquisas da Amazônia (INPA) have concentrated on the area around Manaus. Previous studies in distant areas (e.g., future sites of hydroelectric developments or natural parks) often were restricted to short term visits and served mostly for biogeographic inventories, principally of selected groups. The frequency of such studies have been increasingly biased by rising excursion expenses due to high inflation in the country and critical petroleum supplies.

3. STANDING CROP METHODS, A FIRST ORIENTATION IN ECOSYSTEMS

Tropical ecologists are often requested to do comparative short term studies on soil arthropods of different ecosystems. In this work, a modern soil extraction method was used for the first time to obtain information on the abundance and dominance of the soil fauna in neotropical forests.

Study Area and Sampling Method

Soil arthropods were studied in March 1981 during the main rainy season (cf. Adis 1981, Ribeiro and Adis 1984) in the following previously somewhat investigated forest types near Manaus:

The Dryland Region (Terra Firme). Two sites were at the Reserva Florestal Ducke of INPA: (1) in a primary dense forest on yellow latosol (RA), study area of Penny and Arias (1982) and Adis et al. (1985); (2) in a streamside palmetum forest on sandy soil (RI), study area of Beck (1971) and Franken (1979); the third site was in a secondary forest (cut but unburned capoeira) on yellow latosol at the INPA campus (IN), study area of Prance (1975).

The White Water Region (Várzea). Sampling was carried out during the emersion (= non-inundation) period at Rio Solimões: (1) in a riverine forest on clayey soil (montmorillonite) on Ilha de Curarí (IC), study area of Adis (1981), Adis et al. (1985), and Irmler (1975, 1979); and (2) in an island forest on clayey soil (montmorillonite) on Ilha de Marchantaria (MA), study area of the Tropical Ecology Working Group, Max Planck Institute for Limnology, Plön (Furch et al. 1983, Irion et al. 1983, Klinge et al. 1983, Junk et al. 1983).

The Black Water Region (Igapó). Samples were taken during the emersion period in a riverine forest on soil consisting of clay,

silt, and sand material (TM), study area of Adis (1981), Adis
et al. (1979, 1985), Beck (1976), Erwin (1983), Irion and Adis
(1979), Irmler (1975, 1979), Katz (1981), Worbes (1983), and others.

   The Mixed Water Region.   This region is influenced by black
water and white water rivers (cf. Prance 1979, Adis 1984b).   Sampling
was realized during the emersion period in a forest on clayey soil
located between the Rio Negro and the Rio Solimões (LJ), study area
of Irmler (1979) and Erwin (1983).

   In each forest type, three soil samples (diameter 21 cm, depth
3.5 cm, including litter) were taken at random with a split corer
(steel cylinder with lateral hinges).   The corer was driven into
the soil by a mallet.   Animals were extracted from each sample
following a modification of the method of Kempson et al. (1963).

Results and Value of Data

   In the main rainy season of 1981, abundance of soil arthropods
in Central Amazonian forest types varied between 3,500 and 9,000
ind/m^2 (Fig. 1).   In soils of inundation forests, abundance was
somewhat lower if compared to dryland forests, except on Ilha de
Marchantaria (MA).   In all forest types, Acari and Collembola
represented 60-80% of the soil fauna collected.   However, dominance
of both taxa in the total catch varied significantly among forest
types; for example, 19% was recorded for Collembola from INPA's
secondary dryland forest (IN) and 55% for Collembola from the
riverine white water inundation forest (IC).   Differences even
occurred within the same region: in the Várzea, highest dominance
was found for Acari on Ilha de Marchantaria (MA) with 31% of the
total catch, whereas Collembola dominated (55% of the total catch)
on nearby Ilha de Curarí (IC).   Dominance of other "common" arthropod
taxa varied among forest types as well: only a few Coleoptera
occurred in samples of four forests (IC, RA, MA, IN); the number of
Formicidae was low in two forests (LJ, TM).

   The extraction method used supplies rapid preliminary information
on the soil fauna of neotropical forest ecosystems.   The number of
samples taken for comparison of forest types depends on the restricted
capacity of sophisticated laboratory equipment.   Our study permitted
three samples to be taken per forest within a four week study
duration.   The somewhat high standard deviations in some forests
(LJ, RI, MA) indicate that abundance of soil arthropods and dominance
of taxa may change with increasing sample numbers.   However, differences among forest types are obvious, especially where standard
deviation is less than 30% (TM, IC, RA, IN).   Nevertheless, data
represent only the soil fauna sampled at a certain period, the main
rainy season, and depend somewhat on local microclimatic conditions
as well and cannot be used for general statements.

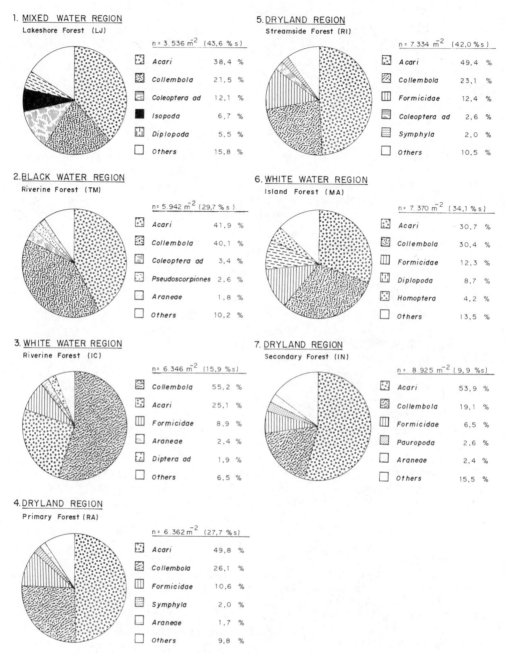

Fig. 1.   Abundance of arthropods with standard deviation (%) and
          dominance of the five most frequent taxa in soil samples
          taken from seven forest types of Central Amazonia in March
          1981 (see text for further explanation).

Standing crop methods, such as the soil extraction method used here, are valid for preliminary studies, particularly orientation projects, which depend mostly on fast results. If an overall comparative ecosystem study has to be accomplished within a short term period (e.g., three months), we recommend that several current standing crop methods be used simultaneously (Janetschek 1982, Phillipson 1971, Southwood 1980) in addition to permanent catching apparatuses (cf. Section 4). Soil extraction by the Kempson method is, above all, suited for long term studies. A minimum of 12 samples of the size cited above should be taken at least once a month (Adis, in preparation).

4. ECOSYSTEM ANALYSES VIA COMBINATION OF SAMPLING METHODS

Insights into the importance and function of arthropods in forest ecosystems can be obtained by determining "key taxa" in different strata. A combination of various collecting methods, depending on the goal to be achieved, enables comparative statements on the composition of arthropod populations in neotropical forest ecosystems, even within short term periods. In the following study, the arthropod faunas of two riverine inundation forests, one in the Black Water Region and one in the White Water Region (TM and IC, respectively), and of a primary forest in the Dryland Region (RA) were compared. Data presented are gathered from investigations realized by various scientists (mostly during the main rainy season) between 1976 and 1982 in the study areas cited above (cf. Section 3). Results permit first statements on the abundance of arthropods and on the dominance of taxa in different strata (i.e., the strata-cenoses of soil/litter, trunk, and canopy). Methods used have been developed only recently, mostly for ecosystem studies in temperate zones (cf. Grimm et al. 1975, Schauermann 1977, Weidemann 1971), and were tested for the first time under neotropical conditions.

Sampling Methods

In each forest type, ground and arboreal photo-eclectors (Funke 1971) were set up as permanent catching apparatuses. Arboreal photo-eclectors (Fig. 2) detect upward migrations of non-flying animals and approaches of flying arthropods to trunks. Such data provide valuable information on activity periods and life cycles (Adis 1981, Funke 1971). Ground photo-eclectors (Fig. 3) detect animals particularly on the ground surface and soil, but also catch inhabitants of other habitats (e.g., trunk and canopy regions). Their samples enable statements on species inventory, dominance structure, activity density, and phenology of arthropods (Funke 1971, 1977, Adis 1981, Thiede 1977). For insects that are epigeous as imagines or that have soil living development stages, ground photo-eclectors can provide data for the calculation of emergence abundance, a basic value for estimating the production of imagines/ m^2 (ha) x year (Funke 1972).

Fig. 2.   An arboreal photo-eclector consists of connected cloth fun-
          nels which form a closed ring around a trunk. This type
          detects upward migrations and trunk approaches of arthropods
          (Funke 1971). It is mounted on *Aldina latifolia* in the
          riverine inundation forest of the Black Water Region (from
          Adis 1977).

          In both inundation forests, three arboreal photo-eclectors
were placed at about 3.6 m height on the trunks of abundant tree
species: in the Black Water Region (TM), two apparatuses each on
Aldina latifolia (Leg.) and one eclector on *Peltogyne venosa* (Leg.);
in the White Water Region (IC), two eclectors on *Pseudobombax
munguba* (Bombac.) and one eclector on *Ficus* sp. (Morac.) (detailed
description in Adis 1981). Data on the ground faunas were collected
in both inundation forests with one ground photo-eclector each,
which were moved every five weeks to new locations (cf. Adis 1981).
All catch results were evaluated from January until March 1976. In
the dryland forest (RA), one arboreal photo-eclector was set up on
Qualea paraensis (Vochys.) and data evaluated from January until
March 1982. Data from ground photo-eclectors were calculated from
catches realized in December 1977 and March 1978 by Penny and Arias
(1982).

Fig. 3. Ground photo-eclectors exposed in the riverine inundation forest of the Black Water Region to capture arthropods of the ground surface and soil.

In addition to the permanent collecting methods, complementary standing crop samples were taken: the soil fauna of each forest was extracted from samples collected in March 1981 (Section 3). Canopy arthropods were obtained by applying a pyrethrum fog into three crowns of selected trees and subsequently collecting the fallen material from suspended cloth sheets (Adis et al. 1985, Adis 1984b). Tree species sampled were *Aldina latifolia* (Leg.) and *Erisma calcaratum* (Vochys.) in the Black Water Region (TM), *Pseudobombax munguba* (Bombac.) and *Ficus* sp. (Morac.) in the White Water Region, and *Eschweilera* cf. *odora* (Lecythid.) as well as *Dipteryx alata* (Leg.) in the Dryland Region (RA). Fogging was executed between 28 July and 2 August 1977, in the early dry season (with respect to rainfall) and when inundation forests were completely flooded.

Results and Value of Data

The relative abundance (Schwerdtfeger 1968) of arthropods (Acari and Collembola disregarded) ranged between 1,000 and 1,500

ind/m^2 in the soil and from 1,000 to 7,000 specimens/m^2 in ground
photo-eclectors (Fig. 4). Between 1,000 and 8,000 animals occurred
per tree trunk and 30 to 60 ind/m^2 were collected from the canopy.
Altogether, the relative abundance of arthropods sampled was higher
in the inundation forest of the Black Water Region (TM), intermediate
in the primary dryland forest (RA), and somewhat lower in the white
water inundation forest (IC). However, lowest arthropod abundance
in soil samples was found in the Igapó (TM) and lowest activity
density on tree trunks in Terra Firme (RA). The dominance of taxa
varied significantly among strata within the forests and among
forest types as well. In the White Water and Dryland Regions,
Formicidae were most dominant in the soil, on the trunk, and in the
canopy. In ground photo-eclectors, emerging adult Diptera were
most frequent. In the Black Water Region, various groups occurred
more abundantly in soil samples, principally adult Coleoptera and
Pseudoscorpiones. In ground photo-eclectors, adults of Coleoptera
and Diptera were most frequent. On trunks, numerous Pseudoscorpiones
and Araneae occurred, whereas Formicidae dominated in canopy samples.

In all forest types investigated, zoophagous groups were
predominant. Numerous imagines of Diptera (especially Sciaridae,
Cecidomyiidae, and Phoridae) collected in ground photo-eclectors
indicate that their larvae may be of great importance for litter
decomposition. Other primary decomposers (e.g., Isopoda, Diplopoda)
were almost absent.

Results presented up to now are based on material sorted to
higher taxon levels only. Profound taxonomic studies would provide
much more information. For example, comparison to genus and species
levels may reveal if non-flying taxa caught on the ground (e.g.,
Araneae, Formicidae, Pseudoscorpiones) are terricolous arthropods
or represent an arboricolous fauna which occasionally occurs on the
forest floor. Taxonomic studies on flying arthropods may show
which genera and species represent strata changers that are living
in the trunk-canopy region after having emerged from the forest
floor (e.g., Diptera, Coleoptera). Species composition of dominant
groups would provide the first information on arthropod diversity
among forest types (cf. Beck 1971).

Future studies of neotropical forest ecosystems should attempt
to compare seasons of the same year. Data could then be compared
by the month and provide the first information on arthropod density
and abundance depending on biotic and abiotic factors (cf. Section 5).
Sampling methods used can be altered or supplemented: if emphasis
is on studies of non-flying arthropods, we suggest the additional
use of pitfall traps on the ground and arboreal photo-eclectors for
downward migrations (Fig. 5); for investigations on the flying
arthropod fauna, flight traps might be used at different heights
and strata (Penny and Arias 1982, Sturm et al. 1970). Each sampling
method, however, is somewhat biased in being selective for certain

arthropod groups (cf. Adis 1979, 1981, Grimm et al. 1975, Thiede
1977, Penny and Arias 1982). Data should therefore be analysed
most critically. For preliminary studies, we recommend the simultaneous use of one arboreal eclector type mounted on the same (dominant)
tree species, one ground photo-eclector, and three pitfall traps.
In addition, extraction of soil samples by the Kempson method
should be realized once a month.

5. POPULATION FLUCTUATIONS SET AGAINST ALTERING ABIOTIC CONDITIONS

    Moist tropical forests traditionally are distinguished by
insignificant temperature changes and seasonal variation in rainfall
(Richards 1976, Walter 1970). Forests of Central Amazonia are
exposed to a rainy season (December until May, 1450-1850 mm precipitation) and to a dry season (June until November, 450-600 mm precipitation). The months of heaviest rainfall generally are January-April;
the driest months usually are July-September (cf. Adis 1981, Falesi
et al. 1971, Ribeiro and Adis 1984). During the dry season, day
temperatures in forest ecosystems are higher and differences between
day and night temperatures are more pronounced (Fig. 6). Population
fluctuations of neotropical arthropods thus may correspond somewhat
with periodically altering climate conditions, mostly accompanied
by biotic changes in the ecosystem. Ecological investigations on
activity dynamics of arthropod populations are subject to long term
studies. Data on the activity density of arthropods on tree trunks
in four neotropical forest types are now available, and can be
compared for the first time.

Study Area and Sampling Method

    Data presented were gathered from arboreal photo-eclectors
(Funke 1971), which record upward migration and trunk approach of
arthropods. Three apparatuses each were set up on tree trunks in
the Black and White Water Regions (TM, IC) from January 1976 to
May 1977 (cf. Section 3, also Adis 1981). One eclector was installed
in a primary forest of the Dryland Region (RA) between December 1981
and December 1982 (cf. Section 3). In addition, investigations
were carried out in a secondary dryland forest (capoeira; TC)
bordering the black water inundation forest (TM). One eclector was
mounted on the trunk of *Jacaranda copaia* (Bigon.), about 50 m from
the Igapó forest. Climatic conditions were recorded continuously
in each study area (cf. Adis 1981).

Results and Value of Data

    Arthropods caught per month per trunk amounted to about 1,500
specimens in the Igapó forest (TM), 700 individuals in the Várzea
forest (IC), 660 specimens in the secondary dryland forest (TC),
and 400 individuals in the primary forest on Terra Firme (RA).

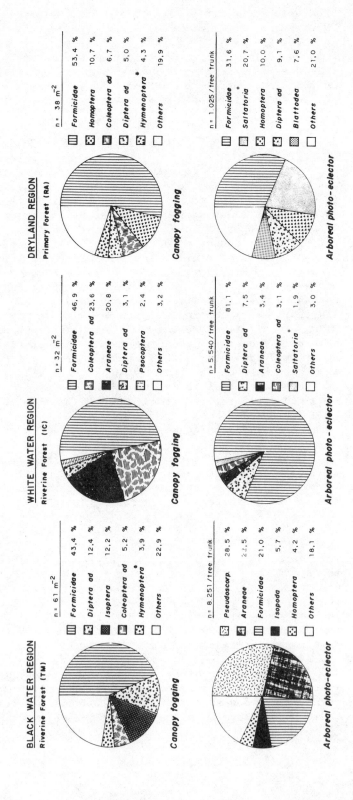

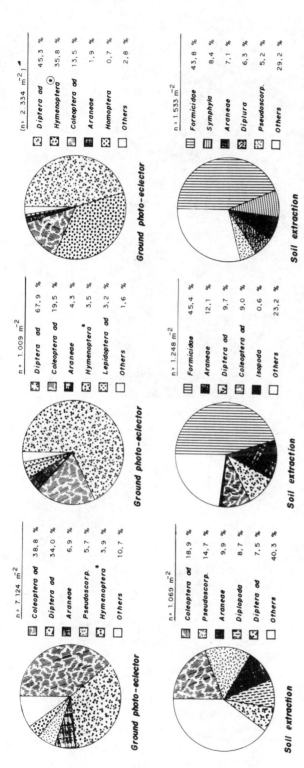

Fig. 4. Abundance and activity density of arthropods and dominance of the five most frequent taxa, sampled with various methods from different strata in three forest types of Central Amazonia (excluding Acari and Collembola). Ground and trunk data were gathered from studies carried out during the rainy season of different years; canopy data were obtained during the early dry season and with riverine forests flooded (see text for further explanation). (+) Caelifera and Ensifera; (*) without Formicidae; (⊛) including Formicidae; (◣) extrapolated from two weekly captures of 27 traps (13 December 1977, 15 February 1978) (modified from Penny and Arias 1982).

Fig. 5.   Arboreal photo-eclector to detect downward migrations and
          trunk approaches of arthropods, mounted on *Aldina latifolia*
          in the riverine inundation forest of the Black Water Region
          (from Adis 1981).

Somewhat more arthropods were caught during the rainy season (TM =
61%, TC = 70%, IC = 73% of the total catch), except in the primary
dryland forest (RA = 43% of the total catch). Formicidae represented
the predominant group of all arthropods sampled (Figs. 7 and 8).
They accounted for 81% of the total catch in the Várzea forest
(IC), 66% in the secondary dryland forest (TC), 43% in the primary
forest on Terra Firme (RA), and 36% in the Igapó forest (TM).
However, ant dominance in samples varied within and between seasons
due to distinct activity patterns. This variation is also true for
the dominance of other taxa. Pronounced trunk ascents of ants were
observed during the dry season and in all forest types investigated.
Additional data from the Igapó indicate that a dry, warm climate
seems to stimulate tree inhabiting, meso-xerophilous species to
become highly active and to migrate between canopy, trunk region,
and forest floor (Fig. 9). With the beginning of the rainy season

and changing microclimate (Fig. 6), their activity densities on
trunks steadily decreased (Figs. 7-9). Another even more pronounced
vertical migration was noted on tree trunks of the Igapó forest:
with the beginning of the rainy season, numerous non-flying (terricolous) arthropods, some of them still juvenile, migrated into the
trunk and canopy region (especially Arachnida and Myriapoda; Fig. 7)
where they remained throughout forest inundation--normally of 5-6
months duration (March-August/September) (Adis 1981, 1984a, 1984b).
Trunk ascents of hygrophilous species are certainly influenced
(regulated) by rising wetness on the forest floor and increasing
relative humidity in the lower trunk region, as shown for Pseudoscorpiones (Fig. 10) and Symphyla (Adis 1981, Adis and Scheller 1984).
Arthropods of the Várzea forest under study (IC) are considered mostly
trunk and canopy inhabitants (Adis 1981). Terricolous groups,
which (as in the Igapó) emigrate in high numbers to the trunk-canopy
area with the beginning rainy season, did not exist in the Várzea
(Fig. 8) (cf. Adis 1981).

Reactions to microclimatic changes other than trunk migrations
are already known for neotropical forest arthropods as well, especially from the Igapó. For example, an aestivation period due to
the dry season was discovered for the arboricole millipede *Epinannolene* n.sp. (Pseudonannolenidae) (Fig. 11) (cf. Adis 1984a). In
ground photo-eclectors, the lowest number of arthropods sampled was
recorded in October. This month of the dry season has the lowest
precipitation and highest insolarity after the forest floor has
emerged and dried (cf. Adis 1981, Ribeiro and Adis 1984). With the
beginning rainy season, activity and emergence of holometabolous
insects, especially Diptera, increased on the forest floor. Based
on annual population fluctuations and activity dynamics of arthropods,
two seasonal aspects ("Aspektfolge"; Tischler 1955) were proposed
for the emersion period of the Igapó forest under study: a dry
season aspect and a rainy season aspect (Adis 1984a, 1984b). Their
validity for other forest ecosystems in the Central Amazon has to
be tested in future long term studies, executed with various sampling
methods (cf. Section 4). Data which Penny and Arias (1982) gathered
from different strata in the primary dryland forest (RA) indicate
that insect populations are somewhat higher in December and April
(i.e., at the beginning and at the end of the rainy season as
well). Thus, the hypothesis that tropical ecosystems with a sufficiently high number of species lack population fluctuations (Saunders
and Bazin 1975) cannot be confirmed for Central Amazonian forests.
However, it remains to be investigated how pronounced population
fluctuations are in the Amazon Basin, when compared to the actual
wet, but more pronounced seasonal tropics (e.g., Panama) (cf. Wolda
1978a, 1978b, 1980, Wolda and Foster 1978).

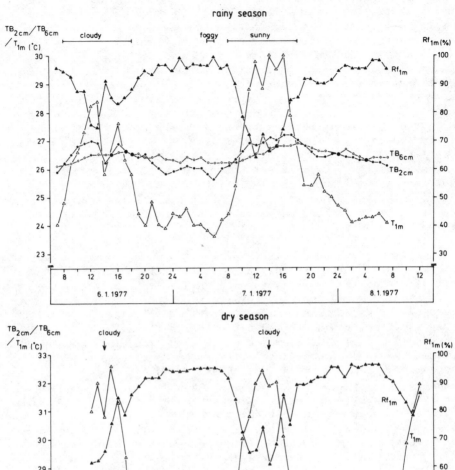

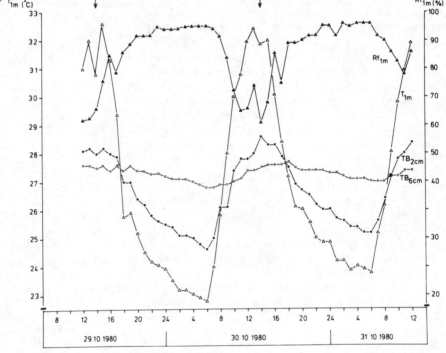

6. THEORIES ABOUT THE DIVERSITY OF AMAZONIAN ARTHROPODS
    (Cited, in part, from Erwin and Adis 1982 [reprinted by permis-
    sion of Columbia Univ. Press, New York, NY USA], Adis 1984a.)

Biological diversification in the tropics is the subject of
permanent and frequently antagonistic discussions. Several theories
have been proposed to explain the incredible diversity of neotropical
plants and animals, mostly in a deductive sense, based on selected
groups. Their general validity still must be confirmed (cf. Prance
1982, Simpson and Haffer 1978, Vanzolini 1973). Many scientists
explain species diversity in the American tropics by the "Haffer
effect" (Haffer 1969, 1982), that is, the theory of tropical forest
fragmentation during northern ice advances and temporary existence
of forest and grassland (savanna) "island" refugia (Prance 1982).
However, other possibilities may exist as supplemental or alternative
theories.

Fittkau (1973) and Klinge (1973) consider high plant and
animal diversity in forests and rivers on Terra Firme to be a
response to extremely low nutrient concentrations in geochemically
impoverished ecosystems. Species richness and simultaneously
occurring high diversity of moist tropical ecosystems do not,
according to Fittkau (1982), reflect high nutrient supply, but
apparently a mode of adaptation to continuous restriction of nutrients
or food substances under otherwise permanently favorable living
conditions. Plants and animals are assumed to act as highly efficient
"nutrient traps." During evolution of multiple "Lebensformen"
(life forms; Tischler 1955), shortages therefore must have been of
significant importance at all times.

Beck (1976, 1983) assumes that soil arthropods of Amazonian
inundation forests represent individual species of the Terra Firme

---

Fig. 6. Microclimate in the riverine inundation forest at Rio Tarumã
         Mirím (Black Water Region) during the rainy season (6-8 Jan-
         uary 1977) and dry season (29-31 October 1980). (▲) relative
         humidity at 1 m height (Rf_{1m}); (Δ) temperature at 1 m height
         (T_{1m}); (●) soil temperature at 2 cm depth (TB_{2cm}); (o) soil
         temperature at 6 cm depth (TB_{6cm}). Maximum difference
         between day and night air temperature (T_{1m}): rainy season
         6.4°C; dry season 9.2°C. Maximum difference between day and
         night soil temperature (TB_{2cm}): rainy season 1.4°C; dry
         season 4.0°C. (Dry season measurements in collaboration with
         graduate students of INPA, Dep. of Entomology.)

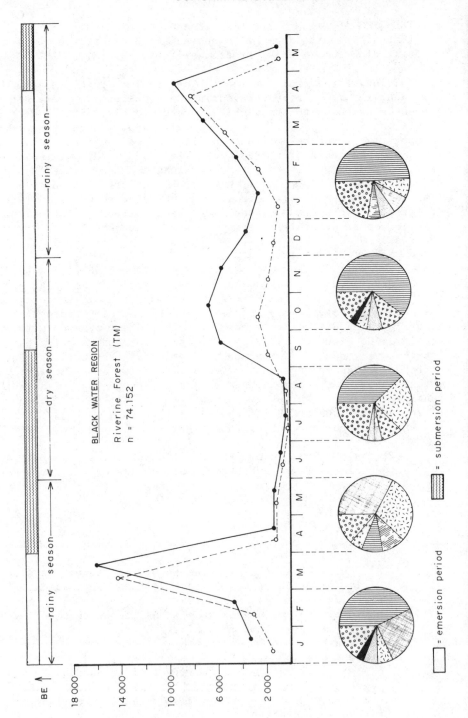

BLACK WATER REGION

Riverine Forest (TM)

n = 74.152

= emersion period

= submersion period

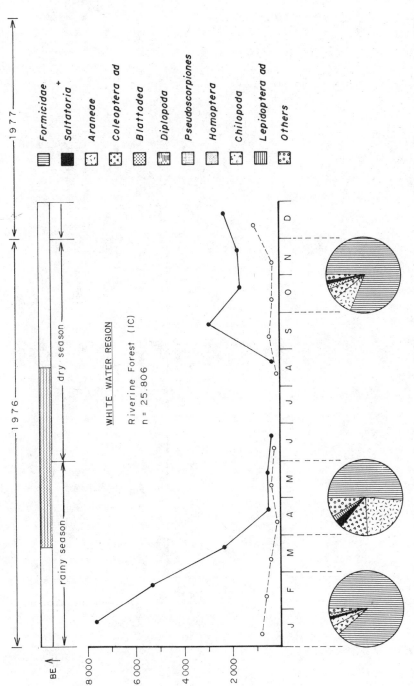

Fig. 7. Trunk ascents of arthropods and dominance of frequent taxa (3 arboreal photo-eclectors, respectively [BE↑]) in a riverine inundation forest of the Black Water Region (January 1976-May 1977) and of the White Water Region (January-December 1977; eclectors were flooded in July). Not included are Acari, Collembola, and Diptera. (+) Caelifera and Ensifera; (*) without Formicidae; (●) total catch; (○) total catch without Formicidae.

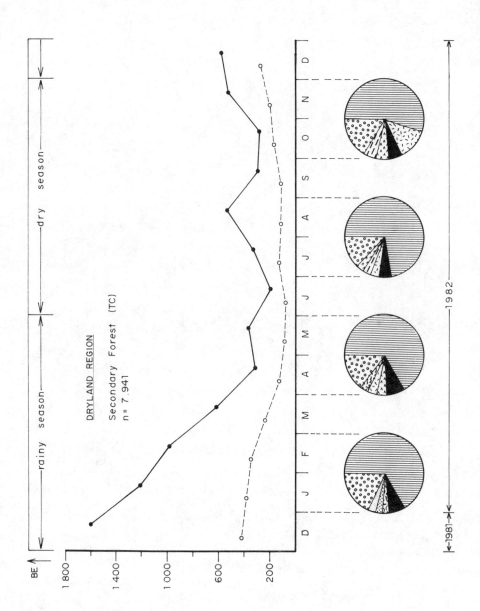

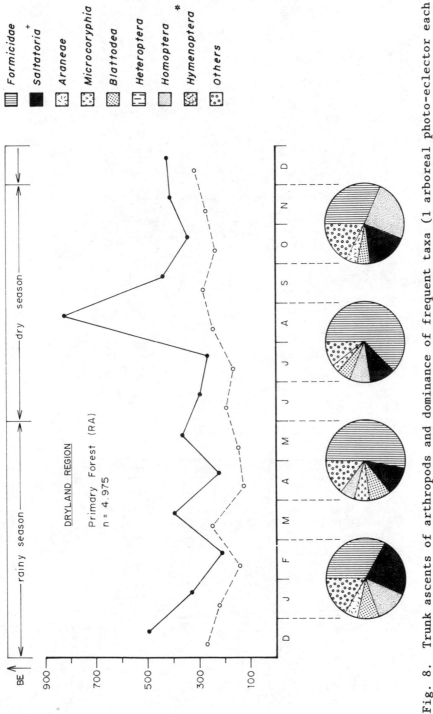

Fig. 8. Trunk ascents of arthropods and dominance of frequent taxa (1 arboreal photo-eclector each [BE↑]) in a secondary and a primary forest of the Dryland Region between December 1981 and December 1982. Not included are Acari, Collembola, and Diptera. (+) Caelifera and Ensifera; (*) without Formicidae; (●) total catch; (○) total catch without Formicidae.

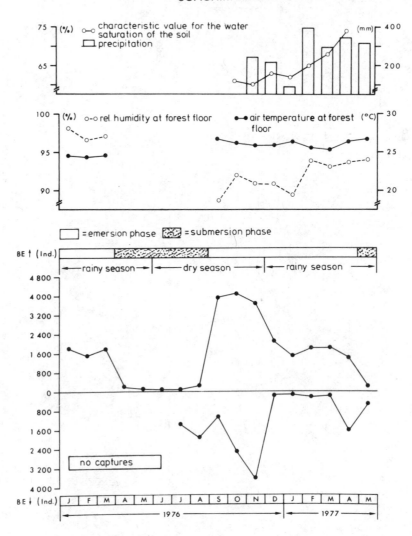

Fig. 9. Trunk migration (upwards/downwards) of Formicidae (3 arboreal
photo-eclectors, [BE↑/BE↓]) in the riverine inundation forest
of the Black Water Region between January 1976 and May 1977
(from Adis 1981).

soil fauna, which--due to pre-adaptation in their way of living and
their reproduction cycles--could immigrate into periodically flooded
forests.

Another possibility was indicated by Erwin and Adis (1982):
Haffer's forest refuge model as stated deals only with Terra Firme
forests, their fragmentation and (or) replacement by savannas

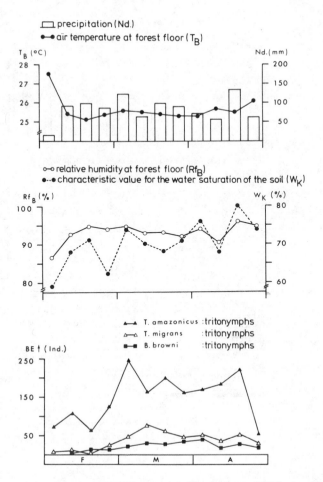

Correlations to abiotic factors (December 1976-April 1977)

	r_1	r_2
T. amazonicus	+0.524[a]	+0.550[a]
T. migrans	+0.545[a]	+0.492
B. browni	+0.702[b]	+0.588[a]

where r_1 = correlation coefficient of trunk ascent and characteristic value for water saturation of the soil (W_K)/week; r_2 = correlation coefficient of trunk ascent and relative humidity (Rf_B)/week; a = significant for $p < 0.05$ ($r > 0.497$; $n = 16$); b = significant for $p < 0.01$ ($r > 0.623$; $n = 16$).

Fig. 10.   Trunk ascent of tritonymphs of *Tyrannochthonius amazonicus*, *Tyrannochthonius migrans*, and *Brazilatemnus browni* (Pseudoscorpiones; 3 arboreal photo-eclectors [BE↑]) in the riverine inundation forest of the Black Water Region between February and April 1977 (from Adis 1981).

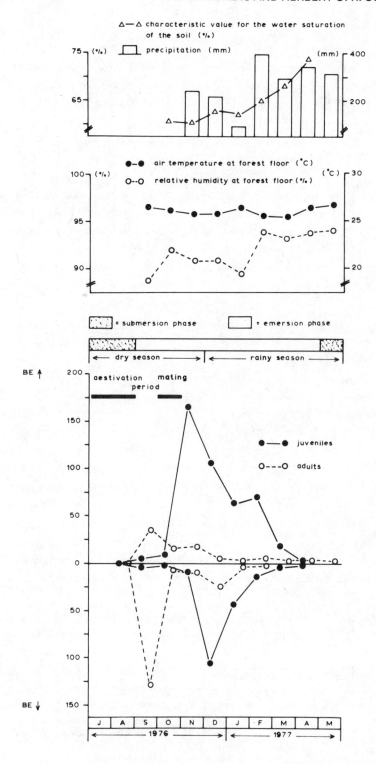

(sensu lato). Apparently there has been no suggestion that riverine forests disappeared along the lengths of the rivers, but rather that savannas were behind such forests as they are today.

For this reason, Central Amazonian inundation forests, espe- cially along black water rivers, are considered to be short term refuges and, additionally, long term evolutionary centers (Erwin and Adis 1982, Adis 1984a). Flood cycles greatly influence specia- tion rate and dispersion of arthropods. Evidence from Rio Tarumã Mirím near Manaus suggests alternation of two types of inundation: annual flooding and continental watertable rise over the last million years. The area examined originally was covered by a Terra Firme forest (Irion and Adis 1979). Periods of high sea level (cf. Fairbridge 1961) caused a back-up of main riverine courses in Amazonia. During these periods, the affluent Rio Tarumã Mirím was high enough that annual water fluctuations of the Rio Negro led to an inundation lasting several months in the forest under study. First inundation was several million years ago (letter dated 25 April 1984 from G. Irion, Institut für Meeresgeologie und Meeresbiologie Senckenberg, Wilhelmshaven, W. Germany). Since that time, various periods of high sea level have occurred, for example, 250,000 years, 170,000 years, and 85,000 years ago (cf. Fig. 12), with annual flooding of 10,000-30,000 years duration. Six thousand years ago, the sea level reached a height of about 5 m below mean sea level. Since then, the examined forest has been inundated again. Flood cycles were responsible for extant vegetation formation in the Igapó. During high sea level the (lower) Rio Negro bank suffered longer episodes of inundation and possible formations of "ria-lakes" (Sioli 1968). Igapó forests were drowned (at least partially) and backed up minor tributaries which subsequently had annual flooding (e.g., Rio Tarumã Mirím) (Fig. 12A). They replaced dryland forests, previously occupying the area. Intermittent dry

Fig. 11.   Phenology and vertical movements on trunks (upwards/down- wards; 3 arboreal photo-eclectors [BE↑/↓BE]) of the arbori- cole millipede *Epinannolene* n.sp. (Pseudonannolenidae) in the riverine inundation forest of the Black Water Region between July 1976 and May 1977 (from Adis 1984a; reprinted by permission of Junk, The Hague, The Netherlands). Aestivation period due to dry season was observed between July and September; mating occurred throughout October in the lower trunk region. Juveniles migrated into the upper trunk-canopy region (especially during dry season), where they were found mostly under bark and occasionally on epiphytes.

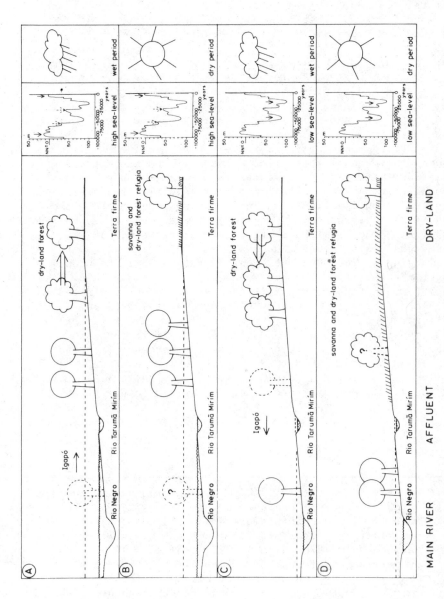

Fig. 12. Postulated formations and shifts of black water inundation forests along main rivers and affluents of Central Amazônia within the last 100,000 years due to changes in sea level and climate (see text for further explanation; sea level according to Fairbridge 1961).

spells during high sea level periods may have caused formation of
savanna and dry land forest refugia (Fig. 12B), as postulated by
Haffer (1969). During low sea level periods, annual inundation
occurred only along main rivers (e.g., Rio Negro); this has been
the situation for at least one million years, probably including
all glacial advances of the Pleistocene and perhaps even since the
Andes took their present configuration. Terra Firme vegetation
spread extensively, forcing Igapó forests along affluents to retreat
to main rivers (Fig. 12C; cf. Irion 1976). Intermittent dry periods
(Haffer 1969) again caused savanna formation with possible dryland
forest refugia and, in addition, "Igapó forest refuges" (Fig. 12D).

Flood cycles have dramatic impact on inundation forest faunas.
This is especially important when two kinds of flooding patterns
manifest themselves through time. In the greater Amazon region,
there could have been periods favoring lakeshore species, then
inundation forest species, then riparian species (cf. Erwin and
Adis 1982). For example, during high sea level and maximum wetness,
flooding could have been minimized because the larger water bodies
essentially were lakes that buffered flooding, major shore lines
were more stable, and the associated flora and fauna were selected
towards species without flood response, while flood response species
shifted up affluents (cf. Fig. 12A). At the other extreme, that
is, periods of low sea level and maximum aridity, flooding would
have decreased due to lack of water at affluents and the biota
would have been composed of species that were non-responsive to
flooding. However, flood responsive species would then have been
concentrated in annually inundated forests along main rivers (Fig.
12D). Both situations cause interruption and isolation of Igapó
forests--in the first case due to large waterbodies, and in the
second due to drying affluents, emerging sandy riverbeds, and
additional regression of dryland forests. Species vagility then
seems to become a crucial factor in sorting out and isolating gene
pools. Thus, small barriers may be highly disruptive to populations
of individuals with low vagility, particularly non-flying soil
arthropods or even carabid beetles (cf. Erwin and Adis 1982). If
we suppose that intense competition and predation promotes rapid
speciation (especially in Igapó forests, where flooding mixes
terrestrial and canopy biotas together during half the year) and
that the isolation of Igapó forests is long enough for rapidly
evolving species, we may have found another key to tropical species
richness. In fact, preliminary studies indicate that (1) Igapó
forests retain endemic species, for example, the pseudoscorpions
Tyrannochthonius amazonicus and *T. migrans* (Adis 1981, Mahnert
1979); (2) the fauna of Igapó forests seems to be different in
composition and activity patterns when compared to adjacent dryland
forests (Fig. 8) (Adis, in preparation); and (3) in Igapó forests
across the Rio Negro and along its tributaries, taxonomically
different species and subspecies are found (e.g., within the Pseudoscorpiones [cf. Mahnert 1979], the Symphyla [letter dated 4 October

1981 from U. Scheller, Storfors, Sweden], and the Opilioues [unpub-
lished identification list dated February 1984 from B. Frefe, Max
Planck Institute, Plön, W. Germany]).

Numerous strategies (adaptations) apparently have been evolved
through time by arthropods of inundation forests to compensate for
periodic losses of their terrestrial habitat. Summing up, organisms
either (1) acquire adaptations for remaining in flooded terrestrial
areas, (2) move to trunk canopy areas, (3) move to adjacent dryland
biotopes during inundation, or (4) do a combination of these things
(cf. Adis 1981, 1982a, 1982b, 1984a, 1984b, Erwin 1983, Erwin and
Adis 1982, Paarmann et al. 1982).

7. RESEARCH QUESTIONS FOR FUTURE NEOTROPICAL STUDIES

Research on Amazonian arthropods raises a series of questions
or approaches for tropical ecologists. Some of these are

(1) How many different ecosystems compose tropical forests in the
     Central Amazon? What are the average size and percentage of
     each ecosystem type in the entire forest system?
(2) How many (endemic) arthropod species does each ecosystem
     contain and what are the reasons for their presence or absence?
(3) What are the average number and biomass of arthropods per
     ecosystem, their yearly fluctuations taken into consideration?
(4) What is the specific function of (each) arthropod species in
     these ecosystems (e.g., primary and secondary consumption,
     decomposition rates) and how many trophic groups or levels
     exist?
(5) How high are production and energy flow within arthropod-cenoses
     of different forest ecosystems?
(6) Are species richness, abundance, and biomass of arthropods
     higher in disturbed (secondary) forest ecosystems and which
     faunal components are changing when compared to primary forest
     ecosystems?
(7) Where is the "critical point" of human impact in Amazonian
     forest ecosystem (e.g., how many trees can be removed before
     the ecosystem "collapses")?
(8) What are the different recommendations necessary for the
     biological management of each forest ecosystem type?

8. FINAL REMARKS

Our knowledge about Central Amazonian forest ecosystems is
still very fragmentary. Nevertheless, considerable progress has
been achieved within recent years (Bergamin 1981). However, because
of various interests, different study directions, and distinct
scientific needs (cf. Salati et al. 1979), information gathered on

Amazonian forests is as diverse as their flora and fauna. Most
research has been highly site specific. Intensive and comparative
studies on different forest types are nearly lacking. Basic research,
however, provides knowledge on the structure and function of ecosystems. It is the foundation, for example, for ecological criteria
to be used in biological management plans for Amazonia (cf. Schubart
1977, Schubart and Salati 1982). Analytical procedures are, as
shown in this paper, still in a methodological development stage.
Nevertheless, the first tools are being tested and applied to
tropical ecosystem studies, especially to multiple and overdue
comparative short term projects. Still another problem is the lack
of trained taxonomists capable of classifying and describing the
arthropod material collected (cf. National Research Council 1980).
In the study carried out by Adis (1981), about 75% of the species
captured were assumed to be new to science. To achieve indepth
insights into the structure and function of ecosystems, integrated
long term studies are required. Priorities should be given to
critical areas indicated by the National Research Council (1980).
At present, the first integrated and long term ecosystem study in
the Várzea region is being carried out on an island of the lower
Rio Solimões (Ilha de Marchantaria) by the Tropical Ecology Working
Group of the Max Planck Institute for Limnology, Plön, West Germany,
in close collaboration with scientists of the Instituto Nacional de
Pesquisas da Amazônia (INPA) in Manaus, Brazil (cf. Section 3).

ACKNOWLEDGEMENTS

     This research was sponsored partially by the Max Planck Institute
for Limnology in Plön, West Germany, the Instituto Nacional de
Pesquisas da Amazônia (INPA) in Manaus, Brazil (OAS project and
Convênio INPA/Max Planck), the German Research Foundation (project
of Prof. Dr. W. Funke, Ulm), the "Studienstiftung des Deutschen
Volkes," and the II. Zoological Institute of the University of
Göttingen, West Germany.

     We are indebted to all colleagues who in some way supported
the realization of this study. Special thanks are due to Irmgard
Adis and our technical staff, especially Vera Bogen, for their
enormous efforts in the field and laboratory. Norman Penny (INPA)
and June H. Cooley (INTECOL) kindly corrected the English manuscript.

REFERENCES CITED

Adis, J. 1977. Programa mínimo para análises de ecossistemas:
     Artrópodos terrestres em florestas inundáveis da Amazônia
     Central. Acta Amazonica 7(2):223-229.
Adis, J. 1979. Problems of interpreting arthropod sampling with
     pitfall traps. Zool. Anz. 202(3/4):177-184.

Adis, J.   1981.   Comparative ecolgical studies of the terrestrial
    arthropod fauna in Central Amazonian inundation-forests.
    Amazoniana 7(2):87-173.
Adis, J.   1982a.   Eco-entomological observations from the Amazon.
    II. Carabids are adapted to inundation-forests! Coleopt.
    Bull. 36(2):440-441.
Adis, J.   1982b.   Zur Besiedlung zentralamazonischer Überschwemmungs-
    wälder (Várzea-Gebiet) durch Carabiden (Coleoptera). Arch.
    Hydrobiol. 95(1/4):3-15.
Adis, J.   1984a, in press.   "Seasonal Igapó-forests" of Central
    Amazonian black-water rivers and their terrestrial arthropod
    fauna. In H. Sioli, ed. The Amazon--limnology and landscape
    ecology of a mighty tropical river and its basin. Monogr.
    Biol., Junk, The Hague, The Netherlands.
Adis, J.   1984b, in press.   Adaptations of arthropods to Amazonian
    inundation-forests. In Ecology and resource management in the
    tropics. Proc. VII Symp. Int. Soc. Trop. Ecol., Bhopal,
    India, 1981.
Adis, J., and U. Scheller.   1984, in press.   On the natural history
    and ecology of Hanseniella arborea Scheller, a migrating
    symphylan from an Amazonian black-water inundation forest
    (Myriapoda, Symphyla, Scutigerellidae). Pedobiologia.
Adis, J., K. Furch, and U. Irmler.   1979.   Litter production of a
    Central-Amazonian black water inundation forest. Trop. Ecol.
    20(2):236-245.
Adis, J., Y.D. Lubin, and G.G. Montgomery.   1985, in press.   Arthro-
    pods from the canopy of inundated and terra firme forests near
    Manaus, Brazil, with critical considerations on the pyrethrum-
    fogging technique. Stud. Neotrop. Fauna Environ.
Anderson, A.B., and W.W. Benson.   1980.   On the number of tree
    species in Amazonian forests. Biotropica 12(3):235-237.
Beck, L.   1971.   Bodenzoologische Gliederung und Charakterisierung
    des amazonischen Regenwaldes. Amazoniana 3(1):69-132.
Beck, L.   1976.   Zum Massenwechsel der Makro-Arthropodenfauna des
    Bodens in Überschwemmungswäldern des zentralen Amazonasgebietes.
    Amazoniana 6(1):1-20.
Beck, L.   1983.   Bodenzoologie der amazonischen Überschwemmungs-
    wälder. Amazoniana 8(1):91-99.
Bergamin, H. F., ed.   1981.   O INPA e o Museu Goeldi nos 30 anos do
    CNPq. Suppl. Acta Amazonica 11(1):1-206.
Braga, P.I.S.   1979.   Subdivisão fitogeográfica, tipos de vegetacão,
    conservacão e inventário florístico da floresta amazônica.
    Suppl. Acta Amazonica 9(4):53-80.
Council of Environmental Quality and U.S. Foreign Ministery.   1980.
    The global 2000 report to the President. U.S. Gov. Print.
    Off., Washington, DC USA.
Erwin, T.L.   1982.   Tropical forests: Their richness in Coleoptera
    and other arthropod species. Coleopt. Bull. 36(1):74-75.
Erwin, T. L.   1983.   Beetles and other insects of tropical forest
    canopies of Manaus, Brazil, sampled by insecticidal fogging.

Pages 59-75 *in* S.L. Sutton, T.C. Whitmore, and A.C. Chadwick, eds. Tropical land in forests: Ecology and management. Blackwell, Oxford, Great Britain.

Erwin, T.L., and J. Adis. 1982. Amazonian inundation forests: Their role as short-term refuges and generators of species richness and taxon pulses. Pages 358-371 *in* G.T. Prance, ed. Biological diversification in the tropics. Proc. Int. Symp. Trop. Biol. Columbia Univ. Press, New York, NY USA.

Fairbridge, R.W. 1961. Eustatic changes in sea level. Phys. Chem. Earth 4:99-185.

Falesi, I.C., T.E. Rodrigues, I.K. Morikawa, and R.S. Reis. 1971. Solos do Distrito Agropecuário da SUFRAMA (Trecho: km 30 - km 79, Rod. BR-174). Instituto de Pesquisas e Experimentacão Agropecuárias da Amazônia Ocidental (IPEAAOc), Sér. Solos 1(1):1-99.

Fearnside, P.M. 1982a. Deforestation in the Brazilian Amazon: How fast is it occurring? Interciencia 7(2):82-88.

Fearnside, P.M. 1982b. Desmatamento na Amazônia Brasileira: Com que intensidade vem ocorrendo? Acta Amazonica 12(3):579-590.

Fittkau, E.J. 1970. Esboco de uma divisão ecologica da Região Amazônica. Pages 365-372 *in* I.M. Idrobo, ed. II. Symposio y foro de biologia tropical amazônica. Assoc. Biol. Trop., Bogotá, Colombia.

Fittkau, E.J. 1971. Ökologische Gliederung des Amazonas-Gebietes auf geochemischer Grundlage. Münster. Forsch. Geol. Paläontol. 20/21:35-50.

Fittkau, E.J. 1973. Artenmannigfaltigkeit amazonischer Lebensräume aus ökologischer Sicht. Amazoniana 4(3):321-340.

Fittkau, E.J. 1982. Struktur, Funktion und Diversitat zentralamazonischer Ökosysteme. Arch. Hydrobiol. 95(1/4):29-45.

Franken, M. 1979. Major nutrient and energy contents of the litter-fall of a riverine forest of Central Amazonia. Trop. Ecol. 20(2):211-224.

Funke, W. 1971. Food and energy turnover of leaf-eating insects and their influence on primary production. Ecol. Stud. 2:81-93.

Funke, W. 1972. Energieumsatz von Tierpopulationen in Landökosystemen. Verh. Dtsch. Zool. Ges. (Helgoland 1971):95-106.

Funke, W. 1973. Rolle der Tiere in Wald-Ökosystemen des Solling. Pages 143-174 *in* H. Ellenberg, ed. Ökosystemforschung. Springer, Berlin, W. Germany.

Funke, W. 1977. Das zoologische Forschungsprogramm im Sollingprojekt. Verh. Ges. Ökol. (Göttingen 1976):49-58.

Furch, K., W.J. Junk, J. Dieterich, and N. Kochert. 1983. Seasonal variation in the major cation (Na, K, Mg and Ca) content of the water of Lago Camaleão, an Amazonian floodplain-lake near Manaus, Brazil. Amazoniana 8(1):75-89.

Grimm, R., W. Funke, and J. Schauermann. 1975. Minimalprogramm zur Ökosystemanalyse: Untersuchungen an Tierpopulationen in Wald-Öksystemen. Verh. Ges. Ökol. (Erlangen 1974):77-87.

Haffer, J. 1969. Speciation in Amazonian forest birds. Science
    165:131-137.
Haffer, J. 1982. General aspects of the refuge theory. Pages
    6-24 in G.T. Prance, ed. Biological diversification in the
    tropics. Columbia Univ. Press, New York, NY USA.
Irion, G. 1976. Die Entwicklung des zentral- und oberamazonischen
    Tieflands im Spät-Pleistonzän und im Holozän. Amazoniana
    6(1):67-79.
Irion, G., and J. Adis. 1979. Evolucão de florestas amazônicas
    inundadas, de igapó - um exemplo do rio Tarumã Mirím. Acta
    Amazonica 9(2):299-303.
Irion, G., J. Adis, W. Junk, and F. Wunderlich. 1983. Sedimento-
    logical studies of an island in the Amazon River. Amazoniana
    8(1):1-18.
Irmler, U. 1975. Ecological studies of the aquatic soil inverte-
    brates in three inundation forests of Central Amazonia.
    Amazoniana 5(3):337-409.
Irmler, U. 1977. Inundation-forest types in the vicinity of
    Manaus. Biogeographica 8:17-29.
Irmler, U. 1979. Abundance fluctuations and habitat changes of
    soil beetles in Central Amazonian inundation forests (Col.:
    Carabidae, Staphylinidae). Stud. Neotrop. Fauna Environ.
    14:1-16.
Janetschek, H., ed. 1982. Ökologische Feldmethoden. Ulmer,
    Stuttgart, W. Germany.
Junk, W.J., G.M. Soares, and F.M. Carvalho. 1983. Distribution of
    fish species in a lake of the Amazon river floodplain near
    Manaus (Lago Camaleão), with special reference to extreme
    oxygen conditons. Amazoniana 7(4):397-431.
Katz, B. 1981. Preliminary results of leaf litter-decomposing
    microfungi survey. Acta Amazonica 11(2):410-411.
Kempson, D., M. Lloyd, R. Ghelardi. 1963. A new extractor for
    woodland litter. Pedobiologia 3:1-21.
Klinge, H. 1973. Struktur und Artenreichtum des zentralamazonischen
    Regenwaldes. Amazoniana 4(3):283-292.
Klinge, H. 1983. Forest structures in Amazonia. Pages 13-23 in
    Ecological structures and problems of Amazonia. Proc. Int.
    Union Conserv. Nature Nat. Resour. Workshop, São Carlos,
    Brazil, 1982.
Klinge, H., K. Furch, E. Harms, and J. Revilla. 1983. Foliar
    nutrient levels of native tree species from Central Amazonia.
    1. Inundation forests. Amazoniana 8(1):19-45.
Mahnert, V. 1979. Pseudoskorpione (Arachnida) aus dem Amazonas-
    gebiet (Brasilien). Rev. Zool. 86(3):719-810.
Myers, N. 1980. Conversion of tropical moist forests. Nat. Acad.
    Sci., Washington, DC USA.
National Research Council. 1980. Research priorities in tropical
    biology. Nat. Acad. Sci., Washington, DC USA.
Paarmann, W., U. Irmler, and J. Adis. 1982. *Pentacomia egregia*
    Chaud. (Carabidae, Cicindelidae) an univoltine species in the
    Amazonian inundation forest. Coleopt. Bull. 36(2):183-188.

Penny, N.D., and J.R. Arias. 1982. Insects of an Amazon forest. Columbia Univ. Press, New York, NY USA.

Phillipson, J., ed. 1971. Methods of study in quantitative soil ecology: Population, production and energy flow. Blackwell, Oxford, Great Britain.

Prance, G.T. 1975. The history of the INPA capoeira based on ecological studies of Lecythidaceae. Acta Amazonica 5(3):261-263.

Prance, G.T. 1979. Notes on the vegetation of Amazonia. III. The terminology of Amazonian forest types subject to inundation. Brittonia 31(1):26-38.

Prance, G.T. 1980. A terminologia dos tipos de florestas amazônicas sujeitas a inundacão. Acta Amazonica 10(3):495-504.

Prance, G.T., ed. 1982. Biological diversification in the tropics. Columbia Univ. Press, New York, NY USA.

Prance, G.T., W.A. Rodrigues, and M.F. da Silva. 1976. Inventário florestal de um hectare de mata de terra firme, km 30 da Estrada Manaus - Itacoatiara. Acta Amazonica 6(1):9-35.

Revilla, J.D. 1981. Aspectos florísticos e fitossociológicos da floresta inundável (Igapó) Praja Grande, Rio Negro, Amazonas, Brasil. M.S. Thesis, INPA, Manaus, Brazil.

Ribeiro, M.N.G., and J. Adis. 1984, in press. Local rainfall--a bias for bioecological studies in the Central Amazon. Acta Amazonica.

Richards, P.W. 1976. The tropical rain forest. Cambridge Univ. Press, Cambridge, Great Britain.

Salati, E., P.I.S. Braga, and R. Figliuolo, eds. 1979. Estratégias para politica florestal na Amazônia Brasileira. Suppl. Acta Amazonica 9(4):1-216.

Salati, E., J. Marques, and L.C.B. Molion. 1978. Origem e distribuicão das chuvas na Amazônia. Interciencia 3:200-206.

Saunders, P.T., and M.J. Bazin. 1975. Stability of complex ecosystems. Nature 256:120-121.

Schauermann, J. 1977. Untersuchungen an Tierpopulationen in den Buchenwäldern des Solling: Die Tiere der Bodenoberfläche und des Bodens. Jahresber. Naturwiss. Ver. Wuppertal 30:104-107.

Schubart, H.O.R. 1977. Critérios ecológicos para o desenvolvimento agrícola das terras firmes na Amazônia. Acta Amazonica 7(4):559-567.

Schubart, H.O.R., and E. Salati. 1982. Natural resources for land use in the Amazon region: The natural systems. Pages 219-249 in S. Hecht, ed. Amazoniana: Agriculture and land use research. Cent. Int. Agric. Trop., Cali, Colombia.

Schwerdtfeger, F. 1968. Ökologie der Tiere. II. Demökologie. P. Parey, Berlin, W. Germany.

Simpson, B.B., and J. Haffer. 1978. Speciation patterns in the Amazonian forest biota. Annu. Rev. Ecol. Syst. 9:497-518.

Sioli, H. 1956. Über Natur und Mensch im brasilianischen Amazonasgebiet. Erdkunde 10(2):89-109.

Sioli, H.   1968.   Zur Ökologie des Amazonas-Gebietes.   Pages 137-170
    in E.J. Fittkau, J. Illies, H. Klinge, G.H. Schwabe, and
    H. Sioli, eds.   Biogeography and ecology in South America,
    vol. I.   Junk, The Hague, The Netherlands.
Southwood, T.R.E.   1980.   Ecological methods with particular refer-
    ences to the study of insect populations.   Chapman and Hall,
    New York, NY USA.
Sturm, H., A. Abouchaar, R. de Bernal, and C. de Hoyos.   1970.
    Distribucion de animales en las capas bajas de un bosque
    humedo tropical de la region Carare-Opon (Santander, Colombia).
    Caldasia 10(50):529-578.
Thiede, U.   1977.   Untersuchungen über die Arthropodenfauna in
    Fichtenforsten (Populationsökologie, Energieumsatz).   Zool.
    Jahrb. Syst. 104:137-202.
Tischler, W.   1955.   Synökologie der Landtiere.   G. Fischer,
    Stuttgart, W. Germany.
Vanzolini, P.E.   1973.   Paleoclimates, relief and species multipli-
    cation in equatorial forests.   Pages 255-258 in B.J. Meggers,
    E.S. Ayensu, and W.D. Duckworth, eds.   Tropical forest ecosys-
    tems in Africa and South America: A comparative review.
    Smithson. Inst. Press, Washington, DC USA.
Walter, H.   1970.   Vegetationszonen und Klima.   UTB, Ulmer, Stuttgart,
    W. Germany.
Walter, H.   1973.   Allegemeine Geobotanik.   UTB, Ulmer, Stuttgart,
    W. Germany.
Weidemann, G.   1971.   Food and energy turnover of predatory arthro-
    pods of the soil surface.   Ecol. Stud. 2:110-118.
Wolda, H.   1978a.   Seasonal fluctuations in rainfall, food and
    abundance of tropical insects.   J. Anim. Ecol. 47:369-381.
Wolda, H.   1978b.   Fluctuations in abundance of tropical insects.
    Am. Nat. 112:1017-1045.
Wolda, H.   1980.   Seasonality of tropical insects.   I. Leafhoppers
    (Homoptera) in Las Cumbres, Panama.   J. Anim. Ecol. 49:277-290.
Wolda, H., and R. Foster.   1978.   Zunacetha annulata (Lepidoptera;
    Dioptidae), an outbreak insect in a neotropical forest.
    Geo-Eco-Trop. 2(4):443-454.
Worbes, M.   1983.   Vegetationskundliche Untersuchungen zweier
    Überschwemmungswälder in Zentralamazonien - vorläufige Ergeb-
    nisse.   Amazoniana 8(1):47-65.

CONSIDERATIONS ON SOME ECOLOGICAL PRINCIPLES

Oscar Ravera

Department of Physical and Natural Sciences
Commission of the European Communities
Joint Research Centre, 21010 Ispra (Varese), Italy

ABSTRACT

In this paper, I discuss certain fundamental concepts about which there is yet no general agreement among ecologists. The text consists of three parts: (1) the ecosystem, (2) the community, and (3) diversity and stability. In (1), I compare the ecosystem and continuum concepts and cite cases supporting each. All ecosystems are subject to energy and material inputs. To establish clear boundaries limiting an ecosystem and community often is very difficult. For this reason, I review a spectrum of examples--from those best corresponding to the ecosystem scheme to those best corresponding to the continuum concept. Between these extremes, are many intermediate cases, which probably form the majority. In (2), I compare the community as a system of interrelated species with the community defined as a unit formed by all species living in the same physical environment. In addition, I discuss the nature and the value of the relationships among different species and different trophic levels. In (3), I focus on the concept and the evaluation of diversity and on the concept of stability and the relationships between diversity and stability. To better understand these concepts, more information and more collaboration between plant, animal, and microbial ecologists are necessary. The future of applied ecology will depend upon the progress made in studies of these fundamental concepts of ecology.

ECOSYSTEM

The "ecosystem" has been defined in several ways. Some authors define it as a unit comprised of a community and the physical

145

environment (habitat) in which it lives. A community consists of a
system of interlinked populations interacting with the physical
environment by interrelationships that become closer with time,
increasing the stability of the system (e.g., Clements and Shelford
1939, Braun-Blanquet 1951, Odum 1973).

According to Margalef (1974), the ecosystem is an abstract
unit expressed by a level of organization whose constitutive elements are individuals belonging to different species. This definition appears to be more realistic than the concept of the ecosystem
as a "microcosm." The most important merit of Margalef's concept
is its abstraction of the physical limits of the ecosystem and its
accent on the relationships between species rather than on the
species themselves.

Some ecologists consider the unit host plants (Janzen 1968,
1973) and inflorescences with the associated fauna (Seifert 1975)
as "ecological islands" (and then ecosystems). Others (e.g.,
Ramonsky 1926, Gleason 1939, Whittaker 1951), adopting the concept
of the "continuum," deny the reality of the ecosystem and community,
and believe that groups of species-populations live in the same
physical environment only because they have the same ecological
needs. The structure of this group of populations, narrowly determined by physical factors, varies gradually along a spatial gradient
of one or more parameters. Whittaker and others have tried successively to unify the concept of ecosystem with that of continuum.

The following examples seem to support both the ecosystem and
continuum concepts.

A river, with its tributaries and catchment basin (which controls the characteristics of the whole hydrographic system), may be
considered as a unit, or each of the tributaries may be considered
as an ecosystem independent from the main river. Often it is even
harder to establish the limits between neighbouring ecosystems.
For example, consider a fairly uniform terrestrial ecosystem (e.g.,
grassland) in which there are two lakes with catchment basins
separated by a small relief of land. If we consider the grassland
as an ecosystem and each of the two lakes as an independent ecosystem,
there are three neighbouring ecosystems; but if we consider each
lake with its catchment basin as an independent ecosystem, there
are only two ecosystems (i.e., the two lakes with their respective
catchment basins) (Fig. 1).

The sea may be considered globally as one ecosystem divided
into subecosystems (e.g., pelagic, bathyal, intertidal, and neritic).
But it is difficult to define a subecosystem in any but an approximate
way. For example, the characteristics of the neritic zone are very
different from those of the intertidal zone, although these two
ecosystems (or subecosystems) are closely connected.

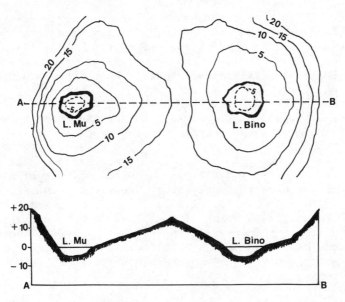

Fig. 1.   Schematic map of an imaginary region (from Ravera 1980).

In meromictic lakes (e.g., Lake Kiwu, Africa), the thermal and chemical stratifications are stable and therefore two superimposed and clearly differentiated ecosystems exist in the same lake. Examples of temporary stratification occur in lakes in the temperate zone; in some deep lakes (e.g., Lake Maggiore, Lake Lugano), full winter circulation occurs at intervals of four or six years, depending on meteorological events.

A typical ecosystem with superimposed layers (strata) is the forest. It is obvious that the physical, chemical, and biological characteristics of the foliage layer are very different from those of the litter, although some birds that nest in the canopy eat insects from the litter and some insects living in the litter find their food on the trunks and branches of the trees.

In groundwater studies, rivers, lakes, and underground springs usually are considered independent ecosystems. Rouch and Bonnet (1976) judged this separation to be artificial and affirmed that a karsic system (in the hydrological definition) is a single ecosystem formed by the rocks, the waters (lotic and lentic) contained in them, and the associated fauna.

Marshes surrounding a lake represent a transitional environment between aquatic and terrestrial ecosystems, just as mangrove formations are intermediary ecosystems between the sea and land, and estuaries are intermediary between rivers and littoral zones.

None of these ecosystems are completely closed because all
receive solar energy and (or) chemical energy from neighbouring
ecosystems and release part of the energy and material received to
the outside. Some ecosystems that do not receive solar energy,
such as subterranean systems and hypolimnetic fresh and marine
waters, must receive chemical energy (in the form of organic matter)
from ecosystems that do receive solar energy. Subterranean systems,
in addition to organic matter, may also receive large amounts of
epigeal organisms such as copepods (e.g., Rouch 1982).

The functioning of an ecosystem is determined by the energy
flow which it receives and the biogeochemical cycles of a series of
substances. Neighbouring ecosystems have close physical, chemical,
and biological relationships with each other: consider the contribution of solid material, solutes, and freshwater from rivers to the
sea, the load of nutrients from the catchment area to the lake, and
the contribution of substances from the sea to the marshes and
coastal lagoons and from the mountain ecosystems situated upstream
to those downstream.

To judge if the limits between two communities are well defined,
we must consider the largest possible number of compartments--not a
limited number as is often done. For example, the limit between
two neighbouring phytocoenoses (e.g., wood and grassland) may be
clearly established if we consider only plants, but there may be
species of birds and insects which are common to both communities
and others limited to either wood or grassland. Wild animal migrations and human directed migrations (transhumance and summer mountain
grazing) always have considerable influence both on the migrant and
on the affected communities (the one left and the one joined). In
fact, a quantity of organic material equal to that contained in the
migrating population is taken away from the community that the
population leaves and enriches the community that the population
enters. Animals that spend parts of their lives in different ecosystems (e.g., aquatic insects) produce consequences similar to
those mentioned for migration.

If we examine populations along a river from the source down-
wards, we observe a gradual succession of species. A clear example
of this succession is schematized in Figure 2 from a study on the
fish of the River Lot, France (Tourenq and Dauba 1978). Such a
spatial succession is due mainly to the gradual variation in the
characteristics of the substratum and the chemical and physical
properties of the water along the river. These variations are
strongly dependent on current velocity (e.g., Ravera 1951). In a
river, a net separation between two neighbouring communities is the
exception rather than the rule and is due to a sudden morphological
alteration of the river (e.g., sharp slope of the profile) or to
the discharge of noxious or eutrophicating substances or calories.
The type of sediment deposited in a given section of a river may

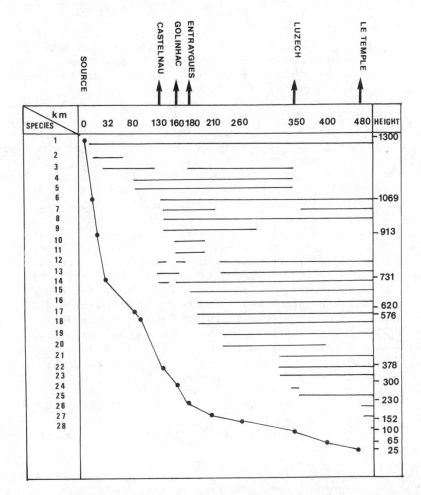

Fig. 2. Distribution of different species of fishes in River Lot:
(1) *Salmo trutta*; (2) *Cottus gobio*; (3) *Phoxinus phoxinus*;
(4) *Barbus barbus*; (5) *Leuciscus cephalus*; (6) *Exos lucius*;
(7) *Lucioperca lucioperca*; (8) *Rutilus rutilus*; (9) *Leuciscus
leuciscus*; (10) *Cobitis barbatula*; (11) *Chondrostoma nasus*;
(12) *Cyprinus carpio*; (13) *Tinca tinca*; (14) *Perca fluvia-
tilis*; (15) *Gobio gobio*; (16) *Scardinius erythrophtalmus*;
(17) *Alburnus alburnus*; (18) *Lampetra planeri*; (19) *Lepomis
gibbosus*; (20) *Chondrostoma taxostoma*; (21) *Micropterus
salmonides*; (22) *Anguilla anguilla*; (23) *Abramis brama*;
(24) *Blicca bjorkna*; (25) *Ameturius nebulosus*; (26) *Mugil*
sp.; (27) *Petromyzon marinus*; (28) *Alona alosa*. (Modified
from Tourenq and Dauba 1978; reprinted by permission of
Centro Comune di Ricerca, Ispra, Italy.)

vary with the season. Thus, a river shows variations in time as well as in space.

Studies on pollen spectra of lake sediments show that often the frequencies of most species living in the same community gradually vary in relation with climatic changes. In this case, a gradient of each individual species in time is evident.

In spite of the "openness" of the ecosystem to energy and material input, sometimes the limits of the community and those of the physical environment are well defined. For example, an alpine lake, excavated in rock and without tributaries and outlet, can be considered as an ecosystem sharply delimited by its banks. Indeed, if it receives solar energy and material from its watershed, the nutrients are locally recycled. A fault and successive erosion may produce two adjacent rocky layers with very different physical and chemical characteristics. It is clear that two differentiated ecosystems will develop on the two different substrata, even if they are controlled by the same climatic conditions. An island, if it is influenced by the continent to which it belongs and the nearest islands, is also a clear example of a well defined ecosystem isolated from the adjacent sea. If the rocky substratum of the sublittoral zone at a certain depth is replaced by sandy sediments, there will be a clear boundary between the two ecosystems with a marked zoning of plants and animals.

Thus, many environments are adequately described as ecosystems (e.g., high mountain lakes), while others seem to fit the continuum concept better (e.g., the side of a mountain from the peak to the base). Consequently, as neither the ecosystem concept nor the continuum concept seem well suited to all the cases studied, both may be considered valid.

We may imagine a series of cases that range from those best corresponding to the ecosystem scheme to those best representing the continuum concept and, between these extremes, we would find numerous examples with intermediate characteristics, probably forming the majority. This scheme could be compared to the intermediate taxonomic forms between the typical forms of two species.

The ecosystem concept offers notable advantages both for the presentation and for the programming of research. On the other hand, to classify natural occurrences according to a scheme that is too rigid is never advisable, even if it is probable that many examples (but not all) will have the characteristics required by the scheme. As a working hypothesis and teaching instrument, we may adopt the concept of ecosystem or of continuum, but we must remember the limitations of both. In my opinion, the critical consideration of the concepts of ecosystem and continuum could further stimulate ecological research. Certainly, more information

on the interactions between neighbouring environments and more
research on the transitional ecosystems are needed.

COMMUNITY

Is the community composed of all the organisms existing in a
given environment (e.g., lake, forest) or is it composed only of
the organisms actually linked by relationships? In the former,
there is a single community in each environment; in the latter,
more than one may exist. Considering one community for each environment is more practical for studying energy flow and biogeochemical
cycles. However, if a community is defined as a complex of organisms
interrelating with each other, it seems more realistic to focus on
the relationships between different species-populations. In my
opinion, the community should be considered more as a system of
species actually interrelated than as a unit composed of all the
species living in the same physical environment and all dogmatically
linked by trophic relationships. In addition, studies on the
partitioning of energy flow among the different food webs (communities) existing in the same environment in relation to their real
structure should produce more useful results than studies on the
pathway of a single energy flow in one community.

The food web is an obvious reality as are many clear examples
of close relationships between different species, such as host-
parasite and predator-prey. On the other hand, some populations
which live in the same area are not necessarily linked to each
other. For example, Terborgh (1971) calculated that the distribution area of only one-third of 261 species of birds was limited by
competition; the remainder was limited by environmental factors.
According to Whittaker (1962), most species living in the same
environment are not united by any relationship--only a few species
are positively (e.g., symbiosis) or negatively (e.g., predation)
linked. Some organisms, such as saprophytes and others, may be
relatively unlinked to other organisms if they have no predator.
For example, oligochaetes, living in a river without predators,
feed on organic material which is released mostly from the land
into the waters. The bulk of algal material that is not preyed
upon by zooplankton or nekton may be used as food by other organisms
(detritivorous, saprophytes) upon the death of the algal cells.
This relationship is between detritivorous organisms and organic
material (allochthonous or autochthonous) and not between organisms
of different species.

In an environment, particularly if it is not homogeneous, more
than one food web may be recognized and the relationships between
the species of different food webs (communities) of the same environment may be very weak. Indeed, the elimination of one or more
important species from a food web may cause its collapse, whereas

other food webs may remain unaffected. Thus, more communities and species relatively independent from one another may co-exist in the same environment.

Two successive trophic levels are connected by prey-predator relationship and the control may be exerted, according to some authors, by the predator (control from above) or, according to White (1978), by the prey (limited from below). This control may be more or less effective. Indeed, the controls exerted by rabbits on Australian vegetation are more severe than those exerted by herbivores of temperate zones on vegetation, because the herbivores consume only a small part of the production (e.g., Golley 1960, Trojan 1968). Likewise, the controls exerted by zooplankton on phytoplankton in oligotrophic lakes are more severe than those exerted by zooplankton on phytoplankton in productive waters. The controls are influenced by the population size and the intrinsic rate of natural increase (IRNI) of the predator in relation to that of the prey. For example, if the population size and IRNI of the prey are far higher than those of the predator, the latter cannot exert effective control on the prey.

In a community, the importance of a species-population also depends upon the number of relationships which link it to other species (e.g., predation, symbiosis). Indeed, the extinction of a species with numerous interlinking relationships will influence community structure more than the elimination of a species with few links. Therefore, in addition to the parameters commonly used to evaluate the importance of each major species-population (e.g., biomass, production, density), the ratio between the number of links associated with each important species and the number of species comprising the community may be a useful index for obtaining a better knowledge of community structure. For example, in a food chain the number of relationships is equal to the number of species minus one, whereas in a food web it is equal to or higher than the number of species (Fig. 3). The problem becomes more complex if the value of the same relationship connecting two species is considered. For example, if a species (A) feeds on only one prey (B), which is preyed upon with the same pressure by two other species (C and D), the value of the relationship A-B amounts to 100% for A, but only about 33% for B.

In studies on the relationship between species-populations, considerable importance has been attributed to competition and predation, whereas positive relationships (e.g., commensalism, symbiosis) have received comparatively little attention (P.W. Price, Department of Biological Sciences, Northern Arizona University, Flagstaff, AZ USA, personal communication, April 1983). Most of the quantitative information on competition and predation has resulted from experiments in which optimum conditions were arranged to emphasize the importance of these relationships (e.g., Underwood 1978). In natural

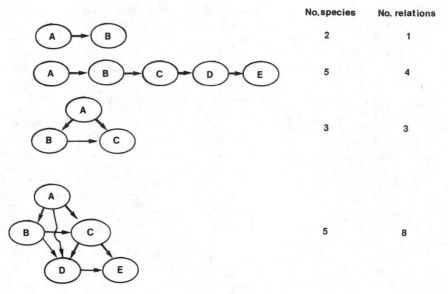

Fig. 3.   Number of species and number of relations in different types
          of communities (from Ravera 1981).

environments, the effects of competition and predation are more
complex than in the laboratory, because generally in the latter
only two species are considered, whereas in the field very frequently three or more species are involved. One of the few clear
examples of a simple prey-predator relationship reported in textbooks
is the hare-lynx periodicity. More recent studies (Fox 1978) have
demonstrated that this pattern is driven by the periodicity of fire
and precipitation. In addition, epidemic diseases may influence
the size of one or both of the populations. In natural ecosystems,
this complexity seems to be the rule rather than the exception
(Watt 1968, 1973).

     Under natural conditions, there are several mechanisms to pre-
vent, or at least to reduce, competition and predation pressure.
Some species may be present during a period of the year when species
upon which they prey or with which they compete are absent or
scarce. In addition, very high specialization in feeding and (or)
in space resources may permit the co-existence of two or more
related species living in the same physical environment. For
example, one species of Diaptomids commonly lives in small lakes;
when two or more species co-exist in the same lake, their body
sizes normally are different. According to Hutchinson (1951),
competition for food is reduced because the larger Diaptomid feeds
on large particles and the smaller on small particles. Bossone and
Tonolli (1954) studied the relationships between three Calanoids

(*Arctodiaptomus bacillifer*, *Acanthodiaptomus denticornis*, and
Heterocope saliens) in a small alpine lake of northern Italy (Lake
Monscera). *A. bacillifer* lives at a higher altitude than *A. denticornis*; consequently, both species co-exist in only a few lakes
situated at an altitude between the lower distribution limit of *A.
bacillifer* and the higher limit of *A. denticornis*. It is noteworthy
that the life cycle of *A. bacillifer* is shorter in lakes with *A.
denticornis* than in those in which the latter species is absent.
The life cycle of *H. saliens* is present during all of the summer,
that of *A. bacillifer* is limited to the first part of this season
followed by that of *A. denticornis*. These cycles limit the co-existence of both Diaptomids and, as a consequence, competition for food
is reduced. The predation pressure of *H. saliens* on *A. bacillifer*
(preferred to *A. denticornis* for its smaller size) increases after
the latter have laid resistant eggs. The mortality of *A. bacillifer*
by predation facilitates the population increase of *A. denticornis*
(Fig. 4). These complex mechanisms permit the co-existence of the
three species in one small lake. Another example of co-existence
and partitioning of the same resource between two species of the same
genus is given by two bees (*Trigona fuscipennis* and *T. fulviventris*)
foraging on shrub of *Cassia* (*Calycophyllum*) *biflora* (Johnson and
Hubbell 1975). *T. fuscipennis* forages on dense clumps of *Cassia*,
preferring the plants which are richer in flowers. *T. fulviventris*
usually visits more widely spaced or isolated plants irrespective

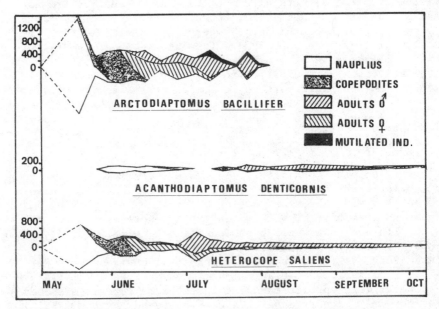

Fig. 4.   Lake Monscera population density (no. ind/m^3) (from Bossone
          and Tonolli 1954; reprinted by permission of Centro Commune
          di Ricerca, Ispra, Italy).

of the number of flowers they show. When this bee forages in
clumps, the few plants visited have fewer flowers than those not
visited. Resource partitioning also seems to be very effective in
bird communities (Cody 1973).

In contrast to these examples, numerous species of reef fish
co-exist successfully with a considerable overlap in their food and
living space requirements (Jones 1968, Choat 1969). The structure
of these communities in isolated colonies of coral is determined by
chance colonization and chance mortality (Sale and Dybdahl 1975).

Thus, considering the very different environments and life
histories of different taxa, it is not surprising that differences
in strategy and organization of the various taxa exist that permit
co-existence in the same area.

DIVERSITY AND STABILITY

Diversity is one of the most important properties of the community. Unfortunately, its measurement commonly involves only a
few taxanomic groups and the value calculated from these groups is
attributed to the whole community. This extrapolation may be
correct if groups of interrelating organisms are considered, such
as phytoplankton, zooplankton, and fish belonging to the same food
chain (Margalef 1968). On the contrary, it seems hazardous to
extend the value of diversity calculated for the planktonic community
to the benthos of a deep lake or the ocean.

There are several methods available to measure diversity, but
ecologists do not agree on which index is the best. May (1975) has
discussed the advantages and disadvantages in adopting the various
evaluations of diversity reported in the literature. Margalef
(Department of Ecology, University of Barcelona, Barcelona, Spain,
personal communication, April 1983) thinks that the best index is
Simpson's and that the concept of diversity must always be related
to space. Hurlbert (1971), in a review of the literature on diversity, concluded that the most commonly used methods for measuring
this property are meaningless. The various causes of error in
diversity evaluation have been analysed by Hughes (1978); therefore,
I do not think it useful to repeat this list here.

Another fundamental property of the community is stability.
Some communities seem to be very stable, others are not. If the
physical environment evolves very slowly, the community structure
is roughly constant and its change is small and gradual. Conversely,
rapid major changes in the environment produce significant modification of the community, including its extinction and substitution by
another with a higher degree of fitness to the new situation. For
example, the comparison between samples of bottom fauna collected

from Lunzer Untersee (Austria) in 1955 and those from the same lake
in 1964 showed that biomass and biocoenotic structure did not
change significantly because of the stability of the physical
environment (Ravera 1966). D'Ancona (1942) has observed significant
change in the biocoenotic structure of the zooplankton in Lake Nemi
(central Italy), probably caused by a great and artificial decrease
of the lake level. The modification of zooplankton structure of
Lake Maggiore (northern Italy) recorded from 1909 to 1958 seems to
be the effect of the increasing enrichment of this lake in recent
years. Indeed, no significant modification was observed from 1909
to 1950 (Baldi 1951), whereas Tonolli (1962) emphasized important
changes from 1950 to 1958. Quantitative and qualitative modifications of the zooplankton structure of Lake Lugano (on the border
between Italy and Switzerland) are the indirect effect of the
progressive nutrient enrichment of this water body since 1940
(Ravera 1977). On the other hand, the modification of community
structure caused by the introduction of new species in an unchanged
physical environment has been observed and, more rarely, stable
communities in changing environments have been reported. Ward
(1975), for example, found a similar macroinvertebrate composition
in a river after 29 years (1945-1974) despite differences in water
flow, temperature, and riparian vegetation. There is evidence that
communities living in temporary tarns and rice fields maintain
their structure during successive years in spite of periodic dry
periods. Accurate studies carried out for several years should
produce useful information about the stability of these environments.
In contrast, communities of reef fish show very weak stability
(Sale and Dybdahl 1975). Any variation in ecological conditions
favouring one species over another leads to variation in community
structure. Only further chance fluctuations in ecological conditions
will return the previous structure. According to Sale and Dybdahl
(1975), if the source areas of fish are large, there may be considerable inertia preventing modification of community structure. This
inertia is the reason why the community seems to be stable but, in
fact, is not.

There is some disagreement among ecologists about the meaning
of stability; Orians (1975) discussed seven different definitions
reported by different authors. According to MacArthur (1955) there
is evidence that stability is related to community structure, and
the degree of stability is the result of the number of trophic
pathways in the ecosystem and the energy flow passing along each
pathway. Several causes may increase the number of trophic pathways
in a community, for example, an increase in the number of species,
the passage of some herbivores or carnivores to an omnivorous diet,
the feeding by species of higher levels on a larger number of
species belonging to lower levels. In contrast, the addition of
species belonging to a new trophic level (e.g., top carnivores)
does not increase the number of pathways. Equitability of the
energy flow along different trophic pathways prevents (or reduces)

the effects due to the variation of population size of one or more
species on the whole community. In addition, equitability promotes
the channeling of energy flow into other pathways. This hypothesis
is logical and extremely interesting, but, unfortunately, the
energy flow along trophic pathways presently cannot be measured
directly and, consequently, McArthur's hypothesis cannot be confirmed.

The stability concept must be related to time; for example, a
community may be stable for 10 years, but not for 10 centuries.
Because the stability value is not absolute, it seems more useful
to evaluate the stability of a given community in relation to that
of other communities.

Diversity depends upon the total number of species forming the
community and their frequencies. Therefore, diversity must be
expressed by a number, whereas at present there is no general
agreement on the best parameters to consider when defining stability.
We might suppose that the degree of constancy of certain important
rates (e.g., production rate, organic matter degradation rate,
respiration rate), over a reasonable period of time, indicates the
level of stability of a community. However, it would be preferable
to consider the constancy of the ratio of certain rates. For
example, an increase in the ratio between production and degradation
of organic matter in a water body would indicate that the ecosystem
was evolving towards a progressive eutrophication. In general, a
heterogeneous physical environment with a high number of species is
more stable than a homogeneous environment with a small number of
species. Most ecologists (e.g., Clements and Shelford 1939, Elton
1958, Patten 1963, Margalef 1968, Odum 1973) agree on a positive
relationship between the degree of diversity and the stability of a
community, but it is not known what increase in diversity is needed
to obtain a significant increase in stability.

The increase in stability of a community, considered as an
effect of diversity, is based largely on the following facts (Elton
1958): (1) very simple systems are not stable, for example, the
artifical ecosystems used in laboratory research and cultivations
(in particular monoculture); if these systems are to maintain their
integrity, they require continuous intervention by man; and (2) the
forests of the tropical zone, rich in species, have a high degree
of stability, whereas subarctic ecosystems, with relatively small
numbers of species, have low stability due to the large fluctuations
in populations.

Maynard Smith (1975) observed that conclusions different from
those of Elton may be drawn from the same facts. For example,
natural communities are more stable than cultivated fields, not so
much because they have a greater diversity, but because their
evolution takes longer and thus there has been enough time to
establish a complex network of interrelations among the different

components of the ecosystem. According to this author, a greater
complexity of the system is not enough to cause greater stability.
Stability is then the product of two selective processes: genetic
feedback and species exclusion. The selective mechanisms modify
the species which are linked by relationships (e.g., parasitism,
competition) in that they maximally favour their probability of
survival, thereby increasing the stability of the community (genetic
feedback). Maynard Smith examined the different consequences
resulting from the introduction of one or more species into an
area. The invader may be eliminated either by competitors already
existing in the area or by adverse environmental conditions, or it
may replace one or more pre-existing species and occupy its ecological
niche, or it may co-exist with the indigeneous species (species
exclusion). According to Maynard Smith, these two selective mechanisms are the cause of diversity and stability in the community,
which is not a randomly formed group of species, but a system of
species selected to co-exist in the best way. Odum (1975) recom-
mended that diversity not be used as an index of ecosystem degradation. Some authors (e.g., Orians 1975, Uhlmann 1978) do not deny
the causal relationship between diversity and stability but believe
that such a relationship has not yet been demonstrated satisfactorily.
Barret (1969), in contrast to the authors who have calculated low
diversity values in polluted environments, observed that after the
distribution of an insecticide, the diversity of arthropods increased
because the dominant species were more sensitive to the insecticide
than the common and rare species. May (1971a, 1971b, 1972) has
demonstrated theoretically that an increase in diversity of a
community reduces its stability. Thus, the relationship between
diversity and stability still is not clear.

Odum (1975) introduced the concept of energy in the relationship between diversity and stability. Ecosystems with low diversity and high stability either have huge energy reserves or energy
is provided continuously and in great amounts (e.g., wet coastal
zones and environments modified by natural events or by humans).
In these ecosystems, a relatively simple community may use the huge
quantity of nutrients available and the large energy flow better
than a complex one. Conversely, stable ecosystems with high diversity, generally receive, in addition to solar energy, a modest
supply of nutrients (e.g., oligotrophic lakes, rain forests). In
these ecosystems, the recycling of nutrients is fundamental and the
community must be very diversified to maintain stability and thus
to use the small amount of energy available in the best way.

Were Odum's statements generally applicable, the practical
consequences would be very important. Because, according to Margalef's theory, ecosystems tend towards a high stability level,
diversity must decrease (or increase) in unstable systems according
to the degree of energy flow received. Consequently, in stable
systems, variation in diversity is an index of variation in size

and (or) quality of energy flow. For example, the discharge of
toxic substances into ecosystems receiving large energy flows
(e.g., eutrophic lakes) or small energy flows (e.g., oligotrophic
lakes) probably decreases their diversity, which increases stability in the former and decreases it in the latter. Initially, a
decrease of energy flow in an ecosystem rich in nutrients (e.g., by
reduction of nutrient load by wastewater treatment) or an increase
where nutrients are scarce (e.g., by nutrient load increase or
thermal enrichment) should produce instability. In both these
ecosystems, stability will be increased, but by an increase of
diversity in the first and a decrease in the latter.

CONCLUSIONS

We need more studies on the measurement of stability and
better agreement on its meaning. The reliability of the above
hypotheses must be demonstrated by research both in the field and
in the laboratory. The results from these studies will be useful
for understanding the functioning of ecosystems better and for
choosing actions to restore degraded environments or to produce
more organic substances (e.g., agriculture, aquiculture).

More information is needed to understand the relationships
between different species-populations and between these and their
physical environment. The quantitative knowledge of these relationships is the basis for manipulating ecosystems.

Research carried out on the pathways of energy flow in different
compartments of the community in relation to the biochemical cycles
of essential elements should produce useful information for a
better understanding of ecosystem dynamics.

Because of the fundamental importance of primary production,
animal ecologists cannot neglect plants in their research, even if
they do not always have satisfactory knowledge of them. On the
other hand, plant ecologists rarely consider the animal populations
which live in the environment they are studying. Moreover, some
aspects of ecology have been treated mainly from a botanical point
of view (e.g., variations in space and time of taxonomic and structural relationships), others mainly from a zoological point of view
(e.g., trophic food webs). There has not been an active exchange
of information among plant, animal, and microbial ecologists. This
lack of communication probably is the principal reason why the same
terms used by these groups of experts sometimes have assumed different meanings. For example, by "community" the animal ecologist
means a unit made up of all the organisms (animal and plant) which
live in the same physical environment, while generally the plant
ecologist means a certain number of plant populations. The plant
ecologist defines "phytocoenosis" as a topographic and edaphic unit

with a homogeneous physionomy, a definition which is not very
different from the animal ecologist's meaning of community, but the
latter includes animals also. A more effective collaboration
between plant, animal, and microbial ecologists is needed. The
future of applied ecology will depend, at least partly, on the
advances in studies of the problems I have mentioned.

REFERENCES CITED

Baldi, E. 1951. Stabilité dans le temps de la biocénose zoo-
    plantique du Lac Majeur. Verh. Int. Ver. Limnol. 11:35-40.
Barret, G.W.P. 1969. The effect of an acute insecticide stress on
    a semi-enclosed grassland ecosystem. Ecology 49:1019-1035.
Bossone, A., and E.V. Tonolli. 1954. Il problema della convivenza
    de *Arctodiaptomus bacillifer* (Koelb.), di *Acanthodiaptomus
    denticornis* (Wierz.) e di *Heterocope saliens* (Lill.). Mem.
    Ist. Ital. Idrobiol. 8:81-94.
Braun-Blanquet, J. 1951. Pflanzernsociologie. Grundzüge der
    vegetationskunde. Springer, Wien, Austria.
Choat, J.H. 1969. Studies on labroid fishes. Ph.D. Dissert.
    Univ. Queensland, Brisbane, Queensland, Australia.
Clements, F.E., and V.E. Shelford. 1939. Bio-ecology. Wiley, New
    York, NY USA.
Cody, M.L. 1973. Competition and the structure of bird communi-
    ties. Princeton Univ. Press, Princeton, NJ USA.
D'Ancona, U. 1942. Relazione sulle ricerche idrobiologiche e
    idrografiche compiute nel Lago di Nemi. Int. Rev. Hydrobiol.
    41:235.
Elton, C.S. 1958. The ecology of invasions by animals and plants.
    Methuen, London, England.
Fox, J.F. 1978. Forest fires and the snowshoe hare-Canada lynx
    cycle. Oecologia 31:349-374.
Gleason, H.A. 1939. The individualistic concept of the plant
    association. Am. Midl. Nat. 21:92-110.
Golley, F.B. 1960. Energy dynamics of a food chain of an old-field
    community. Ecol. Monogr. 30:187-208.
Hughes, B.D. 1978. The influence of factors other than pollution
    on the value of Shannon's diversity index for benthic macro-
    invertebrates in streams. Water Res. 12:359-364.
Hurlbert, S.H. 1971. The non-concept of species diversity: A
    critique and alternative parameters. Ecology 52:577-586.
Hutchinson, G.E. 1951. Copepodology for ornitologists. Ecology
    32:571-577.
Janzen, D.H. 1968. Host plants as islands in evolutionary and
    contemporary time. Am. Nat. 102:592-595.
Janzen, D.H. 1973. Host plants as islands. II. Competition in
    evolutionary time. Am. Nat. 107:786-790.
Johnson, L.K., and S. Hubbell. 1975. Contrasting foraging strate-
    gies and coexistence of two bee species on a single resource.
    Ecology 56:1398-1406.

Jones, R.S. 1968. Ecological relationships in Hawaiian and Johnston Island Acanthuridae (surgen-fishes). Micronesica (J. Coll. Guam) 4:309-361.

Margalef, R. 1968. Perspectives in ecological theory. Univ. Chicago Press, Chicago, IL USA.

Margalef, R. 1974. Ecologia. Omega, Barcelona, Spain.

May, R.M. 1971a. Stability in model ecosystems. Proc. Ecol. Soc. Aust. 6:18-56.

May, R.M. 1971b. Stability in multispecies community models. Bull. Math. Biophys. 12:59-79.

May, R.M. 1972. Will a large complex system be stable? Nature 238:413-414.

May, R.M. 1975. Patterns of species abundance and diversity. Pages 81-120 in M.L. Cody and J.M. Diamond, eds. Ecology and evolution of communities. Belknap Press, Cambridge, MA USA.

Maynard Smith, J. 1975. Models in ecology. Cambridge Univ. Press, Cambridge, England.

McArthur, R.H. 1955. Fluctuations of animals populations and a measure of community stability. Ecology 36:533-536.

Odum, E.P. 1973. Fundamentals of ecology. Saunders, Philadelphia, PA USA.

Odum, E.P. 1975. Diversity as a function of energy flow. Pages 11-14 in W.H. van Dobben and R.H. Lowe-McConnell, eds. Unifying principles in ecology. Junk, The Hague, The Netherlands.

Orians, G.H. 1975. Diversity, stability and maturity in natural ecosystems. Pages 139-150 in W.H. van Dobben and R.H. Lowe-McConnell, eds. Unifying principles in ecology. Junk, The Hague, The Netherlands.

Patten, B.C. 1963. Plankton: Optimum diversity structure of a summer community. Science 170:894-898.

Ramonsky, L.G. 1926. Die Grundgesetzmassigkeiten im aufbau der vegetationsdecke. Bot. Centbl. N.S. 7:453-455.

Ravera, O. 1951. Velocità di corrente e insediamenti bentonici. Studio su una lanca del fiume Toce. Mem. Ist. Ital. Idrobiol. 6:221-267.

Ravera, O. 1966. Stability and pattern of distribution of the benthos in different habitats of an alpine oligotrophic lake: Lunzer Untersee. Verh. Int. Ver. Limnol. 16:233-244.

Ravera, O. 1977. Effects of eutrophication on the zooplankton of a subalpine lake: Lake Lugano. Pages 97-104 in W.K. Downey and G. Ni. Uid, eds. Lake pollution prevention control. Natl. Sci. Counc., Dublin, Ireland.

Ravera, O. 1980. L'ecosistema. Pages 101-159 in VI Seminario sull'evoluzione biologica. Ecologia ed etologia. Contrib. Cent. Linceo Interdiscipl. Sci. Mat. Appl. 51, Accad. Naz. Lincei, Roma, Italia.

Ravera, O. 1981. Alcune considerazioni sull'ecosistema. Pages 35-40 in A. Moroni and O. Ravera, eds. Ecologia. Edizioni Zara, Parma, Italia.

Rouch, R. 1982. The budget karsic system. XIII. Comparison of the
    Harpacticoidea drift at the input and output of the aquifer.
    Ann. Limnol. 18:133-150.
Rouch, R., and L. Bonnet. 1976. Le système karsique du Baget.
    IV. Premières données sur la structure et l'organization de
    la communauté des Harpacticides. Ann. Spéléol. 31:27-41.
Sale, P.F., and R. Dybdahl. 1975. Determinants of community
    structure for coral reef fishes in an experimental habitat.
    Ecology 56:1343-1365.
Seifert, R.P. 1975. Clumps of *Heliconia* influorescences as ecolog-
    ical islands. Ecology 56:1416-1422.
Terborgh, J. 1971. Distribution on environmental gradients:
    Theory and a preliminary interpretation of distributional
    patterns in the avifauna of the Cordillera Vilcabamba, Peru.
    Ecology 52:23-40.
Tonolli, V. 1962. L'attuale situazione del popolamento planctonico
    del Lago Maggiore. Mem. Ist. Ital. Idrobiol. 15:81-133.
Tourenq, J.N., and F. Dauba. 1978. Transformation de la faune des
    poissons dans la rivière Lot. Ann. Limnol. 14:133-138.
Trojan, P. 1968. Energy flow through a population of *Microtus
    arvalis* (Pall.) in an agrocenosis during a period of mass
    occurrence. Pages 267-279 *in* K. Petrusewicz and L. Ryszkowski,
    eds. Energy flow through small mammal population. Pol. Sci.
    Publ., Warsaw, Poland.
Uhlmann, D. 1978. Hydrobiology. Wiley, New York, NY USA.
Underwood, A.J. 1978. An experimental evaluation of competition
    between three species of intertidal Prosobranch Gastropods.
    Oecologia 33:185-202.
Ward, J.V. 1975. Bottom fauna-substrate relationships in a Northern
    Colorado trout stream: 1945-1974. Ecology 56:1429-1434.
Watt, K.E.F. 1968. Ecology and resource management. McGraw-Hill,
    New York, NY USA.
Watt, K.E.F. 1973. Principles of environmental science. McGraw-
    Hill, New York, NY USA.
White, T.C.R. 1978. The importance of a relative shortage of food
    in animal ecology. Oecologia 33:71-86.
Whittaker, R.H. 1951. A criticism of the plant association and
    climatic climax concept. Northwest Sci. 25:17-31.
Whittaker, R.H. 1962. Classification of natural communities.
    Bot. Rev. 28:1-239.

TOWARDS A LANDSCAPE ECOLOGY OF RIVER VALLEYS

Henri Décamps

Centre d'Écologie des Ressources Renouvelables
Centre National de la Recherche Scientifique
29 Rue Jeanne-Marvig
31055 Toulouse Cedex, France

INTRODUCTION

The interest given to detritus in streams of allochthonous origin dates back almost half a century (Thienemann 1912, Jewell 1927), as pointed out by Hynes (1970). Numerous studies have been devoted to this theme, particularly during the seventies, when a multitude of articles concerning the role of dead leaves in the functioning of water courses was published. Hence, there evolved a more holistic concept of lotic ecosystems.

But unlike that of small water courses, research on rivers has practically ignored the necessity of not "divorcing the stream from its valley" (Hynes 1975). The essential question in ecology concerning the interactions between water courses and their terrestrial environment cannot be resolved merely by studies of small watersheds.

It is now time to analyze with new insight the problems arising from river ecology and to develop a landscape ecology of river valleys.

METHODOLOGICAL AND CONCEPTUAL DIFFICULTIES

It has always been difficult to apply the ecosystem concept to rivers. Some limnologists even question the appropriateness of such a concept when they consider the longitudinal diversity of rivers extending several hundred kilometres (Rzoska 1978). A better understanding of the ecological nature of large rivers is

163

certainly necessary. Furthermore, the fact that these systems are
probably among the most anciently perturbed natural systems cannot
be disregarded. For a long time, fluvial valleys have been colonized
by humans. In recent years, dams, often constructed in series on
water courses, have profoundly modified the conditions of flow;
numerous and various waste deposits have transformed lotic flora
and fauna; the relationships between rivers and their inundation
beds have suffered drastic changes.

Moreover, the methods of investigation are incomparably more
difficult in large rivers than in small ones. This problem explains
mainly why so little is known about the ecology of large rivers
(potamon) as compared to that of small streams (rhithron). For a
long time, work on running waters arose from knowledge of what
species were present and how numerous they were (Macan 1974).
Reliable methods allowing the sampling of benthos of large rivers
have long been due. Recently, Elliott and Drake (1981a, 1981b) and
Drake and Elliott (1982) evaluated 14 samplers used for the collection
of macroinvertebrates in deep rivers. Their comparison of seven
grabs, four dredges, and three air lifts constitutes an adequate
basis for future quantitative research on the benthos of profound
rivers.

In spite of these difficulties, rivers seem increasingly ideal
as experimental sites for the testing of ecological theories at
population, community, and ecosystem levels. This explains the
number of books recently published or yet to be published on streams
(Ward and Stanford 1979, Lock and Williams 1981, Likens 1981,
Lillehammer and Saltveit 1984, Fontaine and Bartell 1983, Whitton
1984).

RECENT CONCEPTS ON THE ECOLOGY OF RUNNING WATERS

During the last few years, several North American authors have
drawn ecologists' attention with the "river continuum" and the
"nutrient spiralling" concepts. Longitudinal zonations already had
been proposed for European running waters from studies of fish
(Huet 1949) and invertebrates (Illies and Botosaneanu 1963).
However, even if the idea of longitudinal zonation is recognized as
useful generally, the proposal to define distinct zones precisely
really has not opened a new path in research. The idea of a continuum
from the springs to the mouths, as illustrated by Cummins (1979),
is more adapted to the fact that rivers constitute ecological
clines (Hynes 1970). The nutrient spiralling concept is indeed
linked to the continuum concept, and has been clearly illustrated
by Zhadin and Gerd (1963).

Of major interest, the river continuum and nutrient spiralling
concepts, as they are presented by Vannote et al. (1980) and by

Wallace et al. (1977), place river ecology in an evolutionary perspective. Variations in community types correspond to gradual changes in the physical conditions from upstream to downstream and the structure and function of these communities appear adapted to the most probable locations in the physical system. This variation/ gradation leads to a distribution of the resources as well as to a substitution of species over time and along the river courses from the springs to the mouths. Downstream communities appear to be organized in such a way as to take advantage of the incompletely used organic matter lost from upstream. The adopted strategies thus convey minimum energy losses. During the last few years, numerous publications·in ecology were inspired by these ideas.

However, as noted by Webster and Patten (1979), most of the research has been centered on headwater streams in forested areas (Fisher and Likens 1972, 1973, Boling et al. 1974, Cummins 1974, Fisher 1977). Large rivers, presenting very different conditions from those of small streams, have seldom been studied. But some of their characteristics (drastic changes at certain confluents, amplitude of the main bed, consequences of floods, transverse heterogeneities, etc.) deserve consideration during a discussion on the river continuum and nutrient spiralling concepts. The alternation of high waters and still periods acquires major importance when trying to understand the dynamics of river ecosystems. Not known is the type of control headwater streams exert on the more elevated stream orders, nor if the stability of these elevated streams can be attributed to their resilience as is the case for small streams. Finally, the amplitude of nutrient spiralling and the characteristics which modify these spirallings are unknown for rivers, and the role of sediments has been totally ignored. Recently, Melillo et al. (1983) have shown that the rate of decomposition of the same type of litter is not the same for woody material in a first order stream as it is in a larger stream. Obviously, better understanding of the processes existing in large rivers is needed.

It has to be recognized that regulation of flow by dams has transformed many large rivers. Therefore, the longitudinal continuum is often interrupted and replaced by alternate successions of the lentic and lotic segments. The interruptions of the upstream-down-stream continuum and of the nutrient spiralling flux can, in fact, interfere at various points along the longitudinal profile of water courses. Therefore, experimental systems that test the basic theories of lotic ecology already exist. The serial discontinuity concept has been introduced by Ward and Stanford (1983) to generate a larger theoretical view of perturbed lotic ecosystems. Although simplistic and general, this hypothetical framework provides a structure for designing and interpreting research on altered rivers (Fig. 1). A more precise quantification of the effects of flow regulation on specific examples, taking into account phenomena such as specific abundance, biotic diversity, P/R ratio, etc., would be useful.

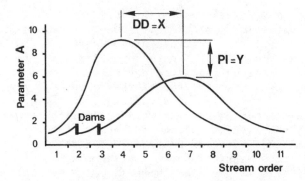

Fig. 1.   Theoretical influence of the presence of dams on a river.
Discontinuity distance (DD) is the shift in distance (X)
of a given parameter. Parameter intensity (PI) is the
difference (Y) of this parameter due to stream regulation.
(After Ward and Stanford 1983; reprinted by permission of
Ann Arbor Science, Ann Arbor, MI USA.)

The interest on regulated streams is attested to by the success
of two international symposiums held at Erie (USA) in 1979 and in
Oslo (Norway) in 1982. In several countries around the world,
research on the ecology of regulated streams is being actively
pursued.

Décamps et al. (1976, 1979) have studied the interactions
governing water quality of the regulated River Lot (France).
Figure 2 shows that flow is the major factor modified by human
activities, and that it influences the whole system by deposition
and re-suspension of sediments and by action on temperature and on
the processes of photosynthesis and biodegradation. Figure 3
summarizes the dynamics of suspended matter in the river. There
exists an everchanging equilibrium among the factors which provoke
re-suspension of organic and mineral, non-algal matter and the
factors which induce both decomposition of the suspended matter and
increasing algal biomass. None of the interactions interfering in
the water mass have been quantified in a dynamic fashion along the
water course. This reflects the gaps existing in our understanding
of the biological processes of rivers. It would be particularly
useful to encompass the role of sediments in the hydrobiology of
rivers. This knowledge seems fundamental when considering the
large quantities of sediments transported by running waters throughout
the world (Meybeck 1982).

In the River Lot, when flow is sufficient, planktonic algae
follow a spirally pathway, which leads them to fairly lengthy
periods of residence in the depths of the aphotic zone, thus limiting
their biomass growth. On the contrary, with weak to very weak

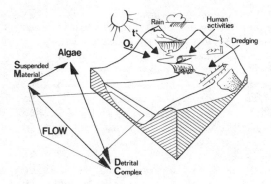

Fig. 2.   Dynamics of suspended matter in the river with flow
           influencing the quantities of suspended material, algae,
           and detrital complex.

flows, the algae remain in the river's euphotic zone and turbidity
tends to decrease, thus increasing net production and diminishing
loss due to respiration.   More than 50% of the river's oxygen
consumption results from bacterial degradation of organic matter
(Capblancq and Décamps 1978).   Thus, a regulated river, such as the
Lot, simultaneously presents high rates of photosynthesis and
biodegradation.   Along with the irregularity of flow, the system's
functioning is sometimes autotrophic and at other times heterotrophic.

THE ROLE OF ALLOCHTHONOUS ORGANIC MATTER

      Many pioneering works, including those of Hynes (1963) and
Minshall (1967), have directed attention to the important role of

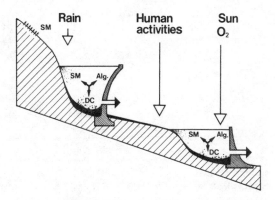

Fig. 3.   Summary of factors influencing the ecology of the River Lot:
           suspended material (SM); detritus complex (DC); algae (Alg).

allochthonous organic matter in streams. This organic matter, originating frequently from leaf litter, is brought to running waters from forested areas and also from the whole hydrographic network because most stream edges are covered with trees and bushes. Pollen and spores also can constitute an important source of import, qualitatively speaking, as has been suggested for the Amazonian river systems where the complex ecological relationships have been studied by Fittkau (1964) and his German colleagues. In addition to these essentially seasonal imports, most rivers receive domestic, industrial, and agricultural effluents.

A few authors have discounted general balances, but it remains difficult to establish a better comparison than that outlined following the International Biological Program (Table 1). The fate of the dead leaves that fall into streams was thoroughly studied for a woodland stream system (Kaushik and Hynes 1971). Young et al. (1978) demonstrated that the distances traveled by large particles of organic matter are not great. But, as noted by Dance (1981), stream order may affect the distance traveled by leaves after their entry into the stream. Greater distances of transport are possible in wider rivers.

Two other areas deserve more emphasis: (1) the conversion of dissolved matter into particulate organic matter, and (2) the variations of the ratio CPOM/FPOM (coarse particulate organic matter/fine particulate organic matter). As decomposition progresses, large particles of detritus are fractionated into smaller ones by the combined action of detritivorous organisms and microbes, which

Table 1. Comparison between general balances (after Le Cren and Lowe-McConnell 1980).

	Allochthonous particulate organic matter, total input $(KJ/m^2 \cdot a)$	Net primary production $(KJ/m^2 \cdot a)$
River Thames below Kennet	1,363	8,163
Bear Brook (USA)	13,495	40
Watershed 10 (USA)	8,962	94
Augusta Creek (USA)	9,695	79

are responsible for the enzymatic hydrolysis of organic matter.
Inversely, bacteria assimilate dissolved organic matter and tend to
produce particles by binding and associating with particulate
matter. There exists, therefore, a tendency to produce particles
of intermediate size. In addition to this, the faecal pellets of
benthic invertebrates are successively taken up in the trophic
networks from upstream to downstream. The role such processes play
in the functioning of rivers, as well as the meaning of the dissolved
organic matter pool arising from continued leaching of the numerous
organic substrates, remains to be specified.

The progress brought about in aquatic microbiology (Wiebe, this
volume) enables further understanding of the mechanisms involved.
The impact of the nutritive quality of the various components of
organic detritus are insufficiently elucidated. As indicated by
Anderson and Sedell (1979), the impact is probably important for
invertebrate populations and, of course, for microbial population
dynamics. Likewise, fungal population dynamics and their role in
leaf decomposition deserve to be studied (Bärlocher and Kendrick
1981). Everything concerning faecal pellets of invertebrates is
worth more attention: their nature, decomposition, and utilization
(Bird and Kaushik 1981). It is also obvious that the quality of
the input (coniferous or mixed forest, for instance) affects stream
productivity. Moreover, the original food value of litter varies
seasonally from October to June (Gosz et al. 1973), and the importance
of denitrification certainly has been underestimated as already
indicated by Owens et al. (1972).

Comparative studies on the relative amounts of organic and
inorganic materials being transported in rivers are lacking. In
the River Lot, for example, most of the particles are included in
the size range between 1 and 20 μm (Table 2).

Table 2.   Suspended matter (mean ± CL) in the River Lot in August
           1975 (after Décamps and Casanova-Batut 1978).

Diameter of particles (μm)	Mineral matter (mg/1)	Organic matter (mg/1)	Turbidity (FTU)
$\phi > 32$	0.77 ± 0.18	0.58 ± 0.10	1.36 ± 0.24
$32 > \phi > 20$	1.10 ± 0.17	0.73 ± 0.11	1.83 ± 0.20
$20 > \phi > 1$	2.55 ± 0.62	4.21 ± 0.62	6.76 ± 0.82

Finally, in agreement with Gelroth and Marzolf (1978), it must be stressed how the type of water course considered plays a major role in the formation of leaf packs, the gathering of the leaf packs in the pools, their introduction into the sediment, etc.

INFLUENCE OF THE VALLEY AND THE CONCEPT OF STREAM CORRIDOR

It is true that one of the most important discoveries of running water ecologists has been the existence of hydrologists, foresters, and soil scientists (Hynes 1975). However, these researchers have not yet realized the enormous input running water ecology can contribute to the better understanding of river systems. Terrestrial ecologists are more aware of the benefits of better understanding of soil and plant components, especially in a historical perspective (Fig. 4).

It is not surprising, therefore, that terrestrial ecologists recently have defined the concept of stream corridor (Forman and Godron 1981). Stream corridors include a vegetation belt accompanying running waters. This belt can cover not only the channel's borders, but also the floodplain and part of the elevation that overhangs them (Schlosser and Karr 1981). Stream corridors can attain widths of many kilometres in certain waterway valleys. They constitute

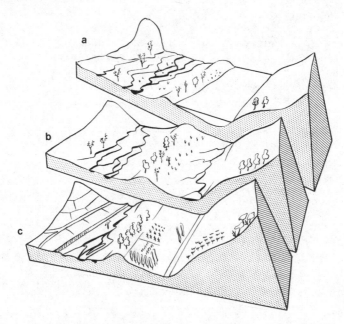

Fig. 4. Evolution of a stream corridor from (a) to (c) with increasing human influence.

one of the major spatial configurations of landscapes as defined by
Forman and Godron (1981): kilometre-wide areas where clusters of
interacting stands or ecosystems are repeated in a similar form.
In fact, stream corridors result from specific geomorphological
processes and disturbances of the component stands. These disturbances include natural events and human activities. Width, degree
of heterogeneity, and connectivity characterize stream corridors,
as does the flux of energy, mineral nutrients, and species among
the constitutive patches (Fig. 5).

This concept seems particularly well adapted to the study of
the ecology of rivers from theoretical as well as applied points of
view. In effect, stream corridors exert a certain control on water
flux and nutrients of river valleys, enabling the delay of outflow
which will eventually influence the systems downstream. They also
form privileged trails of passage for the extension of certain
plant and animal species bound to water.

During the seventies, the effect of riparian vegetation on
natural river systems was considered mainly in a management perspective (Turner 1974, Horton and Campbell 1974, Warren and Turner
1975, Burkham 1976, Dawson 1976, Karr and Schlosser 1978). But in
spite of recent consciousness (Binder et al. 1983, Scheurmann
1983), engineers do not always consider that rivers and streams are
not to be seen merely as ducts and channels.

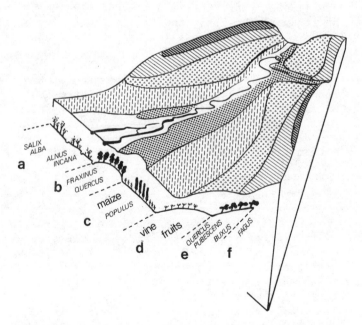

Fig. 5.   Schematic representation of the constitutive parts of the
          Upper Rhone Valley (after Pautou and Bravard 1982).

The propagation of the introduced salt cedar along the river
systems of the western United States presents an example of the
complex adjustments existing in riparian zones (Everitt 1980).
Clearly, the dynamics of riverbank ecology can be understood only
through an interdisciplinary approach linking ecology, hydrology,
geomorphology, pedology, and hydraulic engineering.

Pautou et al. (1979) have related the distribution of forest
groups in the Rhône Valley to the mean depth of the water table.
Species such as *Salix alba* and *Salix cinerea* are present in areas
having shallow water tables and liable to flooding. On the contrary,
Fraxinus excelsior, *Quercus robur*, and *Populus alba* coincide with
water table depths varying from one to three metres. Depending on
their locations, dikes and dams generally either halt the succession
of riparian vegetation or, on the contrary, encourage the development
of hardwoods. In this regard, Carbiener (1978) mentioned the
precariousness of vegetation groups of the Rhine Valley after
modifications of the hydrological regime by the construction of the
Alsace Canal.

As reported by Décamps (1984), modifications of the riparian
vegetation should be considered as part of the evolution of the
whole valley. The upper Rhône Valley constitutes a dynamic system
rejuvenated every year by the processes of erosion and sedimentation
(Pautou and Bravard 1982). Nowadays, water energy of artificial
alluvial plains is becoming canalized towards electricity production,
drainage, and irrigation. Thus, the interactions between the river
and its valley are completely modified, as is the landscape of the
stream corridor.

Studies on the effect of riparian vegetation on the productivity
of rivers such as the work of Mathews and Kowalczewski (1969) have
to be pursued in the future, as well as work in mineral cycling
(Ulehlova et al. 1976, Mitsch et al. 1979). The influence of the
diversity of side arms and tributaries also must be stressed,
particularly when fish populations are considered (Amoros et al.
1982).

Paradoxically, stream corridors simultaneously shelter alluvial
forests (among the most ancient ecosystems still found in their
natural states), cultivated lands, pasture lands, gravel pits, and
urban zones. This diversity reflects an everincreasing use of
these environments. In fact, each river structures its valley into
a heterogeneous space in which numerous human activities are developed. How do the processes of diversification intervene? What are
the consequences of the resulting diversity at the ecological,
economic, and social levels? Does the knowledge of the fluxes of
matter and energy help to understand the interactions among the
different levels of analysis? The interactions among the various
elements forming the mosaic must be considered, as well as the

Fig. 6.   Some alluvial forests in Europe.

interactions among the levels in which they are included.   These
questions must be answered before a real understanding of the
landscape evolution of stream corridors can be obtained.   In this
perspective, the ecology of alluvial forests constitutes a poorly
explored area.   In Europe, it would be advisable to retain unities
of significant dimensions, among which should figure not only the
forests of the Rhine and of the Danube but also those of other
valleys (Fig. 6) as proposed by Yon and Tendron (1980).

CONCLUSION

     Ward and Stanford (1979) were the first to draw attention to
the exploitation of existing "experiments" to develop a more theoretical approach to river ecology.   The functioning of dams and the
regulation of a water course's flow thus have constituted a useful
source of observation and experimentation.   I suggest that these
perturbations as well as many others, such as water transfers,
forestal exploitations, urbanization of valleys, etc., be used more
widely in the study of the ecology of rivers in their valleys.   The
response of rivers to water transfers at an interregional scale
such as those already achieved in North America and Europe or
projected in the USSR, Mexico, and India (Golubev and Biswas 1979)
are still poorly understood.   Similarly, the transformation of
alluvial forests (as shown in Fig. 4 for European countries or as
has actually been undertaken in African and South American countries)

as well as the consequences are far from being fully known. The
ecological problems of tropical rivers, with their known implications
concerning public health, must be considered at the level of the
entirety of each river and its valley (Dasman et al. 1974).

More comparative studies, such as the one on the Rhine and
Rhône (Golterman 1982), are badly needed. New tools also are
necessary, particularly more complex models analyzing spatial
heterogeneity along water courses, interactions between component
systems of fluvial valleys, and their variations with time. The
use of remote sensing with satellites, having greater spatial
resolution, also has to be investigated.

LITERATURE CITED

Amoros, C., M. Richardot-Coulet, J.L. Reygrobellet, G. Pautou, J.P.
    Bravard, and A.L. Roux. 1982. Cartographie polythématique
    appliquée à la gestion écologique des eaux; étude d'un hydrosys-
    tème fluvial: le Haut-Rhône francais. Cent. Natl. Res. Sci.,
    Paris, France.
Anderson, N.H., and J.R. Sedell. 1979. Detritus processing by
    macroinvertebrates in stream ecosystems. Annu. Rev. Entomol.
    24:351-377.
Bärlocher, F., and B. Kendrick. 1981. The role of aquatic hyphomy-
    cetes in the trophic structure of streams. In D.T. Wicklow
    and G.C. Caroll, eds. The fungal community, its organization
    and role in the ecosystem. Marcel Dekkes, New York, NY USA.
Binder, W., P. Jürging, and J. Karl. 1983. Natural river engineering
    --characteristics and limitations. Garten + Landschaft 2:91-94.
Bird, G.A., and N.K. Kaushik. 1981. Coarse particulate organic
    matter in streams. Pages 41-68 in M.A. Lock and D.D. Williams,
    eds. Perspectives in running water ecology. Plenum, New York,
    NY USA.
Boling, R.H., Jr., E.D. Goodman, J.A. Van Sickle, J.O. Zimmer, K.W.
    Cummins, R.C. Petersen, and S.R. Reice. 1974. Toward a model
    of detritus processing in a woodland stream. Ecology 56:141-151.
Burkham, D.E. 1976. Hydraulic effects of changes in bottomland
    vegetation on three major floods, Gila River in southeastern
    Arizona. U.S. Geol. Surv. Prof. Pap. 665-J:1-14.
Capblancq, J., and H. Décamps. 1978. Dynamics of the phytoplankton
    in the river Lot. Verh. Int. Ver. Limnol. 20:1479-1484.
Carbiener, R. 1978. Un exemple de prairie hygrophile primaire
    juvénile: l'Oenantho Lachenalii Molinictum de la zonation
    d'atterrissement rhénane résultant des endiguements du 19è
    siècle en moyenne Alsace. Pages 13-42 in J.M. Gehu, ed. La
    végétation des praires inondables. Cramer, Lille, France.
Cummins, K.W. 1974. Structure and function of stream ecosytems.
    Bioscience 24: 631-641.

Cummins, K.W. 1979. The natural stream ecosystem. Pages 7-24 in J.V. Ward and J.A. Stanford, eds. The ecology of regulated streams. Plenum, New York, NY USA.

Dance, K.W. 1981. Seasonal aspects of transport of organic and inorganic matter in streams. Pages 69-95 in M.A. Lock and D.D. Williams, eds. Perspectives in running water ecology. Plenum, New York, NY USA.

Dasman, R.F., J.P. Milton, and P.H. Freeman. 1974. Ecological principles for economic development. Wiley, London, England.

Dawson, F.H. 1976. Organic contribution of stream edge forest litter fall to the chalk stream ecosystem. Oikos 27:13-18.

Décamps, H. 1984, in press. Biology of regulated rivers in France. In A. Lillehammer and S.J. Saltveit, eds. Regulated streams. 2nd Int. Symp., Oslo, Norway. Univ. of Oslo Press.

Décamps, H., J. Capblancq, H. Casanova, A. Dauta, H. Laville, and J.N. Tourenq. 1981. Ecologie des rivières et développement: l'expérience d'aménagement de la vallée du Lot. Pages 219-234 in J.C. Lefeuvre, G. Long, and G. Ricou, eds. Ecologie et développement. Cent. Natl. Res. Sci., Paris, France.

Décamps, H., J. Capblancq, H. Casanova, and J.N. Tourenq. 1979. Hydrobiology of some regulated rivers in the southwest of France. Pages 273-288 in J.V. Ward and J.A. Stanford, eds. The ecology of regulated streams. Plenum, New York, NY USA.

Décamps, H., and T. Casanova-Batut. 1978. Les matières en suspension et la turbidité de l'eau dans la rivière Lot. Ann. Limnol. 14:59-84.

Décamps, H., J.C. Massio, and J.C. Darcos. 1976. Variations des teneurs en matières minérales et organiques transportées dans une rivière canalisée, le Lot. Ann. Limnol. 12:215-237.

Drake, C.M., and J.M. Elliott. 1982. A comparative study of three air-lift samplers used for sampling benthic macro-invertebrates in rivers. Freshwater Biol. 12:511-533.

Elliott, J.M., and C.M. Drake. 1981a. A comparative study of seven grabs used for sampling benthic macroinvertebrates in rivers. Freshwater Biol. 11:99-120.

Elliott, J.M., and C.M. Drake. 1981b. A comparative study of four dredges used for sampling benthic macroinvertebrates in rivers. Freshwater Biol. 11:245-261.

Everitt, B.L. 1980. Ecology of saltcedar. A plea for research. Environ. Geol. 3: 77-84.

Fisher, S.G. 1977. Organic matter processing by a stream-segment ecosystem: Fort River, Massachusetts, U.S.A. Int. Rev. Gesamten Hydrobiol. 62:701-727.

Fisher, S.G., and G.E. Likens. 1972. Stream ecosystem: Organic energy budget. Bioscience 22:33-35.

Fisher, S.G., and G.E. Likens. 1973. Energy flow in Bear Brook, New Hampshire: an integrative approach to stream ecosystem metabolism. Ecol. Monogr. 43:421-439.

Fittkau, E.J. 1964. Remarks on limnology of Central-Amazon rain-forest streams. Verh. Int. Ver. Limnol. 15:1092-1096.

Fontaine, T.D., and S.M. Bartell, eds. 1983. Dynamics of lotic
    ecosystems. Ann Arbor Sci., Ann Arbor, MI USA.
Forman, R.T.T. 1984, *in press*. Corridors in a landscape: Their
    ecological structure and function. Ekologia (CSSR).
Forman, R.T.T., and M. Godron. 1981. Patches and structural
    components for a landscape ecology. Bioscience 31:733-740.
Gelroth, J.V., and G.R. Marzolf. 1978. Primary production and
    leaf litter decomposition in natural and channelized portions
    of a Kansas stream. Am. Midl. Nat. 99:238-243.
Golterman, H.L. 1982. La géochimie du Rhin et du Rhône et l'impact
    humain. Hydrobiologia 91:85-91.
Golubev, G.N., and A.K. Biswas. 1979. Interregional water transfers:
    Projects and problems. Water Supply Manage. 2:59-211.
Gosz, J.R., G.E. Likens, and F.H. Bormann. 1973. Nutrient release
    from decomposing leaf and branch litter in the Hubbard Brook
    Forest, New Hampshire. Ecol. Monogr. 43:173-191.
Horton, J.S., and C.J. Campbell. 1974. Management of phreatophyte
    and riparian vegetation for maximum multiple use values. U.S.
    Dep. of Agric. For. Serv. Res. Pap. RM-117:1-23.
Huet, M. 1949. Apercu des relations entre la pente et les popula-
    tions piscicoles des eaux courantes. Schweiz. Z. Hydrol.
    11:333-351.
Hynes, H.B.N. 1963. Imported organic matter and secondary productiv-
    ity in streams. Int. Congr. Zool. 16:324-329.
Hynes, H.B.N. 1970. The ecology of running waters. Univ. of
    Toronto Press, Toronto, Ontario, Canada.
Hynes, H.B.N. 1975. The stream and its valley. Verh. Int. Ver.
    Limnol. 19:1-15.
Illies, J., and L. Botosaneanu. 1963. Problèmes et méthodes de
    las classification et de la zonation écologique des eaux
    courantes, considérées surtout du point de vue faunistique.
    Mitt. Verh. Int. Ver. Theor. Angew. Limnol. 12:1-57.
Jewell, M.E. 1927. Aquatic biology of the prairie. Ecology 7:289-
    298.
Karr, J.R., and I.J. Schlosser. 1978. Water resources and the
    land water interface. Science 201:229-234.
Kaushik, N.K., and H.B.N. Hynes. 1971. The fate of the dead
    leaves that fall into streams. Arch. Hydrobiol. 68:465-515.
Le Cren, E.D., and R.H. Lowe-McConnell. 1980. The functioning of
    freshwater ecosystems. IBP 22. Cambridge Univ. Press, Cambridge,
    England.
Likens, G.E., ed. 1981. Flux of organic carbon by rivers to the
    oceans. U. S. Dep. of Energy CONF - 8009140, Nat. Tech. Inf.
    Serv., Springfield, VA USA.
Lillehammer, A., and S.J. Saltveit, eds. 1984, *in press*. Regulated
    streams. 2nd Int. Symp., Oslo, Norway. Univ. of Oslo Press.
Lock, M.A., and D.D. Williams, eds. 1981. Perspectives in running
    water ecology. Plenum, New York, NY USA.
Macan, T.T. 1974. Running water. Mitt. Int. Ver. Limnol. 20:301-
    321.

Mathews, C.P., and A. Kowalczewsky. 1969. The disappearance of leaf litter and its contribution to production in the River Thames. J. Ecol. 57:543-552.

Melillo, J.M., R.J. Naiman, J.D. Aber, and K.N. Eshelman. 1983. The influence of substrate quality and stream size on wood decomposition dynamics. Oecologia 58:281-285.

Meybeck, M. 1982. Carbon, nitrogen and phosphorus transport by world rivers. Am. J. Sci. 282:401-450.

Minshall, G.W. 1967. Role of allochtonous detritus in the trophic structure of a woodland spring brook community. Ecology 48:139-149.

Mitsch, W.J., C.L. Dorge, and J.R. Wiemhoff. 1979. Ecosystem dynamics and a phosphorus budget of an alluvial cypress swamp in southern Illinois. Ecology 60:1116-1124.

Owens, M., J.H.N. Garland, I.C. Hart, and G. Wood. 1972. Nutrient budgets in rivers. Symp. Zool. Soc. London 29:21-40.

Pautou, G., and J.P. Bravard. 1982. L'incidence des activités humaines sur la dynamique de l'eau et l'évolution de la végétation dans la vallée du Haut-Rhône francais. Rev. Geogr. Lyon 57:63-79.

Pautou, G., J. Girel, B. Lachet, and G. Ain. 1979. Recherches écologiques dans la vallée du Haut-Rhône francais. Doc. Cartogr. Ecol. 22:1-63.

Rzoska, J. 1978. On the nature of rivers. Junk, The Hague, The Netherlands.

Scheurmann, K. 1983. The design of river and stream beds. Garten + Landschaft 2:94-98.

Schlosser, I.J., and J.R. Karr. 1981. Water quality in agricultural watersheds: Impact of riparian vegetation during base flow. Water Res. Bull. 17:233-240.

Thienemann, A. 1912. Der Bergbach des Sauerland. Int. Rev Gesamten Hydrobiol. Hydrogr. Suppl. 4:1-125.

Turner, R.M. 1974. Quantitative and historical evidence of vegetation changes along the Upper Gila River, Arizona. U.S. Geol. Surv. Prof. Pap. 655-H:1-20.

Ulehlova, B., E. Klimo, and J. Jakrlova. 1976. Mineral cycling in alluvial forest and meadow ecosystems in southern Moravia, Czechoslovakia. Int. J. Ecol. Environ. Sci. 2:15-25.

Vannote, R.L., G.W. Minshall, K.W. Cummins, J.R. Sedel. and C.E. Cushing. 1980. The river continuum concept. Can. J. Fish. Aquat. Sci. 37:130-137.

Wallace, J.B., J.R. Webster, and W.R. Woodal. 1977. The role of filter feeders in flowing waters. Arch. Hydrobiol. 79:506-532.

Ward, J.V., and J.A. Stanford, eds. 1979. The ecology of regulated streams. Plenum, New York, NY USA.

Ward, J.V., and J.A. Stanford. 1983. The serial discontinuity concept of lotic ecosystems. Pages 29-42 in T.D. Fontaine and S.M. Bartell, eds. Dynamics of lotic ecosystems. Ann Arbor Sci., Ann Arbor, MI USA.

Warren, D.K., and R.M. Turner. 1975. Saltcedar seed production,
    seedling establishment, and response to inundation. J. Ariz.
    Acad. Sci. 10:131-144.
Webster, J.R., and B.C. Patten. 1979. Effects of watershed perturba-
    tion on stream potassium and calcium dynamics. Ecol. Monogr.
    49:51-72.
Whitton, B.A., ed. 1984, in press. Ecology of European rivers.
    Blackwell, Oxford, England.
Yon, D., and G. Tendron. 1980. Etude sur les forêts alluviales en
    Europe, éléments du patrimoine naturel international. Rap.
    Cons. Eur. 65782:1-68.
Young, S.A., W.P. Kovalak, and K.A. Del Signore. 1978. Distance
    travelled by autumn-shed leaves introduced into a woodland
    stream. Am. Midl. Nat. 100:217-222.
Zhadin, V.I., and S.V. Gerd. 1963. Fauna and flora of the rivers,
    lakes and reservoirs of the U.S.S.R. Israel Prog. Sci. Transl.,
    Jerusalem, Israel.

AND NOW?   ECOSYSTEM RESEARCH!

Hermann Remmert

Fachbereich Biologie
Philipps-Universität
D-3550 Marburg/Lahn, FRG

INTRODUCTION

With Lindeman's great discovery a new epoch began. Ecology,
up to that time regarded as the last nature reserve for lovers of
butterflies and birds, suddenly was able to cope with modern scientific disciplines. Exact measurement and exact quantification of
results were now possible, as well as the inclusion of these results
into one grand framework. So ecology was accepted, especially when
the problems of the ecosphere became obvious. The huge success of
the International Biological Program (IBP) and the popularily of
ecology even among molecular biologists are not the least results
of Lindeman's discovery.

But after the end of the IBP, with so many well produced
analyses of ecosystems available, disillusionment among ecologists
spread worldwide. In fact, what ecosystem research had brought was
by no means ecosystems analysis. There were many more, obviously
decisive parameters in ecosystems which simply had been overlooked.
Energy transfer and flux of matter, studied extensively during the
IBP, are only one (very important) facet of ecosystems.

In the following, I shall try to emphasize some further parameters which in my view are highly important and should be considered
along with energy and matter flow in further studies.

LONG TERM RESEARCH: IS THE OSCILLATING SYSTEM THE "NORMAL" SYSTEM?

Virtually all our ecosystem research has implied that ecosystems
under uniform conditions (disregarding seasonal fluctuations)

179

remain constant over long time spans. But what is meant by constant
and by long? Cycles of many small mammal populations in the northern
hemisphere show that the concept must be qualified. Recently,
moreover, ecological research has provided a great deal of evidence
that wide fluctuations can occur in large systems, even with very
long periodicity (Fig. 1). Because all ecological papers, including
those of the IBP, have considered only short time spans, this fact
is not too well documented. For lack of anything more reliable, I
therefore shall consider a number of speculative suggestions.

In the region of the Neusiedler See, on the border between
Austria and Hungary, there is comparatively little precipitation.
The lake exists only because water flows into it from the relatively
low surrounding mountains, where precipitation is higher. The lake
is surrounded by a broad belt of reeds, which is steadily expanding;
the water is very shallow and thus provides an excellent place for
Phragmites communis to grow. Precipitation in the region of the

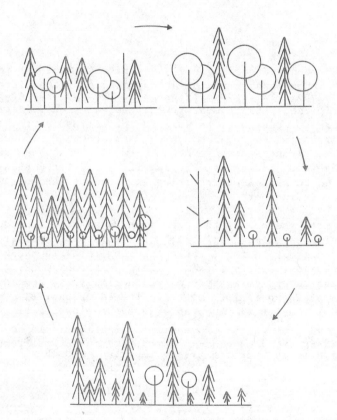

Fig. 1. Cycle of a virgin forest in Austria (combined after different simultaneous stages of succession; severely simplified).

lake is much less than the amount of water consumed by the vegetation. At times in the past, the lake has not existed; towards the end of the last century it was not there. The following hypothesis might be possible: At present the reeds are expanding; after a while they will occupy almost the entire lake and then will use up considerably more water than can be supplied by precipitation and streams; they will pump the lake dry. When that happens, the reeds will die off; the dry basin formed will begin to fill with water that flows into it; lake will be re-formed, and reeds will spread again.

Since this hypothesis was proposed for the Neusiedler See, doubts have been loudly voiced, but even doubters firmly attest to the likelihood of fluctuating lakes in arid regions. The soil in a given ecosystem therefore may be the soil made by this ecosystem or by another very different phase of this ecosystem, and thus one specific phase may dictate the plant composition and, consequently, the flow of energy and matter in another phase.

The situation in the North American taiga is similar. Stands of spruce and pine are intermingled to form a mosaic. Pines grow very rapidly, and allow a great deal of light to reach the ground. Beneath the pines, spruce grow and gradually suppress the pines, as they cast almost unbroken shade on the ground. Eventually the pines are replaced by a very dense stand of spruce. Once this has happened, insect pests can spread and reproduce at a high rate. They can destroy the entire stand of spruce, and pine will grow again in its place. This ecosystem "taiga" could be subdivided into two temporal ecosystems, pine forest and spruce forest. In reality, such a subdivision, usually much more complicated, seems to be possible for all forests. The complicated mosaic of different plant (tree) communities in Itasca State Park (Minnesota, USA) can be explained partly in this way. Beech forests of central Europe give way to birch in their terminal phase, and then to elm, ash, and cherry, which eventually are replaced again by beech. The same holds for pine and spruce forests in central Europe. The rapid decline of many animals such as black grouse or capercallie is due partly to the fact that these "in-system successions" are now prohibited by man.

The most spectacular, but often overlooked, example is the lack of natural regrowth in many Sequoia forests of California. Such a natural regrowth had stopped before the arrival of the white man. Very probably, even in these forests with very old trees, a comparable cycle of different species composition can be postulated.

As indicated in the examples of capercallie and black grouse, these in-system successions exert very great influence on the animals of the system and, of course, on the soil processes, which are extremely different under the canopies of different tree species. As different needles and leaves have very different characteristics

of decomposition, and as different densities of trees lead to very
different groundwater levels, we can postulate a cycle of groundwater
level in synchrony with the stage of in-system succession. The
large herbivores of the steppe of east Africa possibly overexploit
their habitat. In the long run, such overusage kills the essential
food plants. The animals, in turn, will die out and the plants can
re-establish themselves. Petrides (1974) has suggested that there
is a very long cycle in which grass steppe (with its typical large
mammals) alternates with thornbush steppe (with other large mammals).
Similar relationships may be found in the humid regions (oases) in
central Iceland, where rapid reproduction of cotton grass (*Eriophorum*)
accompanied by whooper swans may alternate in a regular cycle with
sedge and pinkfooted geese (A. Gardasson, personal communication
cited in Remmert 1980b). There may be approximately 70 year cycles
of reindeer in Spitzbergen, involving all the lichens, grasses,
herbs, and forbs, as well as the soil formation.

Very probably these views hold also for limnic and marine
systems. From the 60- to 100-year-old studies of benthos communities
in the North Sea, it has been possible to demonstrate that in the
German Wadden Sea area a community once dominated by bivalves is
now dominated by polychaetes (Reise 1982). Even after a rather
short period, Gerlach (1981) was able to demonstrate dramatic
changes in the bottom fauna of the North Sea. The famous outbreak
of starfish on the Great Barrier Reef leads to similar assumptions.

Of course, all these views are not new. They have been demonstrated by Richards (1952) for tropical rain forests and by Reichle
(1973) for temperate forests, as well as by Ellenberg (1978) for
many European vegetations type, *Calluna* being the best known example.
But in our great ecosystem studies, these views have not found
their place; there is no real approach demonstrating the dynamics
of these systems. In fact, there seems to be no study demonstrating
an ecosystem at equilibrium. These views have to be incorporated
into any ecosystem analysis.

Almost all these examples depend on much speculation. But we
need more information about this phenomenon. Might it be that such
an "in-system oscillation" is the very clue to constancy, stability,
and resilience? Of course, nobody knows at the moment, because the
time required for such studies is so long. Our perceptions from
the IBP work might be wrong, or at least give a wrong impression of
the systems, if we do not develop long term programs.

Moreover, animals, which appear to have little influence on
their systems and which consume only about 10% of the primary
production, may have an enormous impact if we look at them with a
different attitude. They may be responsible for very long oscillations and thus for constancy and stability.

Just as humankind has developed high frequency technology (instead of weak direct current technology) to avoid disturbances, and just as animals have evolved digitalization of signals in neurons which are not easily disturbed (instead of the analog signal of the receptor membrane), in-system cycles might be the method by which ecosystems maintain resilience to outside disturbances.

ANIMALS ARE AMPLIFIERS

Lindeman's recipe was to study the flow of matter and energy through the system. If we compare this method with modern physiology, it is as if we had studied feeding, digestion, and defecation of an animal, claiming afterwards we had made an analysis of the animal, that we understood the animal. Nothing would be farther from the truth, or more ridiculous. Even if we had examined everything correctly, we would have overlooked one most important system of the animal--that of receiving and processing information and reacting to the processed information. We would have overlooked the hormonal system, the nervous system, and the sensory system. A sense organ such as the eye causes negligible amounts of energy and matter to flow through the organism, but it is responsible for flight, for avoiding obstacles, for finding food--it determines the very existence of the whole organism. It works like a switch, or better, like an amplifier in an electronic system. Is it possible to compare ecosystems and ecosystem research to physiology and physiological systems? I think so. In the following, I will give examples for my view and will try to demonstrate the importance of physiological studies to further ecological research.

The highly diverse flora and fauna of tropical forests and savannas have been studied intensely over the last decades. I have only to recall some of the volumes of *Ecological Studies*. But the great diversity of these systems has, in fact, been neglected in all these studies. At most we have a list of species, a sentence on their remarkable diversity. We have nothing on the mechanisms that are responsible for the evolution and the maintenance of this diversity, nor of its significance to the whole system. One complex related to diversity should be stated here: the complex of pollination and seed dispersal by animals, which is responsible for the maintenance of diversity. What is animal pollination? It is virtually the same as a sensory organ of an animal: with a minimum of energy and matter transfer, animal pollination, together with seed dispersal, maintains the diversity of the system (as was shown by Regal 1977). Comprehensive papers and volumes on the tropical forests have completely omitted this key factor only because it did not fit into the Lindeman concept. Without animal pollination and seed dispersal, many of our systems would collapse. Yet most ecosystem analyses do not regard pollination and seed dispersal as

a necessary part of ecosystem research: anyone who has tried to
introduce such a study into a highly integrated program of ecosystem
research knows the troubles he gets from it. The result is that we
do not know very much about pollination and seed dispersal by
animals. We know a lot about nectar production, we know a lot
about animals responsible for pollination and seed dispersal, we
know a lot about mechanisms of plants to produce seeds when there
are no animals to pollinate. But we do not know very much about
the investments of plants into their signals to attract pollinators,
we do not know very much about the distances traveled by pollinators
(only a few isolated hints demonstrate that tropical bees easily
travel more than 23 km from their home bases to pollen sources),
and we know almost nothing about the necessity plants have for
pollinators. Almost all plants seem to be capable of self fertilization if there are no pollinators, and thus even the necessity of
pollinators is often discussed. There are virtually no quantitative
data--rape is known to do well by self fertilization, but the
yield rises by about one-fourth when pollinators are present--and
nobody seems to have exact data on how pollination is necessary
genetically.

This is only one example, and I could have demonstrated the
same with equal importance and equal ignorance for the Scandinavian
fjells, the Negev desert, the North American prairies, or the
German mixed mountain forests.

But rather let us proceed to another complex, which everyone
knows as well: the significance of the beaver for the ecosystems
of the northern hemisphere. Virtually all the discussions, papers,
and books on climax vegetation and on the potential natural vegetation of central Europe are wrong, as we can see in Canada or in the
big U.S. national parks. An almost flat area with winding streams
and completely uniform geological bedrock can be changed into a
remarkably diverse mosaic of different ecosystems by a beaver
population (Fig. 2). With beaver dams, ponds alternate with meadows,
and the soil formation in the pond areas and on the pond bottoms is
completely different from the wooded areas (more active humus
formation). After a dike is broken, the ponds become open meadows,
then gradually the forest returns, first with *Alnus* and poplar,
followed by hardwood and (or) pine or spruce. Thus, the landscape
becomes, under the influence of beavers, a highly diverse system
with ponds, meadows, streams, and both young and ripe old forests
of different species. The story of the uniform forest covering
ancient central Europe is a fairy tale. Only in the mosaic of
ponds, meadows, and forests have the European animals found a home.
Capercallie and black grouse need open areas with fens as well as
spruce, pine, birch, and willow. The modern view of all these
systems as different plant communities on different soils is correct,
but all these plant communities develop on the same place in an
extremely rapidly changing mosaic determined by one single animal,
the beaver.

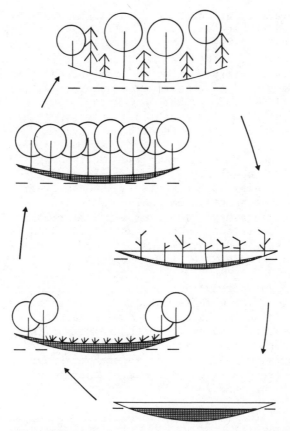

Fig. 2.   Cycle of a beaver influenced area: beaver lake (middle and
          bottom illustrations) with growing layer of organic material;
          beaver meadow (second from bottom) with humus layer from
          converted lake bottom layer with extremely magnified nitro-
          gen fixation and with incoming new trees (second from top);
          original forest (top) without humus layer.   The dashed line
          marks the groundwater level.

     The huge mass of energy transformed and matter transported in
such a system renders the small number of beavers negligible.
There may be about 20 beavers along 10 km of a large stream.   We
may well describe the beaver as a "switch" in our system; a switch
decides which plant community exists and for how long.

     A further example occurs with the reed *Phragmites australis* (=
communis).   The reed reacts to heavy predation pressure by cater-
pillars of Archanara (Lepidoptera) by producing many very thin

shoots which are insufficient for the Archanara (the caterpillar feeds inside the shoots). A mosaic of stands with few strong and thick shoots per unit area (with Archanara) and stands with many thin shoots (without Anchanara) results in areas where the reeds grow on land (the caterpillar moves from one shoot to the other on the ground; therefore, Archanara is missing if *Phragmites* grows in water) (Vogel 1984). The question remains: Why is there a regular, large scale synchronous fluctuation in the larch-larch budworm system and a mosaic in the reed-Archanara system?

There are many similar examples. Mattson and Addy (1975) have added many ideas (a summary is given by Remmert 1980a). The view of the slight significance of animals in ecosystems, as indicated by almost all modern ecosystem analysis, is wrong. The dung beetles introduction to Australia, for example, dramatically demonstrates the enormous significance of that species in savannas. But I do not know of a single ecosystem analysis of steppe and savanna areas that gives dung beetles due attention.

In many plant-herbivore systems, it has been shown that plants produce repellents or more resistant cell walls after attacks by animals. The animal withdraws and attacks another plant. Many plant-herbivore cycles seem to be the result of these induced changes in plants. The animal sets a signal, which induces the plant to react: the quality, not so much the quantity of the attack, is decisive. With a minimum of energy and matter flow, important results are induced.

The energy transfer may be lower still: some insects, such as Danaid butterfly males, need pyrrolizidin alkaloids as precursors of the pheromones with which they attract females. They either leach these from dead, dry plants or they scratch with their tarsae on the leaves of special plants so that a few small droplets of plant fluid become available. This absolutely negligible transfer of energy from very special plants results in possibly serious mass outbreaks of insect pests (Boppré 1983, Boppré and Schneider 1982).

The reindeer introduced to South Georgia eat only a negligible proportion of the aboveground primary production, but in doing so they have destroyed the natural vegetation of vast areas (Leader-Williams 1980, Vogel 1984). They eat the shoots during winter and spring, which, although quantitatively little, kills the plants. By the same mechanism, roe deer and red deer in European forests cause vast damage. In a normal energy flow diagram, this effect is not demonstrable; it is possible, however, to obscure the fact that this effect is not only energy flow, but also amplification.

These are all very old facts. What we need in ecosystem research is to include them along with the systematic search for other such phenomena. A very small cause has very great effects, and the great ecosystem research programs have neglected rather than included these amplifications into their programs. Moreover, what holds for animals probably holds for fungi in the soil as well.

In aquatic systems, ecosystem research and modeling are much older than in terrestrial systems. Over the years, aquatic ecosystem research has reached a degree of refinement of methods and clear results that has not been achieved in terrestrial ecology. This success often leads to the assumption that amplifier functions of aquatic organisms are missing, which is highly unlikely. Because aquatic systems are so much older than terrestrial systems, coevolution should have proceeded much farther, and as a result of coevolution such amplifier functions should have been achieved. Probably the impossibility of close observation under water has led to these assumptions; modern methodology should make such observations possible in the future. The existence of cleaner fish, the famous story of the bitterling and the clams in freshwater, and the recent evolution of marine chemical ecology make the existence of phenomena comparable to those in terrestrial ecosystems highly probable.

None of our great ecosystem analyses gives an answer when the fate of the Amazonian rain forests is discussed, but many tree species need seed transport by special fish, and the seed need some time in the fish gut. Other seeds even seem to need the jaguar for seed dispersal. In central Europe, we exert huge efforts to preserve eagles (*Aquila pomarina, A. clanga*), ospreys (*Pandion haliaetos*), capercallie, and black grouse. But they need rather fast changing terrestrial systems, a mosaic of successions. Herons and cormorants preferably nest in huge, dead trees which are found along winding rivers and in rather young beaver ponds. In nesting in dense forests, they are only seeking a bad refuge. None of our great ecosystem studies gives any help in nature preservation, because this view does not fit into the Linderman recipe.

The famous ecosystem analyses do not help us very much in our current environmental crisis or in predicting constancy, stability, or resilience of a system. This was not foreseeable, but now we do see it and it would be unwise not to say so. What we see now is that in many systems switches and amplifiers play a significant role and their destruction may really destroy our systems. What we do not see is how to find, without investigating the 2 million different animals and plants (also bacteria and fungi--to mention only mycorrhizae), the really important ones. And what is important? Is systematic search for the most important organisms possible?

It is impossible now to say how much primary production animals consume. We know fairly well that grazing by animals makes a high primary production possible; an extreme example is shown in the Nakuru lake in Kenya, as was demonstrated by Vareschi (1977, 1978,

1979). This is also true for terrestrial plants; even they increase their production under grazing stress (see, e.g., Fig. 12/2 in Harper 1977). In reality, as all the studies of predator-prey systems have shown, the predator guild keeps the standing stock of the prey at a low level, that is, in the log phase of reproduction (high production). We know quite a lot about all this in anecdotical form. What we need is a combination of all these things--long term research, Lindeman mode research, and amplifier research--in one single or in several selected systems.

PHYSIOLOGICAL ECOLOGY

Knowledge of the physiological ecology of the different members of the system is necessary. We are extremely poor at predicting the future of our systems. Why did the English sparrow and European starling populations explode under the conditions of North America, but why did no animal or plant from the New World show comparable growth in Europe? We claim that many animals and plants became extinct in Europe during the Pleistocene because of the different topography of the two continents (east-west mountains in Europe versus north-south in America). This means that the European fauna and flora would be impoverished and that extinct species would flourish if repatriated; but nothing of the kind has happened. Instead, the Monterey pine (*Pinus radiata*), a very weak competitor restricted to a minute area in California, now outcompetes all the eucalypts in eastern Australia (Chilvers and Burdon 1983). We should be able to predict. A thorough investigation concerning the physiological basis of ecological adaptation is necessary, as has been demonstrated by Evenari (1982) for the Negev desert. An inclusion of the physiological and biochemical adaptations necessary and possible in any ecosystem is absolutely essential for any prediction (Hochachka and Somero 1973, Townsend and Calow 1981, Krebs and Davies 1978, Remmert 1980a, 1980b). Many approaches to ecosystems have no predictive value, because the authors did not include autecological and physiological considerations into their concepts. In the future, this should no longer be possible.

CONCLUSION

I have often advanced these views and know two questions that always arise.

(1) *So you want to revert from quantifying to simple natural history, you want to go back about 100 years, neglecting the achievements of modern ecology?*

Knowing an animal properly can be hard science, and we can understand our ecosystems better if we pursue this approach with the most modern methods now available. The old method of observing animals halted, because the methodology available at that time made further insights impossible. With modern methods in the field (e.g., telemetry, infrared measurement of body temperature even in insects), many exciting insights are possible. Moreover, ecology in developing countries can do much additional and more valuable work.

(2) *How do you intend to quantify your results, that is, quantify the significance of your animal to a system? We have to quantify, as matter transfer and energy transport have demonstrated. How do you attain the great and grand framework in which matter and energy transfer operate by pursuing different approaches for one single ecosystem?*

Physiologists do not analyze an organism by one single method. No one can quantify the significance of eyes for a given animal, and yet sensory physiology probably is the premier topic in animal physiology all over the world today. Quantification has to be done here in a different way. We can quantify the stimuli that are necessary for such an amplifier, and we can quantify how much amplification results from one single stimulus.

Receptor physiology studies the reaction of a receptor organ and of an organism to a signal. This signal is qualitatively defined, not quantitatively defined like energy or matter flow. In ecology this has not been fully understood, and the result is the attempt to measure the reactions of organisms to signals as if they were quantitatively defined. This approach simply is wrong. Pollination should be regarded as transfer of information, as a qualitative signal, and not as transfer of energy. The same holds for animals feeding on plants: they do not simply transfer energy, but they give a signal to the plant, to which the plant responds by changing its chemical composition and its structure in a fairly large area around the wound. The result is a cycling system, as in the larch-budworm moth system (Fischlin 1982).

The neglecting of sensory physiology in ecology has led to another problem. The optimal foraging theory says that the organism that selects optimal food with a minimum of energy is superior to other organisms. But, as the success of human beings (and many other examples) demonstrates, very often the "brainiest" can outcompete the most economic species in the struggle for existence.

We should give up the approach of trying one single mathematical formula for the description of an ecosystem with all its possible reactions. We should learn from physiology that one single model for any living system, organism, or ecosystem is absurd. We should realize that the way we went for a few decades was good, important, and necessary, but was only one of the many ways needed to understand

ecosystems. We can obtain results only by accepting that this one
way now has been exhausted.

Of course, much more also remains to be done. But the time of
fast progress, of striking new results, is over. It is time to be
open to further developments. We should accept that the most
complex things on our earth, ecosystems, were not analyzed fully by
the short term International Biological Program. Long term studies
are necessary to make understanding and prediction possible.

REFERENCES CITED

Boppré, M. 1983. Scratching up leaves--a strategy to gather
     secondary plant substances in danaine butterflies (Lepidoptera:
     Danainae). Oecologia 59:414-416.
Boppré, M., and D. Schneider. 1982. Insects and prolizidine
     alkaloids. Pages 373-374 in Proc. 5th Symp. Insect-plants-
     relationships. Pudoc, Wageningen, The Netherlands.
Chilvers, G.A., and J.J. Burdon. 1983. Further studies on a
     native Australian eucalypt forest invaded by exotic pines.
     Oecologia 59:239-245.
Ellenberg, H. 1978. Vegetation Mitteleuropas mit den Alpen.
     Ulmer, Stuttgart, W. Germany.
Evenari, M. 1982. Ökologisch-landwirtschaftliche Forschungen im
     Negev. Tech. Hochsch., Darmstadt, W. Germany.
Fischlin, A. 1982. Analyse eines Wald-Insekten-Systems: Der
     Subalpine Lärchen-Arvenwald und der graue Lärchenwickler.
     Ph. D. Dissert., E.T.H., Zurich, Switzerland.
Gerlach, S.A. 1981. Marine pollution. Springer, Berlin, W.
     Germany.
Harper, J.L. 1977. Population biology of plants. Academic Press,
     London, England.
Hochachka, P.W., and G.N. Somero. 1973. Strategies of biochemical
     adaptation. Saunders, Philadelphia, PA USA.
Krebs, J.R., and N.B. Davies. 1978. Behavioural ecology--an
     evolutionary approach. Blackwell, Oxford, England.
Leader-Williams, D. 1980. Population dynamics and mortality of
     reindeer introduced to South Georgia. J. Wildl. Manage.
     44:640-657.
Mattson, W.J., and N.D. Addy. 1975. Phytophagous insects as
     regulators of forest primary production. Science 190:515-522.
Petrides, C.A. 1974. The overgrazing cycle as a characteristic of
     tropical savannas and grasslands in Africa. Proc. 1st Int.
     Congr. Ecol., Pudoc, Wageningen, The Netherlands.
Regal, P. 1977. Ecology and evolution of flowering plants dominance.
     Science 196:622-629.
Reichle, D.E. 1973. Analysis of temperate forest ecosystems.
     Springer, Berlin, W. Germany.

Reise, K. 1982. Long-term changes in the macrobenthic invertebrate
    fauna of the Wadden Sea: Are polychaetes about to take over?
    Neth. J. Sea Res. 16:29-36.
Remmert, H. 1980a. Ecology. Springer, Berlin, W. Germany.
Remmert, H. 1980b. Arctic animal ecology. Springer, Berlin, W.
    Germany.
Richards, P.W. 1952. The tropical rain forest: An ecological
    study. Cambridge Univ. Press, Cambridge, England.
Townsend, C.R., and P. Calow. 1981. Physiological ecology: An
    evolutionary approach to resource use. Blackwell, Oxford,
    England.
Vareschi, E. 1977. Biomasse und FreBrate der Zwergflamingos in
    Lake Nakuru (Kenia). Verh. Dtsch. Zool. Ges. 70:247.
Vareschi, E. 1978. The ecology of Lake Nakuru (Kenya). I. Abundance
    and feeding of the lesser flamingo. Oecologia 32:1-10.
Vareschi, E. 1979. The ecology of Lake Nakura (Kenya). II.
    Biomass and spatial distribution of fish. Oecologia 37:321-335.
Vogel, M. 1984, *in press*. Introduced reindeer and their effects
    on the vegetation and the epigeic invertebrate fauna of South
    Georgia. Oecologia.

NUTRIENT IMMOBILIZATION IN DECAYING LITTER: AN EXAMPLE OF CARBON-NUTRIENT INTERACTIONS

Jerry M. Melillo

The Ecosystems Center
Marine Biological Laboratory
Woods Hole, Massachusetts 02543 USA

John D. Aber

Department of Forestry
University of Wisconsin
Madison, Wisconsin 53703 USA

ABSTRACT

Carbon-nutrient interactions are identified as important topics for ecological research in the 1980s. The linkages between carbon and nutrients during plant litter decay are the focus of this paper. A recently developed model for estimating the maximum net immobilization potential in decomposing litter is reviewed. The factors that control the magnitude of nitrogen immobilization are considered using the model. Key research questions about carbon-nitrogen interactions during plant litter decay are identified.

INTRODUCTION

During the past two decades, there has been impressive progress in the study of whole ecosystems. Research at the ecosystem level has had three main themes: development of organic matter and nutrient budgets for "reference" ecosystems; experimental manipulation of ecosystems to determine the effects of disturbance (e.g., clearcutting, fire, eutrophication) on nutrient cycling patterns; and detailed process studies designed to elucidate factors that control carbon and nutrient dynamics within and among ecosystems. As ecosystem research proceeds, we are beginning to recognize that

an understanding of the mechanisms that control the flux rates of
an element requires consideration of element interactions. The
study of element interactions is certain to be a major theme of
ecological research in the next decade.

   Element interactions fall into two general categories: carbon-
nutrient interactions and nutrient-nutrient interactions. An
example of a carbon-nutrient interaction is the stimulation of net
primary productivity in a forest by added nitrogen (Miller and
Miller 1976). And an example of a nutrient-nutrient interaction is
the stimulation of nitrogen fixation by the addition of phosphorus
(Griffith 1978). Carbon-nutrient interactions are the central
concern of this paper.

   The way in which nutrient cycles are linked to the magnitude
of net carbon fixation, the allocation of fixed carbon, and the
amount of net carbon storage in ecosystems is particularly interesting
to us. Based on a review of the literature, we developed a simple
conceptual model to identify these carbon-nutrient linkages (Fig. 1).
We envision a series of multiple-setting switches in the carbon
flow pathway that are, in part, under the control of the nutrient
status of plants. In turn, the nutrient status of plants is related
to nutrient availability in the soil.

   The depth of our knowledge about the various carbon-nutrient
linkages along the carbon flow pathway is uneven. For the autotrophic
component of ecosystems, plant physiologists have shown that there
is a close relationship between the nutrient (e.g., nitrogen)
status of a plant and its net photosynthesis (Natr 1975). However,
we know little about how the nutrient status of a plant influences
the allocation of fixed carbon. More specifically, we do not have
answers to the following important questions: Does the nutrient
status of plants influence the relative allocation of carbon between
perennial and deciduous plant parts? Is the allocation of fixed
carbon between leaves (needles) and fine roots (and associated
mycorrhizae) controlled by plant nutrient status? If nitrogen is
an important controller of carbon allocation, does the form of
available nitrogen (ammonium vs. nitrate) influence the allocation
pattern? Can we manipulate the magnitude and site (plant vs. soil)
of net ecosystem production by altering nutrient availability? How
do carbon and nutrient allocation patterns influence future nutrient
availability at a site?

   Carbon-nutrient interactions have important implications for
decomposition processes and thus nutrient availability. We know
that the initial litter quality (i.e., carbon-nutrient ratios and
the relative mix of carbon compounds) influences decay rates and
nutrient dynamics of decomposing litter. We are now at the point
of exploring, in detail, the various mechanisms through which these
influences are exerted.

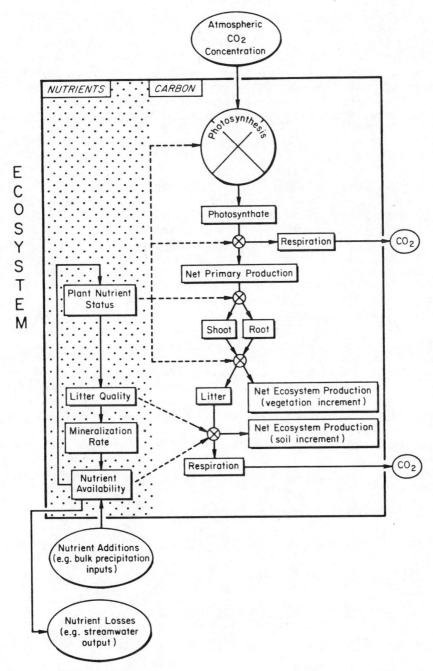

Fig. 1.   Linkages between nutrient cycles and the magnitude of net
          carbon fixation by plants, the allocation of fixed carbon
          within plants, and the net carbon storage in ecosystems
          (after Melillo and Gosz 1983).

An important aspect of decomposition is that nutrients bound
in dead organic matter are ultimately released in an inorganic
form, available for uptake by plants. For example, most of the
nitrogen transferred from the vegetation to the soil in plant
litter is bound in organic molecules such as amino acids, proteins,
coenzymes, nucleic acids, and chlorophyll. Through the decay
process this organically bound nitrogen is released as ammonium, an
organic form of nitrogen that can be taken up by plants.

Three distinct phases often can be recognized in the nitrogen
dynamics of decomposing plant litter (Fig. 2). First, there is a
brief period of leaching during which as much as 25% of the litter's
initial nitrogen mass can be lost. The leaching phase is followed
by a period of net immobilization during which the absolute amount
of nitrogen in the decomposing litter can increase to more than
twice its initial amount. And finally, there is a net release or
net mineralization of nitrogen from the decomposing litter. This
three phase model also can be used in many instances to describe
phosphorus dynamics during litter decay.

The absolute increases of nitrogen (N) and phosphorus (P)
observed during the immobilization phase of decay require the
addition of these nutrients to the decomposing tissue from the
surrounding environment. Decomposing plant litter in the immobilization phase can therefore reduce the amount of nitrogen and phosphorus available to support plant growth. Agriculturalists and foresters
have long recognized the potential for competition for nitrogen
between decomposers and primary producers. For example, Viljoen
and Fred (1924) observed that the incorporation of woody material
into agricultural soils reduced crop (oat) growth rate, and they
concluded that nitrogen immobilization by microbes was the mechanism

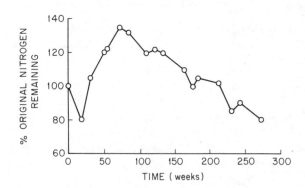

Fig. 2.   Percentage of original nitrogen mass in Scots pine needle
          litter remaining as a function of time. Data from a decom-
          position study conducted in a Scots pine forest in central
          Sweden (Staaf and Berg 1982).

responsible for the reduction. Recently, Thayer (1974) suggested
that the annual cycle of nutrient (N and P) concentrations in a
North Carolina estuarine system regulated phytoplankton production
and that the variation in nutrient concentrations resulted, in
part, from shifts in the equilibrium between microbial immobilization
and re-mineralization associated with the decay of marsh macrophytes.

Some early work was done toward the quantification of the
maximum nitrogen immobilization potential (Nmax) of specific litter
materials and attempts were made to link Nmax to initial litter
quality (Richards and Norman 1931). Through a series of decomposition
studies in laboratory microcosms, Nmax values were established for
some common agricultural waste materials (e.g., Hutchinson and
Richards 1921, Richards and Norman 1931, Acharya 1935). But the
interpretation of the results of these experiments was sometimes
confused, either because the experiments were not run long enough
or because the researchers did not fully understand the relationship
between exogenous nutrient supply and Nmax. No clear link between
Nmax and initial litter quality was established (Richards and
Norman 1931). Recently, Aber and Melillo (1982a) developed a
method of calculating Nmax and they have shown (Aber and Melillo
1982a, Melillo et al. 1983) that, for a given set of environmental
conditions, Nmax can be linked to initial litter quality.

In this paper, we will (1) describe our method of calculating
Nmax, (2) discuss the factors that control the magnitude of nitrogen
immobilization during litter decay, (3) demonstrate the applicability
of our approach for analyzing phosphorus and sulfur (S) immobilization
during litter decay, (4) explore various applications of our analytical approach, and (5) define areas of new research that will give
us a better understanding of the nutrient immobilization during
litter decay.

A METHOD FOR ESTIMATING MAXIMUM NITROGEN IMMOBILIZATION

Our method of calculating maximum net nitrogen immobilization
(Nmax) for a litter material during decay is based on the following
model:

The organic matter and nitrogen dynamics of a single type of
decomposing litter can be described as a simple inverse linear
function. This function is derived by plotting the percent
organic matter remaining on the y axis and the nitrogen concentration in the remaining material on the x axis.

The relationship is described as follows:

$$OM_t = Intercept + (Slope \cdot N_t) \qquad\qquad (1)$$

where OMt = % original organic matter remaining at time t, and Nt = the nitrogen concentration in the remaining tissue at time t. To demonstrate this relationship we will use data from a study of the decay of red pine needles conducted at the Harvard Forest in Petersham, Massachusetts (Melillo and Aber, personal communication). In this study, litter bags containing fallen red pine needles were placed on the forest floor, subsets of bags collected periodically over three years, and the materials analyzed for mass loss and nitrogen content after each collection. Table 1 contains the data for % original mass remaining and % N in the remaining tissue. These data are plotted against time in Figures 3A and 3B. The more or less continual decline in litter mass represents both physical losses of material (i.e., leaching losses) and losses due to microbial catabolism. The increase in nitrogen concentration with time is a consequence of two factors: (1) more rapid carbon loss than nitrogen loss, and (2) an absolute accretion of nitrogen from the surrounding environment (i.e., immobilization). In Figure 3C, we have omitted time and plotted % original mass remaining against the % N in the remaining material (litter-microbe-microbial products complex). The relationship is best described by an inverse linear regression model with a coefficient of determination (r^2) of 0.93.

We have found the inverse linear function to be useful for describing the organic matter and nitrogen dynamics of a wide range

Table 1. Changes in mass and nitrogen concentration in decomposing red pine needles (Melillo and Aber, personal communication).

Time (mo)	Organic matter remaining (%)	Nitrogen remaining (%)
0	100	0.43
2	92	0.46
5	96	0.45
7	85	0.51
9	78	0.59
11	72	0.71
13	68	0.77
15	63	0.72
17	59	0.78
20	67	0.85
22	63	0.96
24	49	1.09
29	47	1.13
36	32	1.30

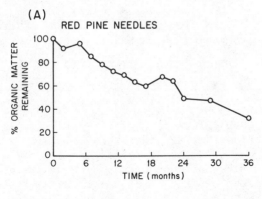

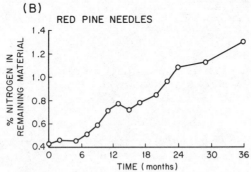

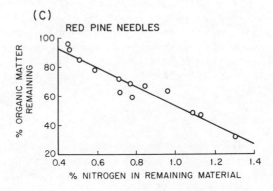

Fig. 3.   (A) Percentage of original red pine needle litter mass
         remaining as a function of time (mo).   Data from a decompo-
         sition study conducted in a red pine plantation at the Har-
         vard Forest, northcentral Massachusetts (Melillo and Aber,
         personal communication).   (B) Nitrogen concentration in
         remaining litter as a function of time (mo).   (C) Percentage
         of original litter mass remaining expressed as a function
         of the nitrogen concentration in the remaining material.

of litter types decomposing under a variety of environments, including cotton grass blades decomposing in the Alaskan tundra (Fig. 4A),
birch wood chips decomposing in a small Quebec stream (Fig. 4B),
and flax straw decomposing in laboratory microcosms at Rothamsted
(Fig. 4C).

Calculating Nmax Using Information from the Inverse Linear Regression Model

The slope and y intercept of the inverse linear function
describing the organic matter and nitrogen dynamics of a given
litter material during decay can be manipulated in a series of
equations to yield an estimate of the maximum amount of exogenous
nitrogen immobilized (Nmax). The general form of the equation for
calculating Nmax is

$$\text{Nmax} = (\frac{\text{Intercept}^2}{-4 \text{ Slope}} - 100 \text{ No}) \cdot 0.1 \tag{2}$$

where No = the initial nitrogen concentration of the material and
Nmax is expressed as mg N immobilized/g initial material. Details
of the derivation of the equation are given by Aber and Melillo
(1982a).

A comparison of observed versus estimated Nmax values for five
plant materials decomposing in laboratory microcosms at Rothamsted
(Richards and Norman 1931) appears in Table 2. The relative order
of Nmax values among materials estimated using Equation 2 is the
same as the relative order of observed Nmax values (highest Nmax in
willow peel, lowest Nmax in retted flax straw). For all materials
except willow peel, the observed Nmax values are very close to the
estimated ones. For willow peel, we think that the experiment was
terminated before Nmax was reached, thus causing the observed value
to be lower than the estimated value.

The Nitrogen Equivalent

Using the simple inverse linear regression model, we can also
calculate the "nitrogen equivalent" (Richards and Norman 1931);
that is, the amount of nitrogen immobilized per unit of litter mass
lost. The equation for calculating the nitrogen equivalent (Neq)
is

$$\text{Neq} = \frac{(\frac{\text{Intercept}^2}{-4 \text{ Slope}} - 100 \text{ No}) \cdot 0.1}{100 - (\text{Intercept} \cdot 0.5) \cdot 10} \tag{3}$$

where Neq is expressed as mg N immobilized/mg OM lost. We are
using this equation to help us to understand the mechanism(s) of

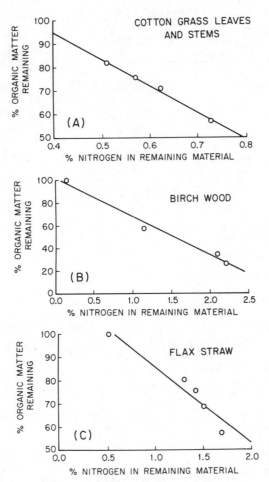

Fig. 4. (A) Percentage of original cotton grass leaf litter mass remaining expressed as a function of the nitrogen concentration in the remaining material. Data from a decomposition study conducted in tussock tundra in northcentral Alaska (personal communication, G. Shaver, Ecosystem Center, Marine Biological Laboratory, Woods Hole, MA USA). (B) Percentage of original birch wood chip material remaining expressed as a function of the nitrogen concentration in the remaining material. Data from a decomposition study conducted in a first order stream in Quebec, Canada (Melillo et al. 1983). (C) Percentage of original flax straw remaining expressed as a function of the nitrogen concentration in the remaining material. Data from a decomposition study conducted in laboratory microcosms at the Fermentation Department, Rothamsted Experimental Station, Harpenden, Herts, England (Richards and Norman 1931).

Table 2.   Observed and estimated maximum net nitrogen immobilization
(mg N/g initial material) for five decomposing plant
materials (Richards and Norman 1981).

Material	Nmax observed	Nmax estimated
	(mg N/g initial material)	
Flax straw (normal)	5.7	5.4
Flax straw (retted)	1.2	1.8
Flax fiber	3.4	4.0
Willow peel	10.3	16.9
Oat straw	8.0	8.2

exogenous nitrogen accumulation in decomposing tissue. Initially,
it was thought that the absolute increase in nitrogen was due to
microbial growth; however, recent evidence from direct counts of
microbes on decaying litter suggests that the fungi and bacteria
account for only a few percent of the increase in nitrogen. Now,
several research groups have proposed the hypothesis that nitrogen
accumulation occurs when "reactive" compounds, produced as microbial
enzymes deploymerize the detritus substrate, condense with nitrogen
containing compounds.

Duration of the Immobilization Phase

The time required to reach Nmax can be calculated knowing the
decay rate (k) of the litter and the amount of organic matter
remaining at Nmax. A number of models have been developed and can
be used to estimate decay rates (Swift et al. 1979). One of the
most frequently used models is the exponential decay model of Jenny
et al. (1949):

$$\ln\left(\frac{OMt}{OMo}\right) = -kt \tag{4}$$

where OMt and OMo = the masses at time t and time 0, respectively.
Aber and Melillo (1982a) have shown that the amount of organic
matter remaining at Nmax can be calculated as follows:

$$OM \text{ at } Nmax = Intercept \cdot 0.5 \tag{5}$$

where Intercept = the y intercept of the inverse linear regression
model of the organic matter and nitrogen dynamics during decay of a
given material. Substituting Equation 5 into Equation 4 yields

$$\ln((\text{Intercept} \cdot 0.5) \cdot 0.01) = -kt \tag{6}$$

and we can solve for t, the time required to reach Nmax.

% OM Remaining vs. % P or % S

The relationship between % organic matter remaining and % P or
% S in the remaining tissue also appears to conform to the inverse
linear regression model. Figure 5 shows the relationship for
phosphorus in *Typha latafolia* leaves decomposing in the laboratory.
And Figure 6 shows the relationship for sulphur in Scots pine
needles decomposing in the field. The inverse linear model apparently
describes the dynamics of nutrients (N, P, S) that are critical to
microbial activity and that are often in short supply relative to
demand.

FACTORS CONTROLLING NUTRIENT IMMOBILIZATION

From a series of experiments conducted by us and our analysis
of the literature on nutrient dynamics during the decay of plant
litter, we have added to our understanding of the inverse linear
regression model and gained insight into the factors controlling
nutrient immobilization. We briefly review some of our findings
below:

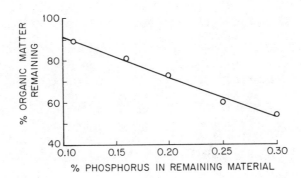

Fig. 5. Percentage of original *Typha* leaf litter mass remaining
expressed as a function of the phosphorus concentration in
the remaining material. Data from a decomposition study
conducted in laboratory microcosms (Melillo, Shaver, Fownes,
and Aber, personal communication).

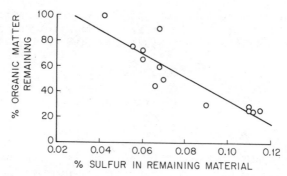

Fig. 6.   Percentage of original Scots pine needle litter mass remain-
          ing expressed as a function of the sulfur concentration in
          the remaining material. Data from a decomposition study
          conducted in Scots pine stand in central Sweden (Staaf and
          Berg 1982).

Initial Litter Quality and Nitrogen Immobilization

Our method of linking initial litter quality and Nmax is most
easily explained by example. We will use data from a study of
woody litter decay conducted in a small stream in eastern Quebec
(Melillo et al. 1983). Wood chips of five tree species were used--
Alnus rugosa, Betula papyrifera, Populus tremuloides, Picea mariana,
and *Abies balsamea.* The initial lignin and nitrogen concentrations
varied among the five woody materials (Table 3). The woody materials
were incubated in a first order stream and after 3, 9, and 16
months, subsets of each litter type were collected and analyzed for
mass loss and nitrogen content. We then plotted % organic matter
remaining against the % N in the remaining material for each type.
The correlation coefficients, slopes, and intercepts describing the
resulting inverse linear functions are given in Table 3.

For the five materials studied, we found one index of initial
litter quality, initial lignin concentration, to be highly correlated
($r^2 = 0.89$) with the slope of the inverse linear relationship
between organic matter disappearance and % N in the remaining
material (Figure 7). The equation describing the relationship
between the slopes of the inverse linear functions and the initital
lignin concentrations of the material is

$$\text{Slope} = -28.48 - (0.91 \cdot \text{Lo}) \tag{7}$$

where Lo = initial lignin concentration.

We can rewrite Equation (1) for the initial condition as follows:

$$\text{Intercept} = 100 - (\text{Slope} \cdot \text{No}). \tag{8}$$

Table 3. Initial litter quality parameters and correlation coeffi-
          cients, slopes, and intercepts of the inverse linear func-
          tions relating mass loss and the nitrogen concentration in
          the remaining material for each of five wood chip types
          decomposing in a first order stream in Quebec (Melillo et
          al. 1983).

| Material | Initial | | Inverse linear function (OM vs. N) | | |
	Lignin (%)	Nitrogen (%)	r	Slope	Intercept
Alder wood	13.1	0.32	-0.89	-39.18	106.29
Birch wood	8.2	0.14	-0.99	-33.71	102.03
Aspen wood	12.0	0.08	-0.88	-43.46	104.09
Spruce wood	25.9	0.06	-0.91	-49.20	104.93
Fir wood	24.6	0.04	-0.97	-52.96	105.18

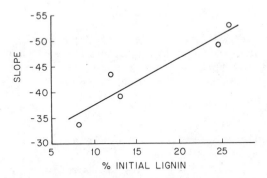

Fig. 7. Slopes of the inverse linear functions relating weight loss
         and nitrogen concentration in the remaining material
         expressed as a function of the initial lignin concentrations
         of the materials. Five wood chip types were used--alder,
         aspen, birch, fir, and spruce. Data from a decomposition
         study conducted in a first order stream in Quebec, Canada
         (Melillo et al. 1983).

Substituting Equations (7) and (8) into Equation (2) yields

$$Nmax = \frac{[100-((-28.48 - (0.91 \cdot Lo)) \cdot No)]^2}{-4(-28.48 - (0.91 \cdot Lo)} - 100 \cdot No \cdot 0.1 \quad (9)$$

which expresses total nitrogen immobilization (mg N/g initial
tissue) by any material decomposing in the first order stream as a
function of the material's initial nitrogen and lignin contents.
We have used Equation (9) to calculate maximum nitrogen immobilization
amounts for a set of hypothetical litter materials decomposing in
the first order stream (Table 4). These results are consistent
with general principles of microbial ecology which can be stated as
follows: (1) substrates broken down more easily will result in a
higher degree of nitrogen immobilization than substrates difficult
to decompose such as lignified tissues (Fenchel and Blackburn
1979); and (2) nitrogen poor detritus will immobilize more nitrogen
than nitrogen rich detritus (Hutchinson and Richards 1921).

Exceptions to both of these rules can be found in the literature.
For example, in a laboratory study of decomposition of plant materials, Richards and Norman (1931) found that of the materials they
studied, the material with the highest initial nitrogen content
also exhibited the greatest nitrogen immobilization potential.
Recent analyses of forest litter decay studies (Berg and Staaf
1980, Melillo et al. 1982, Aber and Melillo 1982) suggest that
nitrogen immobilization in decaying plant litter is positively
correlated with initial lignin content of the material. Reasons
should be sought for these departures from the "conventional wisdom"
of microbial ecology.

Table 4.   Estimates of Nmax for a set of hypothetical litter materials
           that have a range of initial nitrogen and lignin concentra-
           tions and that are decomposing in a first order stream in
           eastern Quebec. The calculations were made using Equation
           (5).

| Initial nitrogen concentration | Initial lignin concentration | | |
| | 10% | 15% | 20% |
(%)	(mg N immobilized/g initial tissue)		
0.02	6.55	5.83	5.26
0.02	5.69	4.98	4.40
0.40	4.80	4.10	3.54

Exogenous Nutrient Supply

The slope of the inverse linear function describing the relationship between % organic matter remaining and % N in the remaining
tissue is sensitive to large changes in exogenous nitrogen supply.
Under high levels of nitrogen the slope is shallow, while under low
levels of nitrogen the slope is steep. In a laboratory study of
Spartina patens leaf litter that was decaying under three levels of
exogenous nitrogen supply, we found the slopes of the inverse
linear regression models to range from -98.4 in the low nitrogen
treatment to -61.5 in the high nitrogen treatment (Jensen et al.
1978).

We also know that a similar response occurs with phosphorus
and evidence for this is presented in Figure 8. Alder wood chips
decomposing under low phosphorus supply exhibited a slope of -576
when % organic matter remaining is plotted against % P in the
remaining tissue, while alder wood chips decomposing under high
phosphorus supply exhibited a slope of -245 (Melillo, Naiman, and
Aber, personal communication). The estimated Pmax (maximum net
phosphorus immobilization) for the alder wood chips decomposing
under low phosphorus supply was 0.3 mg P/g initial material. And
the estimated Pmax for the alder wood chips decomposing under high
phosphorus supply was 0.9 mg P/g initial material. For a given
initial litter material, then, the higher the exogenous phosphorus
supply the greater the Pmax. A similar set of generalizations can
be made for nitrogen.

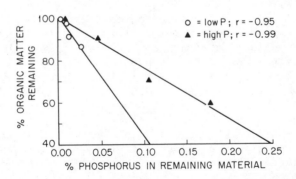

Fig. 8.   Percentage of original alder wood chip mass remaining
          expressed as a function of the phosphorus concentration in
          the remaining material. The lower line (steep slope)
          resulted from decay under low (1x) P enrichment. The upper
          line (shallow slope) resulted from decay under high (50x) P
          enrichment. The same initial material was used in each.
          Data from a decomposition study conducted in laboratory mi-
          crocosms (Melillo, Naiman, and Aber, personal communication).

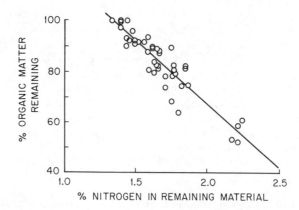

Fig. 9. Percentage of original red pine needle litter mass remaining
         expressed as a function of the nitrogen concentration in the
         remaining material. Data from a decomposition study using a
         litter of the same initial quality that was placed at six
         sites across North America (Melillo, Aber, Gosz, Meentemeyer,
         Shaver, van Cleve, and Vitousek, personal communication).

Temperature and Moisture

For a given litter material, changes in temperature and moisture
appear to alter the rate of nitrogen immobilization but not the
Nmax potential. Red pine needles of uniform initial quality decaying
at six forests across North America had the same Nmax potentials,
that is, the slopes of the inverse linear functions were the same
across sites (Fig. 9). A similar pattern (Fig. 10) can be found
for Scots pine needles of uniform initial quality decaying at 20
sites along a north-south transect in Sweden. In both instances,
there were differences in decay rates among sites. At sites with
rapid decay rates, the litter "materials" had "moved" the farthest
along the line defined by the inverse linear regression model by
the end of the study, while at sites with slowest decay rates, the
"materials" had "moved" the least distance along the line defined
by the inverse linear regression model by the end of the study.
These results suggest that initial litter quality is the dominant
factor influencing the magnitude of Nmax, while the environment
(e.g., temperature and moisture) is the dominant factor influencing
decay rate.

Oxygen Tension

Under anaerobic conditions, the relationship between % organic
matter remaining and % N in the remaining material conforms to the
inverse linear model. Data from Acharya (1935) can be used to
illustrate this point. He studied rice straw decay under aerobic

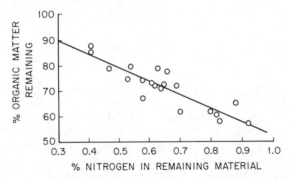

Fig. 10.   Percentage of original Scots pine needle litter mass
           remaining expressed as a function of the nitrogen concen-
           tration in the remaining material. Data from a decomposi-
           tion study using litter of the same initial quality that
           was placed at 20 sites along a north-south transect in
           Sweden (personal communication, B. Berg and H. Staaf, Depart-
           ment of Microbiology, Swedish University of Agricultural
           Sciences, Uppsala).

and anaerobic conditions. Materials decomposing under both conditions
conformed to the inverse linear regression model (Fig. 11).   The
material undergoing anaerobic decay had a steeper slope and thus a
lower Nmax. This suggests that nitrogen immobilization under
anaerobic conditons is less than nitrogen immobilization under
aerobic conditions.

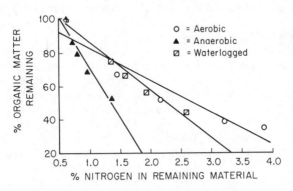

Fig. 11.   Percentage of original rice straw mass remaining as a
           function of the nitrogen concentration in the remaining
           material. Data from a decomposition study conducted in
           laboratory microcosms at the Fermentation Department,
           Rothamsted Experimental Station, Harpenden, Herts, England
           (Acharya 1935).

APPLICATIONS

We can envision a variety of applications for the inverse
linear regression model. Several of our ideas are presented below.

Air Pollution and Decomposition

The impact of air pollutants on decomposition processes in
natural ecosystems such as forests is an increasing concern. The
chronic, low level additions of heavy metals and the continual
inputs of acid precipitation to forest ecosystems in the northeastern
U.S., in Scandinavia, and in the Federal Republic of Germany have
received considerable attention in newspapers and popular magazines
as well as in the scientific literature. We have used our approach
to analyze decomposition data from one study of the impact of air
pollutants on the decay dynamics of leaf litter and the results are
most interesting.

We examined the data from a leaf litter decomposition study
conducted along a heavy metals gradient emanating from a zinc
smelter in Palmerton, Pennsylvania (Strojan 1978). Concentrations
of the heavy metals zinc, iron, lead, cadmium, and copper were one
to two orders of magnitude higher in the surface soil 1 km east of
the zinc smelter than in the surface soil at the control site 40 km
east of the smelter. First year mass loss and nitrogen dynamics
were determined for sassafras leaves and a mixture of chestnut
oak/red oak leaves at three sites along the gradient--1 km, 6 km,
and 40 km from the smelter. The inverse linear functions describing
the relationship between % organic matter remaining and % N in the
remaining tissue for the two litter types at all three sites for
all collection times are plotted in Figures 12A and B. We found
that a single function described the organic matter and nitrogen
dynamics of a particular substrate during decay, regardless of
decomposition site. Heavy metal concentrations at the sites influenced the rate at which the two decomposing litter types moved
along their respective lines, with the materials decomposing 1 km
from the smelter moving the shortest distance along the lines in 12
months and the materials decomposing 40 km from the smelter moving
farthest along the lines in 12 months. These results lead us again
to the conclusion that initial litter quality is the dominant
factor influencing the magnitude of Nmax, while the environment
(e.g., temperature and moisture) is the dominant factor influencing
decay rate.

Agroecosystem Management

An increasing number of farmers are changing to production
methods that demand less soil tillage and maintain crop residues
near the soil surface. Motives for this change include the desire
to (1) reduce soil erosion losses, (2) enhance soil water conserva-

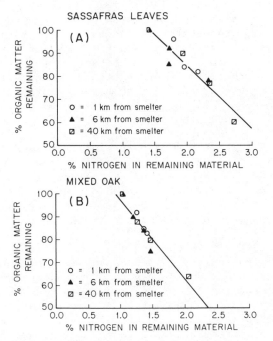

Fig. 12. Percentage of original sassafrass (A) and oak (B) leaf
litter mass remaining as a function of the nitrogen concentration in the remaining material: o = material at site 1
km from the smelter; ▲ = material 6 km from the smelter;
and ⊘ = material 40 km from the smelter (Strojan 1978).

tion, and (3) lower the requirements for fossil fuel and labor
associated with crop production.

     Reduced or no-till farming practices change the soil environment
and soil biology relative to what is found in soils where conventional
methods of moldboard plowing are used. Indications of nitrogen
deficiency and (or) yield limitations in crop plants with no-till
have lead several researchers to conclude that higher levels of
fertilizer nitrogen often are required for no-till than are required
for plowed soils (Bakermans and de Wit 1970, Bandel et al. 1975,
Dornan 1980). We suggest that our approach for quantifying the
magnitude and timing of nitrogen immobilization could be used in
reduced and no-till agroecosystems to develop optimal fertilization
regimes.

Species Composition, Productivity, and Resource Availability in
Forest Ecosystems

     Finally, we believe that one of the most important tasks for
ecologists in the next decades will be the synthesis of information

from the fields of plant and microbial physiology, population and community ecology, and ecosystem studies. This synthesis should lead to a clearer understanding of how ecosystems work and insight into how they will respond to a variety of disturbances. An example of the type of synthesis we envision was begun by us in FORTNITE, a computer model of organic matter and nitrogen dynamics in forest ecosystems (Aber and Melillo 1982b). The model is comprised of a forest production module and decomposition module that are linked to each other. The forest production module simulates the growth, reproduction and mortality of a number of tree species. Species specific growth functions are developed from four factors: available light, growing degree days, actual evapotranspiration, and available nitrogen. Nitrogen availability is an output of the decomposition module. The calculation of nitrogen availability uses the simple inverse linear regression model described earlier.

We have used FORTNITE to simulate the regrowth of northern hardwood forests following various disturbances (Aber et al. 1982). As plant species change during regrowth, so do the amount and quality of litter input to the decomposition module. The combination of variable quantities and qualities of litter influence the rate of decay and the balance between the amount of nitrogen immobilized and the amount of nitrogen available for forest growth. The decomposition module has been used to make estimates of the amount of nitrogen immobilized by various litter types over the first 60 years of regrowth after clearcutting (Aber et al. 1978). In early stages of recovery (Fig. 13) from clearcutting, the decomposition of logging debris may make immobilization a dominant process in the nitrogen regime of the ecosystem. Subsequently, the nitrogen is released from the logging debris through the mineralization process.

The removal of logging debris, as is practiced in whole tree harvesting regimes, reduces the importance of nitrogen immobilization early in forest growth. This makes nitrogen availability greater for both uptake by plants and leaching losses from the ecosystem in the first few years following harvest. If nitrogen losses were large due to slow colonization by plants, then the future productivity of the site would be substantially reduced (Aber et al. 1982).

FUTURE RESEARCH

We have presented a new, relatively untested method for analyzing and interpreting nutrient (N, P, and S) dynamics during litter decay. The method must be tested and its strengths and weaknesses identified. If appropriate, the method should be incorporated into a whole ecosystem view of the relationships between plant nutrient status, litter quality, decay dynamics, and nutrient availability.

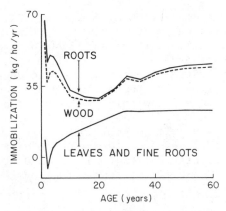

Fig. 13.   Role of different forest litter compartments in the net
           nitrogen immobilization dynamics (N IMMOB) as a function
           of time since clearcutting (Aber et al. 1978).

     Many of the specific research questions that must be asked
relate to the mechanism(s) responsible for the simple, yet very
powerful relationship between % organic matter mass remaining and %
N in the remaining tissue. As we indicated earlier, although some
of the nitrogen accumulation is most certainly related to an increase
in microbial biomass, we think that the complexing of nitrogen in
extracellular products of decay may account for a large fraction of
the nitrogen accumulation during the immobilization phase. Is this
indeed the primary mechanism of "immobilization"? If it is, how
does it relate to humus formation? Are microbial exoenzymes "immobilized" and de-activated as they become trapped in the condensed
macromolecules? And can physical disturbances such as that which
results from plowing or alternate wetting and drying and freezing
and thawing release these "trapped" enzymes with the result being a
burst in decomposition activity? These questions may require
collaboration among ecologists, microbiologists, and biochemists.

LITERATURE CITED

Aber, J.D., D.B. Botkin, and J.M. Melillo. 1978. Predicting the
     effects of different harvesting regimes on forest floor dynamics
     in northern hardwoods. Can. J. For. Res. 8:306-315.
Aber, J.D., and J.M. Melillo. 1982a. Nitrogen immobilization in
     decaying hardwood leaf litter as a function of initial nitrogen
     and lignin content. Can. J. Bot. 60:2263-2269.
Aber, J.D., and J.M. Melillo. 1982b. FORTNITE: A computer model
     of organic matter and nitrogen dynamics in forest ecosystems.
     Univ. of Wisconsin Res. Bull. R3130.

Aber, J.D., J.M. Melillo, and C.A. Federer. 1982. Predicting the
     effects of rotation length, harvest intensity, and fertilization
     on fiber yield from northern hardwood forests in New England.
     For. Sci. 28:31-45.
Acharya, C.N. 1935. Studies on the anaerobic decomposition of
     plant materials. III. Comparison of the course of decomposition
     of rice straw under anaerobic, aerobic and partially aerobic
     conditions. Biochem. J. 29:1116-1120.
Bakermans, W.A.P., and C.T. de Wit. 1970. Crop husbandry on naturally
     compacted soils. Neth. J. Agric. Sci. 18:225-246.
Bandel, V.A., S. Dzienia, G. Stanford, and J.O. Legg. 1975. N
     behavior under no-till vs. conventional corn culture. I:
     First-year results using unlabeled N fertilizer. Agron. J.
     67:782-786.
Berg, B., and H. Staaf. 1980. Leaching, accumulation and release
     of nitrogen in decomposing forest litter. Pages 163-178 in
     P.E. Clark and T. Rosswall, eds. Terrestrial nitrogen cycles.
     Ecol. Bull. (Stockholm) 33.
Dornan, J.W. 1980. Soil microbial and biochemical changes associated
     with reduced tillage. Soil Sci. Soc. Am. J. 44:765-771.
Fenchel, T., and T.H. Blackburn. 1979. Bacteria and mineral cycling.
     Academic Press, New York, NY USA.
Griffith, W.F. 1978. Effects of phosphorus and potassium on nitrogen
     fixation. Pages 80-94 in Phosphorus for agriculture: A
     situation analysis. Potash/Phosphate Inst., Atlanta, GA USA.
Hutchinson, H.B., and E.H. Richards. 1921. Artificial farmyard
     manure. J. Minist. Agric. 28:398-411.
Jenny, H., S.P. Gessel, and F.T. Bingham. 1949. Comparative study
     of decomposition rates in temperate and tropical regions.
     Soil Sci. 68:419-432.
Jensen, C.L., J.M. Melillo, and J.D. Aber. 1978. The effect of
     nitrogen decomposition of Spartina patens. Biol. Bull. 155:446.
Melillo, J.M., J.D. Aber, and J.F. Muratore. 1982. Nitrogen and
     lignin control of hardwood leaf litter decomposition dynamics.
     Ecology 63:621-626.
Melillo, J.M., and J.R. Gosz. 1983. Interactions of biogeochemical
     cycles in forest ecosystems. Pages 177-222 in B. Bolin and
     R.B. Cook, eds. The major biogeochemical cycles and their
     interactions. Wiley, New York, NY USA.
Melillo, J.M., R.J. Naiman, J.D. Aber, and K.N. Eshleman. 1983.
     The influence of substrate quality and stream size on wood
     decomposition dynamics. Oecologia 58:281-285.
Miller, H.G., and J.D. Miller. 1976. Effects of nitrogen supply on
     net primary production in Corsican pine. J. Appl. Ecol.
     13:249-256.
Natr, L. 1975. Influence of mineral nutrition on photosynthesis
     and the use of assimilates. Pages 537-556 in J.P. Cooper, ed.
     Photosynthesis and productivity in different environments.
     Cambridge Univ. Press, Cambridge, England.

Richards, E.H., and A.G. Norman. 1931. The biological decomposition of plant materials vs. some factors determining the quality of nitrogen immobilized during decomposition. Biochem. J. 25:1769-1778.

Staaf, H., and B. Berg. 1982. Accumulation and release of plant nutrients in decomposing Scots pine needle litter. Long-term decomposition in a Scots pine forest II. Can. J. Bot. 60:1561-1568.

Strojan, C.L. 1978. Forest leaf litter decomposition in the vicinity of a zinc smelter. Oecologia 32:203-212.

Swift, M.J., O.W. Heal, and J.M. Anderson. 1979. Decomposition in terrestrial ecosystems. Univ. of California Press, Berkeley, CA USA.

Thayer, G.W. 1974. Identity and regulation of nutrients limiting phytoplankton production in the shallow estuaries near Beaufort, N.C. Oecologia 14:75-92.

Viljoen, J.A., and E.B. Fred. 1924. The effect of different kinds of wood and of wood pulp cellulose on plant growth. Soil Sci. 17:199-208.

STABILITY AND DESTABILIZATION OF CENTRAL EUROPEAN FOREST

ECOSYSTEMS--A THEORETICAL, DATA BASED APPROACH

Bernhard Ulrich

Institut für Bodenkunde und Waldernährung
Universität Göttingen
Büsgenweg 2, D-3400 Göttingen, FRG

ABSTRACT

Based on a thermodynamic model of a forest ecosystem, stability is defined as a system in quasi-steady state. The steady state is defined by the balance between ion uptake (phytomass production) and ion mineralization (secondary production) and by the balance between input and output. In steady state, the ion cycle of the system is closed. A de-coupling (opening) of the ion cycle leads to net production or consumption of protons, which causes changes in the chemical soil environment. Resilience is defined as the ability of the system to maintain the H^+/OH^- balance. Several buffering mechanisms operate in soils and organisms to maintain this balance. The various buffer rates are critical in regulating system resilience. The natural climatic variation, the necessity to re-establish system elements continuously, and the biomass utilization by man lead to a de-coupling of the ion cycle and thus produce chemical stress. A sequence of ecosystem states exist: aggradation, stability with high resilience, humus disintegration, stability with low resilience, buildup of decomposer refuge, and podzolization. From the data known about rates of proton loading and proton buffering, it must be concluded that acid deposition shifts forest ecosystems from stability ranges into destablization phases (transition states), even at low rates of deposition.

DEFINITION OF THE FOREST ECOSYSTEM

A system can be described by its elements and the interactions among them. In ecosystems, the system elements are organisms such as plants, animals, and microorganisms and structural units such as

217

ecological niches or soil types and humusforms. The structural
units constituting organisms also are system elements; the smallest
units of structure considered as system elements are the binding
forms of chemical elements.

Boundaries of neighbouring forest ecosystems are delineated
using the principle of homogeneity: one ecosystem touches another
if the plant or soil associations differ between the two. For
lasting stability, reproduction of species is an indispensable
feature of ecosystems. The various stages of reproduction therefore
are part of the homogeneous forest ecosystem. This may seem strange
if one considers even-aged forests, but it is an essential requirement
for an ecosystem. In a stable, even-aged forest ecosystem, the
different age classes represent different stages in a strictly
cyclic development.

The boundary between ecosystem and atmosphere follows the
plant and soil surfaces. The boundary of the lithosphere is drawn
at the surface of soil minerals so that the surface bound material
(ions and organic matter) belongs to the ecosystem, whereas the
silicate lattice belongs to the lithosphere and thus to the environment of the ecosystem. Thus, the cation release due to silicate
weathering is treated as an input into the ecosystem.

A FOREST ECOSYSTEM MODEL

Ecosystems are open systems, which means they exchange matter
and energy with their environment. Systems of this kind follow the
thermodynamics of irreversible processes. Therefore, one should
consult this field of science for appropriate ecosystems models.
According to Prigogine (1947), open systems in steady state minimize
entropy production. It seems strange that this principle should
apply to ecosystems which use only a few percent of radiation
energy input for photosynthesis. But one has to realize that about
40% of the energy input is used for transpiration (i.e., for transportation), and the rest is used for heat regulation. This model
has received some attention and it seems worthwhile to develop it
further. It is subject to limitations, however: a steady state
is, in principle, unachievable due to the inconstancy of the environment (variability of the climate, change of the lithosphere by
weathering processes) and due to the inconstancy of the system
elements (these elements have a limited life duration and must be
replaced continuously). The fact that forest ecosystems have been
stable for millions of years (e.g., in the tropics since the Tertiary
Age, except that species composition changed due to evolution and
due to changes in the lithosphere caused by weathering) suggests,
however, that forest ecosystems can approach a steady state very
closely--so close that one would not be able to detect the difference
by the methods available. One may say, therefore, that forest
ecosystems can reach a quasi-steady state.

THE ION CYCLE IN THE ECOSYSTEM IN STEADY STATE

All processes included in ion turnover can be reduced to four fluxes: input (e.g., from the atmosphere or by weathering of soil minerals), output (e.g., leaching, denitrification), uptake from soil solution into biomass (by plants and microorganisms), and release of ions during decomposition (mineralization). For an ecosystem in which input and output are equal for each element, a quasi-steady state can be achieved (i.e., in the temporal and spatial sense, all stores in the system are constant). This constancy covers phytomass, total biomass, soil organic matter, and the various chemical elements and their different binding forms. In such an ecosystem, the rate of assimilation must be equal to the rate of respiration, or, if one also includes the turnover of elements other than carbon (C), hydrogen (H), and oxygen (O), the rate of phytomass production must be equal to the rate of mineralization:

$$CO_2 + H_2O + X^+Y^- \quad \underset{\underset{\text{mineralization}}{\text{energy}}}{\overset{\text{phytomass production}}{\rightleftharpoons}} \quad CH_2OX^+Y^- + O_2.$$

In the above equation, X^+Y^- symbolizes the cations and anions of the elements other than C, H, and O; they are taken up from the soil solution and returned again during mineralization.

To account for the ionic status, the precursors of the ions present in soil solution and the binding forms in plants must be known. The following binding forms in the soil matrix and in plants are assumed:

(1) Nitrogen (N): only organic binding form with zero charge. Fixed or exchangeable ammonium ions (NH_4^+) in soil and nitrate ions (NO_3^-) in plants are negligible.
(2) Sodium (Na), potassium (K), magnesium (Mg), calcium (Ca), manganese (Mn), aluminum (Al), iron (Fe): cations. Charges are balanced by inorganic or organic anions.
(3) Sulfur (S), phosphorus (P), cloride (Cl): anions. Charges are balanced by cations (an oversimplification for S [S containing amino acids], but the error introduced is negligible; P in the form of bound pyrophospate creates OH^- dissociation groups during its breakdown).

With the exception of N assimilation and denitrification, the input and output of all elements occur in ionic form. The precursor of assimilated N is uncharged N_2, the reaction products of denitrification are also uncharged compounds (N_2, N_2O). The ion cycle within the ecosystem can be reduced to two processes: ion uptake and ion mineralization. As the following equations show, these processes are connected with proton production (H^+) and consumption (OH^-), respectively:

$$N: \quad R\text{-}NH_2 + H_2O \; \underset{\text{uptake}}{\overset{\text{mineralization}}{\rightleftarrows}} \; R\text{-}OH + NH_3 \begin{cases} NH_4^+ + OH^- \\ NO_3^- + H^+ \end{cases}$$

$$S: \quad R\text{-}SH + H_2O \; \underset{\text{uptake}}{\overset{\text{mineralization}}{\rightleftarrows}} \; R\text{-}OH + H_2S \rightarrow SO_4^{2-} + 2H^+$$

$$P: \quad R\text{-}O\text{-}PO(OH)_2 + H_2O \; \underset{\text{uptake}}{\overset{\text{mineralization}}{\rightleftarrows}} \; R\text{-}OH + PO_4^{3-} + 3H^+$$

$$\text{Cations:} \quad \left[R\text{-}C \overset{O}{\underset{O}{\Big\|}} \right]_2 Ca + 2H_2O \; \underset{\text{uptake}}{\overset{\text{mineralization}}{\rightleftarrows}} \; 2R\text{-}H + CO_2 + Ca^{2+} + 2OH^-.$$

If the N accumulated in plants stems is derived from N_2 (after assimilation) or soil organic N (after mineralization), its transfer to plant organic N is not connected with any net change in charge either in the soil or in the plant; that is, it is not connected with any net proton transfer.

In steady state, ion uptake and ion mineralization balance each other, thus no net proton turnover occurs in the system. Clearly, however, ion uptake and ion mineralization do not always occur at the same place and at the same rate in the ecosystem. This means that not only the retention and disposition of nutrients, but also the buffering of H^+/OH^- ions determine the resilience of the ecosystem. In both respects, the main weight of buffering lies in the soil--the reaction vessel of the ecosystem. If organisms become involved in buffering, adverse consequences are likely (e.g., growth reductions) because their buffering abilities vary a great deal (determining tolerance).

The ion cycle in the ecosystem is controlled by the following principles:

(1) Mass conservation. From this principle it follows that mass balances are a proper way to check steady state conditions and to follow changes of the ecosystem.
(2) Maintenance of electrical neutrality in the flux. From this principle it follows that cation/anion balances of fluxes are appropriate for measuring transfers.
(3) Tendency to maintain the H^+/OH^- balance in the compartments (root, plant, soil, mineral soil, organic top layer). From this principle it follows that net proton production or consump-

tion in a compartment may be a critical variable to realize changes in the ecosystem. Net proton turnover in a compartment can be calculated from the cation/anion balance of the storage changes in this compartment.

RESILIENGE IN THE ION CYCLE

The central aspect of resilience is the H^+/OH^- balance of the compartments of the ecosystem. This statement rests on the knowledge that changes in the chemical soil state greatly influence binding forms, stores, and turnover rates of nutrients.

Plants and decomposers produce CO_2 by respiring at high rates. If no other bases or acids are available, the ecosystem approaches the chemical state of the system H_2O/CO_2 at CO_2 pressures close to or above that of air. Such systems will achieve pH values between 5.0 and 5.6. Therefore, values in this range in the liquid phase of soils indicate the absence of soluble bases or acids, that is, the neutrality range of the ecosystem. This neutrality range probably is the optimal state (i.e., allows maximal diversity both for producers and decomposers) that is possible in the ecosphere. At pH values above 5.6, basicity (e.g., in the form of bicarbonates or carbonates) is accumulated in soils or plants. At pH values below 5, acidity is accumulated. Between pHs 5 and 3, which is characteristic for acid soils, mineral acidity exists as cation acids. The most relevant cation acids are ionic species of Mn, Al, Fe, and heavy metals. Some of these act as micronutrients, but all are toxic at low concentrations. The toxicity depends not only on the kind of chemical element, but also on the ion species of the element. A better understanding of cation acid toxicity (e.g., Al toxicity) depends upon research progress in analytical chemistry.

The H^+/OH^- buffer systems existing in soils are summarized in Figure 1. For the definition of pH ranges given, see Ulrich et al. (1979) and Ulrich (1981a, 1983b); the pHs given apply to the equilibrium soil solution (ESS) or to $pH(H_2O)$. The buffer reactions given represent the principle, not the true reaction chain. The description holds only for ecosystems in steady state; unstable ecosystems in transient states may show features of different buffer ranges (e.g., mull in combination with an acid mineral soil). Under the influence of biomass utilization and (or) acid deposition, the upper soil is often more acidic than the deeper horizons (often a characteristic for ecosystems in transient states).

Several aspects of buffer mechanisms are especially relevant to resilience. One feature is the limited buffer rate of silicate weathering and release of Al ions from silicates. If the acid load exceeds the buffer rate, the system switches over to the buffer system operating at a lower pH, irrespective of buffer capacity

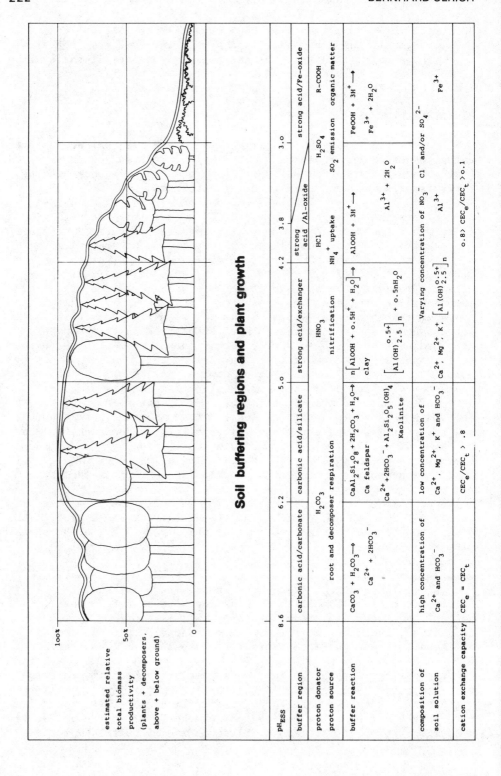

Soil buffering regions and plant growth

pH_{ESS}	8.6	6.2	5.0	4.2	3.8	3.0
buffer region	carbonic acid/carbonate	carbonic acid/silicate	strong acid/exchanger	strong acid/Al-oxide	strong acid/Fe-oxide	
proton donator proton source	H_2CO_3 root and decomposer respiration		HNO_3 nitrification	HCl NH_4^+ uptake	H_2SO_4 SO_2 emission	R-COOH organic matter
buffer reaction	$CaCO_3 + H_2CO_3 \rightarrow$ $Ca^{2+} + 2HCO_3^-$	$CaAl_2Si_2O_8 + 2H_2CO_3 + H_2O \rightarrow$ Ca feldspar $Ca^{2+} + 2HCO_3^- + Al_2Si_2O_5(OH)_4$ Kaolinite	$n[AlOOH + 0.5H^+ + H_2O] \rightarrow$ clay $[Al(OH)_{2.5}^{0.5+}]_n \cdot 0.5nH_2O$	$AlOOH + 3H^+ \rightarrow$ $Al^{3+} + 2H_2O$	$FeOOH + 3H^+ \rightarrow$ $Fe^{3+} + 2H_2O$	
composition of soil solution	high concentration of Ca^{2+} and HCO_3^-	low concentration of Ca^{2+}, Mg^{2+}, K^+ and HCO_3^-	Varying concentration of NO_3^-, Cl^- and/or SO_4^{2-} Ca^{2+}, Mg^{2+}, K^+, $[Al(OH)_{2.5}^{0.5+}]_n$	Al^{3+}	Fe^{3+}	
cation exchange capacity	$CEC_e = CEC_t$	$CEC_e/CEC_t > .8$	$.8 > CEC_e/CEC_t > .1$		$o.8 > CEC_e/CEC_t > o.1$	

estimated relative total biomass

productivity (plants + decomposers, above + below ground)

100%

50%

0

	150 kmol H^+ per 1 % $CaCO_3$	25 kmol H^+ per 1 % silicate	7 kmol H^+ per 1 % clay	150 kmol H^+ per 1 % clay	>2 kmol H^+
buffer capacity per ha and dm soil depth					
buffer rate per ha and year	>2 kmol H^+	0.2 - 2 kmol H^+	small (0.2 kmol H^+ ?)	large as long as $\left[Al(OH)_{2.5}^{0.5+}\right]_n$ is available, otherwise small	>2 kmol H^+
leaching of nutrients	mainly Ca^{2+}	very limited	leaching of exchangeable Ca^{2+}, Mg^{2+}, K^+	leaching complete	
composition of exchangeable cations	Ca^{2+} dominating	balanced	$\left[Al(OH)_{2.5}^{0.5+}\right]_n$ dominating	Al^{3+} dominating	Fe^{3+} and H^+ dominating
Humusform	Mull	Mull and Mull-like Moder	typical moder	rawhumus-like moder	rawhumus
cation/anion balance in plant uptake	anion surplus	balanced	cation surplus	large cation surplus	ion uptake ceases
limitations for plant growth	small limitation by wide Ca^{2+}/K^+ ration and anion surplus in uptake	no limitations by chemical soil factors	suffering of non-tolerant plants by Al-toxicity exclusion of calcicole species	suffering of all plants by Al-toxicity reduction of growth in calcifuge species including beech.	suffering of all plants by Al-Fe-toxicity only plants rooting in top organic layer survive

Fig. 1. H^+/OH^- buffering systems in soils of ecosystems in steady state.

left over. Therefore, highly loaded soils in the Fe buffer range,
as indicated by podzolization, typically still possess remarkable
stores of unweathered primary silicates and clay minerals. Only
calcium carbonate ($CaCO_3$) and Fe oxides show dissolution rates high
enough to buffer any possible load as long as the buffer substances
are distributed evenly in the fine earth.

Another important feature of buffering mechanisms related to
resilience is the reaction product. Buffering converts a stronger
acid to a weaker one. With a $CaCO_3$ buffer, carbonic acid is formed;
with a silicate buffer, silicic acid (or their salts, the clay
minerals). Both acids have low solubilities, approaching zero in
the pH range 7 to 5. It is obvious from the final reaction products,
CO_2 and SiO_2, that protons are transferred finally to the weakest
acid existing: water. Therefore, only these two buffers neutralize
acidity. The release of cation acids below pH 5 indicates that the
load exceeds the possible neutralization rate of the system. Early
in acidification, the cations acids released may be completely
accumulated in the soil (e.g., as interlayer Al-hydroxy complexes
and exchangeable Al ions) with no acidity leaving the system.
Soils can behave like chromatographic columns (with shortcuts
possible, however).

The release of alkali and earth alkali cations from silicates
under proton consumption is the only buffering process in the pH
range somewhat above 6 (this value depends on CO_2 pressure, which
is variable) and 5. The process is not limited to this pH range,
however, and may proceed at a somewhat increased rate at lower pH.
The increase in the rate is, by far, not proportional to the increase
in H^+ concentration, however; it may be a factor of two at the
most. From the data presently available (Ulrich 1981a, Johnson
et al. 1982, Bache 1983, Mazzarino et al. 1983), one can conclude
that most soils developed from sedimentary rocks have a buffer rate
of less than 1 keq H^+/ha·yr down to 1 m soil depth, whereas higher
rates are characteristic for basic magmatic rocks. The ongoing
silicate weathering in acidified soils explains why soils and
ecosystems can recover: if the rate of proton loading becomes less
than the rate of proton buffering by base cation release due to
silicate weathering, acidity accumulated in soil is consumed. This
means that the amount of cation acids is reduced equivalent to the
accumulation of alkali and earth alkali cations (especially Ca and
Mg), and the remaining cation acids are transferred to weaker acids
(e.g., to a higher degree of polymerization in Al polymeric ion
species).

Figure 2 demonstrates the role of cation exchange. The exchange
of Ca ions with protons is a rapid and reversible reaction. It can
be reversed if protons are consumed by some other reaction (e.g.,
reacting with $CaCO_3$ or by some process in ion uptake or mineralization). Protonized clay, however, is not stable and changes slowly

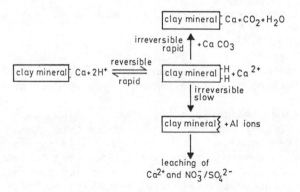

Fig. 2.   The role of cation exchange.

in an irreversible reaction releasing Al ions.   The process is
temperature dependent and may need weeks to months before occurring
extensively.   The cations (Ca^{2+}) exchanged are leached in combination
with the anion of the acid.   The cation exchange $Ca^{2+} \rightleftharpoons 2H^{+}$ is a
very important short term buffer.   As long as this reaction occurs,
the soil can buffer acidification pushes resulting from de-coupling
of the ion cycle in a very convenient way for organisms: the
acid/base status of the soil solution remains almost unchanged, but
the nutrient concentration in the soil solution increases.   But
trouble starts if the ecosystem is not capable of reversing the
exchange reaction before protolysis of the clay begins.   Organic
cation exchangers, that is, a higher store of soil organic matter,
reduces this danger and increases the resilience of the system
considerably.

DE-COUPLING OF THE ION CYCLE: ACIDIFICATION PUSHES

As a consequence of the relative high optimal temperatures of
soil microorganisms, soil temperature determines the soil organic
matter regime.   Low mean soil temperature and a long period with
temperatures below 8°C are the basis for high soil organic matter
stores, whereas opposite conditions favour mineralization and thus
result in low soil organic matter stores in steady state.

With temperature limited decomposition, which is typical for
Central and North Europe, a characteristic de-coupling of the ion
cycle occurs if nitrification exceeds nitrate uptake.   Because
nitrification produces a strong acid (HNO_3), whereas nitrate uptake
consumes the same acid, this de-coupling leads to an acidification
push.   In soils with a high percentage of exchangeable Ca, the
acidification may show up only as a decrease of exchangeable Ca and
an increase of exchangeable H; in soils depleted of exchangeable
Ca, the pH may be lowered also.

Three time spans of de-coupling need to be distinguished:

(1) Short term: seasonal de-coupling, seasonal acidification push. This occurs especially in spring when the soil is heated up and mineralization starts, but ion uptake is limited before leaf burst, or in autumn after re-wetting of a dry soil.

(2) Middle term: climatic acidification push. In soils in the exchange buffer range (Fig. 2), which have already lost considerable fractions of exchangeable Ca, a year or several years substantially warmer than average can lead to longer lasting de-coupling of the ion cycle by favouring mineralization and nitrification. Stimulation of microbial activity resulting from soil acidification has been shown in field and laboratory experiments with acid rain (Tamm and Wiklander 1980, Hovland et al. 1980, Strayer et al. 1981, Killham et al. 1983). The decoupling may last for several years, it may be stopped if in cool years mineralization is slowed down and nitrification is depressed to a rate not exceeding nitrate uptake. A climatic acidification push always leads to the release of Al ions in the soil and the leaching of Ca and NO_3 ions from the root zone.

(3) Long term: humus disintegration. If the exchangeable Ca is depleted, the ecosystem may be unable to return to steady state; that is, it will be unable to close the cycle of mineralization and ion uptake again. In this case, the organic matter and the organic N accumulated in the mineral soil is slowly lost. The process ends if the soil organic matter content approaches low values. It is evident that this process happens, because non-hydromorphic acid soils of stable forest ecosystems show low humus contents in mineral soil, whereas soils in the silicate buffer range may have appreciable contents. The latter ones are the precursors of the former ones. This process was not identified and investigated until recently (for some ideas how buildup and breakdown of humus are influenced, see Ulrich 1981c). The process can be followed by balancing input and output: the nitrate output exceeds nitrate input if the process is going on. A soil rich in organic matter with a narrow C/N ratio has a tremendous acidification potential: 14 kg N can produce up to 1 kmol H^+; N stores in soils may exceed 10,000 kg N/ha before passing through humus disintegration.

ACIDIFICATION EFFECTS OF BIOMASS UTILIZATION

In Table 1, the annual uptake rates and ion balances are given for different tree compartments of a beech (*Fagus silvatica*) stand in the Solling. The tree compartments correspond to different degrees of phytomass utilization: bole (wood + bark), bole +

Table 1.   Annual rates of ion uptake in various tree compartments
of a beech stand (*Fagus silvatica*).

	Bole	Bole + branches	Leaves	Total overstory + understory
Cation sum	0.69	0.83	1.64	4.34
Anion sum	0.11	0.16	0.35	0.83
Cation/anion balance	+0.58	+0.67	+1.29	+3.51

branches, leaves, total overstory and understory. The cation/anion
balance shows that as long as only the woody parts of the overstory
are used and exported from the ecosystem, the soil acidification is
in a range (0.6 kmol H^+/ha·yr) that can be buffered by "better"
soils through base cation release during silicate weathering. If
the N rich parts are continuously exported from the ecosystem
(e.g., by litter utilization, which played a role during the last
centuries), soil acidification increases strongly. Even the utilization of total overstory and understory represents a realistic
picture of some forest areas in Central Europe during the last
centuries. All forest soils, where such utilizations were practiced,
were strongly acidified when modern forestry started.

STABILITY AND DESTABILIZATION PHASES

In Figure 3, a system of stability and destabilization phases
is given for non-hydromorphic soils and post-glacial conditions of
Central Europe. This system is based on the buffer ranges shown in
Figure 1 and the tendency of soil acidification caused by climatic
variation and biomass utilization.

Aggradation Phase

The term "aggradation phase" is used here for ecosystems in
which the organic matter storage in mineral soil is increasing
through accumulation of organic matter with small C/N ratios (~10/1).
The humusform is any subtype of mull. With this phase, soil and
ecosystem development starts at a freshly exposed surface of loose
sedimentary rocks (e.g., after glaciation). As an example, in a
soil where 100,000 kg organic matter/ha has accumulated, the carbon
content of organic matter is 50%, the C/N ratio is 10, and the

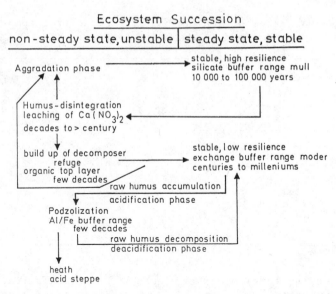

Fig. 3.   System of stability and destabilization phases for non-
          hydromorphic soils of Central Europe.

cation exchange capacity (CEC) is 4 eq/kg C. The CEC indicates a
production of 200 kmol H^+/ha. If the aggradation phase lasts for
1,000 years, the mean annual proton production will be 0.2 kmol
H^+/ha·yr. This production can be buffered easily by silicate
weathering and, therefore, will not result in soil acidification.
The N accumulated originates either mainly from N_2 (legumes) or
from nitrate input with rain. Assuming a nitrate input of 4 kg
N/ha·yr for a period of 1,000 years, the nitrate will contribute
40% of the N accumulated and its transfer to organic bound N will
have consumed 286 kmol H^+/ha. A small nitrate input, as can be
expected under natural conditions, may contribute substantially to
the buffering of protons resulting from biomass and soil organic
matter accumulation.

     Under climatic conditions favouring mineralization, the buildup
of a soil organic matter store will be very limited and may approach
zero. The weathering of silicates by carbonic acid will then
result in the formation of bicarbonates, which means that the soil
will start to accumulate basicity. As long as precipitation exceeds
evapotranspiration, this basicity can be leached and the soil will
remain "neutral," in the pH range 5 to 6. If, however, evapotranspiration exceeds precipitation, the consequence will be alkali-
zation of the soil. Therefore, long term agriculture is possible
only under the condition that precipitation plus irrigation exceed
evapotranspiration.

Stability with High Resilience (Stability Range I)

If the soil organic matter accumulation reaches a steady state (constant store), the aggrading system passes into a stable system with high resilience. The soil can stay either in the carbonate buffer range or in the silicate buffer range, so two different states can be distinguished. The absence of toxic cation acids allows all species of primary and secondary (decomposers and consumers) producers to exist and thus to compete with each other. For the climatical and hydrological conditions given, one can expect great diversity, compared with ecosystems on acidified soils. The absence of toxic cation acids means further that roots tend to contact soil material maximally where water and nutrients are stored and that bacteria play the dominant role in the decomposers.

A stable forest ecosystem with high resilience (stability range I) should, therefore, under the conditions of Central Europe, be characterized by the following properties:

(1) the system is composed of relatively many species, which are structured in layers;
(2) the soil is deeply rooted, and the roots are homogeneously distributed;
(3) the decomposers are characterized by the activity of earthworms;
(4) the soil is in the silicate (or carbonate) buffer range throughout and shows no depth gradient in the soil chemistry;
(5) soil organic matter is accumulated throughout the whole rooting zone, and, due to the activity of soil burrowing animals, the soil is crumbly.

Examples can be found which demonstrate that untouched forest ecosystems in Central Europe would exhibit these properties.

In the higher altitudes of subalpine mountains, it is possible that a biological depth gradient exists in soil due to frost action. This gradient should result in the formation of an organic top layer which is colonized mainly by arthropods, whereas the mineral soil may still be characterized by earthworm activity. The biological depth gradient implies a chemical depth gradient; that is, the organic top layer is more acid than the mineral soil. The climatical limitations imposed on decomposers thus should lower the resilience of the ecosystem--forest ecosystems in higher altitudes or farther north are more easily subjected to plastic strain by climatic or human induced stress.

Even in ecosystems with the soil staying in the silicate buffer range, there is some loss of cations (mainly Ca^{2+}), accompanied by HCO_3^-. This leaching from soil is responsible for the salt content of soft groundwater. To keep the ecosystem stable, this loss has to be balanced by silicate weathering. A long lasting

de-coupling of the ion cycle is to be expected if, in the course of
natural ecosystem development, the weatherable silicates in the
root zone are exhausted. If the output (leaching) of Ca and Mg
becomes larger than the input (from the atmosphere and by weathering),
the ecosystem passes into a non-stationary transition state which
is characterized by decreasing stores of exchangeable Ca and Mg.
The soil acidifies, the chemical soil state passes from the silicate
into the cation exchange buffer range. The time span needed for
this process can be estimated from the buffer capacity and the
buffer rate in the silicate buffer range of the soil. With the
values given in Figure 1 (cf. Ulrich 1983b), the buffer capacity
amounts to 750 keq/ha·m per 1% silicate content in the soil. With
a buffer rate of 0.4 kmol H^+/ha·yr for 1 m soil depth, found by
Mazzarino et al. (1983) for a soil of the last interglaciation
period ("Dasburg" soil), a silicate content of 1% would last for
around 2,000 years. If this buffer rate is accepted as typical for
the natural development of untouched ecosystems on comparable soils
in an interglaciation period, then soils with silicate contents
above 5% would still be in the silicate buffer range or swinging
between the silicate and cation exchange buffer ranges, provided
there was no acidifying influence by man's activities. With a
silicate content of 50% in the parent rock, the ecosystem may stay
in this stability range for 100,000 years. During this period, the
lithospheric environment of the ecosystem will change slowly by the
transfer of primary silicates into secondary clay minerals. As
long as the weathering rate of the silicates remains constant, this
will increase the resilience of the system by enlarging the exchangeable Ca and Mg stores.

Humus Disintegration (Destabilization Phase I)

A very long lasting de-coupling of the ion cycle should result
in loss of the organic matter that was accumulated in the whole
rooting zone, including the subsoil, during the aggradation phase
and which was kept at a constant level during stability range I.
This type of de-coupling is called "humus disintegration" (Ulrich
1980). The ecosystem is in a non-stationary transition state which
is necessarily limited in time. The time period under question may
be decades to centuries, however.

As an example, I will consider the same soil as before, assuming
that the total soil organic matter is mineralized, N is being
nitrified, and denitrification is zero. In such a case, there
would be proton consumption by mineralization of the Ca saturated
acidic groups amounting to 200 kmol/ha. On the other hand, nitrification produces 360 kmol H^+/ha. The balance of both partial processes
yields a proton production of 160 kmol H^+/ha.

If one assumes that this process takes place within 100 years,
the mean annual rate would be 1.6 kmol H^+/ha. This amount exceeds

the possible buffer rate in the silicate buffer range for many
soils, thus increasing soil acidity. Existing data indicate that
in the early stages of humus disintegration, N rich compounds are
preferably mineralized. This mineralization leads to an increase
in the C/N ratio of the remaining soil organic matter. Probably
humus disintegration also increases the mobility of clay and thus
can lead to clay migration (leesivation) in the soil.

Under natural conditions, this process seems to be bound to
strong climate changes which leads to the destruction of the forest
ecosystem. Examples are the changes from interglacial to glacial
periods. The destruction of forest ecosystems by man may trigger
the same process. Sustained damage to primary producers is also
caused by man. Clearcutting, grazing, and shifting agriculture may
result in sustained reduced primary production. Part of this
reduction is in the amount of understory root and overstory leaf
litter produced. It further changes the microclimate of the soil,
that is, the climatical conditions for the decomposers. Both
effects, the reduction in litter production and the increase in
soil temperature, operate to lower organic matter stores in the
mineral soil. They thus may initiate humus disintegration. The
process can be started by the acidity produced (alkalinity not
returned) by the de-coupling of the ion cycle due to biomass utilization.

One might assume that all forest ecosystems in Central Europe
have passed through the aggradation phase and reached stability
range I after the last glaciation period. With the exception of
parent material very low in silicate content and shallow soils,
most forest ecosytems then should still be in stability range I.
This is not the case. Most forest subsoils are in the cation
exchange or Al buffer range; they show no activity of soil burrowing
animals, and the roots are inhomogeneously distributed. These
soils must be assumed to have passed through the phase of humus
disintegration centuries or millenniums ago. It must further be
assumed that most often this process has been human initiated
through biomass utilization.

It seems that all forest ecosystems in Central Europe that
have not been subjected to humus disintegration before have now
switched over into this phase. There is no reason for this other
than acid deposition. Acid deposition buffered at the leaf surface
is transferred to the soil close to the roots via the regulation of
ion uptake (Ulrich 1983a). Acid deposition, therefore, is especially
suited for acidifying the soil in the deeper rooting zone. A
reduction of exchangeable Ca may, in combination with a climatic
acidification push, initiate humus disintegration and prevent the
process from being stopped again.

In the beginning stages of humus disintegration, the ecosystem resembles the stable ecosystem in the silicate buffer range except for low pH values, especially in the deeper rooting zone: the humusform still can be mull, the rooting still deep and homogeneous, and the soil structure still crumb-like. Due to the continuous surplus of nitrification compared to nitrate uptake, tree growth may be excellent and plants indicating high N supply may appear in the shrub and herb layers. The Al ions released from clay minerals are bound to soil organic matter, reducing their toxicity. The feature typical for the process in the deeper rooting zone is the leaching of nitrate in combination with Ca and Mg. The leaching of nitrate indicates the continuous net nitrification (i.e., formation of HNO_3) in the rooting zone. The leaching of Ca and Mg indicates acidification, that is, the loss of basic cations and their replacement by Mn and Al ions (cation acids). The process of humus disintegration can be stopped at any point, if the rate of proton loading (HNO_3 formed exceeds HNO_3 uptake + acid deposition) becomes smaller than the rate of proton consumption by base cation release during silicate weathering. That this condition is reached under the influence of acid deposition is very improbable, however. Also liming may be ineffective if the earthworm activity is not high enough to transport substantial amounts of buffering substances to the subsoil.

Due to its long duration and the excellent growth of Al tolerant trees species, the process is very difficult to recognize and nobody is aware of its dramatic end: the change of the chemical soil state into the Al and Fe buffer ranges with all of the accompanying consequences for plant growth and decomposer activity. Many stands showing fir die back seem to be subjected to humus disintegration (Ulrich 1981b). The same seems to be true for some forest stands in subalpine mountains.

Buildup of a Decomposer Refuge (Destabilization Phase II)

If soil acidification has led to cation acid toxicity such that bacterial activity in the soil is strongly limited (except in the rhizosphere and the interior of dead roots), soil burrowing animals also are missing and the ecosystem tends to carry out litter decomposition in a top organic layer separated from mineral soil. The lower part of the top organic layer may be rooted. In this case, not only leaf but also root litter contributes to its formation. The accumulation of a top organic layer means the buildup of a decomposer refuge after the mineral soil has become toxic due to the presence of cation acids. It is started by the retardation of leaf litter decomposition and accompanied by the loss of ground vegetation. As a consequence, a fermentation layer (OF horizon) is formed (humusform: F-mull). If the OF increases and roots stretch between the OF and the uppermost mineral soil horizon, the accumulation of a well decomposed humified horizon

(OH) starts (humusform: mull-like moder). The accumulation can be stopped at any stage, provided that conditions allow decomposition and mineralization to continue at the same rate as litter production (to a steady state). This is usually achieved after developing a full OL (litter horizon)-OF-OH profile (humusform: moder). The mineral soil, at least its uppermost horizons, stays in the Al buffer range or swings between the exchange buffer range and the Al buffer range.

The buildup of an organic top layer is accompanied by proton production in the root zone. In the beech ecosystem, for example, the rate of proton production can be estimated from the data presented in Table 1. If litter decomposition is completely stopped, the proton production in soil corresponds to the cation/anion balance of the leaf litter and amounts to 1.3 kmol H^+/ha·yr. All values between zero and this maximum value are possible, depending upon the fraction of litter that is accumulated. This means that the ecosystem can again reach a steady state, characterized by an acid soil and moder as humusform. The system can, however, also switch over to podzolization; if the buffer rate in the Al buffer range is too small compared to proton production, proton buffering will then be taken over by Fe oxides.

Stability with Low Resilience (Stability Range II)

As soon as the mineralizaton rate approaches the rate of litter production, proton production ceases and the ecosystem can again reach a quasi-steady state. The chemical soil state may then slowly return to the cation exchange buffer range. The rate of this recovery process depends upon the rate of proton consumption by silicate weathering (base cation release). By this process, cation acids are transferred to uncharged compounds and base cations produced, so the percentage of exchangeable Ca and Mg can slowly increase. Finally, a steady state may be achieved where the base cation percentage fluctuates around a mean value corresponding to the effects of climatic acidification pushes and de-acidification phases.

This stability range is characterized by a much lower resilience than stability range I, with the soil staying in the silicate buffer range. The lack of soil burrowing animals causes soil compartments to develop with different chemical states. Between these compartments, chemical gradients exist. Such compartments are the organic top layer and the mineral soil, the different horizons within these layers, and the interiors and surfaces of aggregates within these horizons.

The extension of roots into the organic top layer may be greatly advantageous for avoiding spatial de-coupling of proton production and proton consumption, but it makes the plants suscep-

tible to drought periods. Within the mineral soil, the roots grow
along the surfaces of aggregates. The humic substances formed
during root decomposition therefore accumulate at the aggregate
surfaces close to the roots and are not mixed within the total soil
mass. This condition considerably increases the risk of cation
acid (Al) toxicity in a strong seasonal acidification push. Nitric
acid forms only in a very small soil volume, which is close to the
roots and which has very limited buffer capacity. The frequency of
the formation of toxic cation acids injuring microorganisms and
roots therefore can vary considerably among ecosystems.

Most forest ecosystems in Central Europe have passed through
humus disintegration and buildup of a decomposer refuge centuries
or millenniums ago and should be in this stability range with the
soil somewhere in the cation exchange buffer range. Almost all of
these forests are subjected to biomass utilization by modern forestry.
As discussed already, the proton production caused by bole utilization
in the beech ecosystem under consideration is 0.6 kmol H^+/ha·yr.
In most forest ecosystems, this internal proton load may be balanced
by silicate weathering, and thus these ecosystems can maintain a
steady state under the influence of biomass utilization with careful
forest management. This buffering ability can be checked by long
term cation/anion balances. But silicate weathering may not be
able to buffer the acidification effect of biomass utilization. If
the latter occurs, the ecosystem will change slowly during tree
generations with productivity decreasing. Such cases may be widespread but hidden due to the overwhelming influence of acid deposition.

In most managed forest ecosystems, no buffer ability by silicate
weathering remains to neutralize acid deposition. This means that
acid deposition decreases resilience and then destabilizes almost
all forest ecosystems. The higher the rate of acid deposition, the
faster the effects will show up. But even low rates of acid input
will lead finally to destabilization of forest ecosystems. It is
our responsibility, here and now, to decrease the emission of acid
forming gases.

Podzolization Phase

Podzolization usually is considered as a soil process, but it
is essentially an ecosystem process. The source of protons again
is the de-coupling of the ion cycle by the accumulation of litter
and litter residues on top of the mineral soil. It differs from
the phase of buildup of a decomposer refuge in that the soil stays
in the Fe buffer range. Due to the presence of stronger acids, the
stress to the trees may be stronger than in the Al buffer range.
Because proton production is bound to the accumulation of the
organic top layer, podzolization ceases if the organic top layer
reaches a steady state. One should therefore carefully distinguish

between active and inactive podzols; the latter shows the features of podzolization (greyish A horizons and accumulation of Fe and organic matter in a deeper horizon) but the process is not going on. Acid tolerant trees (like Norway spruce and Scots pine) can withstand the high acid strength of podzolization for a limited time period, around two decades probably. If the process then stops, the trees will survive and the forest cover will remain. This is the time span which the buildup of a decomposer refuge needs on a strongly acidified soil and thus is what the trees have experienced during evolution. If due to the action of man (e.g., by using biomass, by founding plantations, or by acid deposition) the active podzolization continues over longer periods of time, the trees will die back and the forest will change into some other ecosystem (e.g., a heath). Incipient podzolization (bleaching without a B horizon) is now very common in central European forests as a consequence of acid deposition. If acid deposition continues, the forest eventually will die back. There is no way out except the reduction of acid deposition.

RESILIENCE DUE TO ORGANISMS

The problem of tree die back on acid soils leads back to resilience of ecosystems. As already stated, the main weight of buffering lies on the soil and not on organisms. This is especially true for ecosystems of high diversity in stability phase I. In ecosystems on acid soils, however, organisms have abilities to buffer acidity. The acidity appears in the form of cation acids and one speaks about tolerance to Mn and Al toxicity. From an ecosystem point of view, the mechanisms acting first in acid stress are located in the soil. If these mechanisms are overstressed, they produce cation acids and the rhizosphere and mycorrhiza become involved. If the latter cannot maintain neutrality and become injured, the acid soil solution can enter the free space of the root cortex. Here the apoplasm comes into play; it may possess a substantial buffer capacity in the form of Ca ions bound to acidic groups in the cell walls. The tolerance mechanisms can make use of three types of chemical reactions:

(1) Cation exchange: the cation acids are bound to surfaces; in solution they are replaced by Ca (restriction: limited buffer capacity);

(2) Precipitation of cation acids, e.g., as phosphates or silicates (restriction: limited resources of anions, P deficiency);

(3) Precipitation of cation acids as hydroxides, partially in the cortex (restriction: limited resources of basicity in an acid environment);

(4) Chelation of cation acids: the formation of metal-organic compounds masks the toxic ion; if the complexes are stable enough, they may pass through the symplasm without doing harm.

Acid tolerant tree species seem able to protect the symplasm very effectively. Their problem is, however, the acidification of the apoplasm. Bauch and Schröder (1982) have shown that injured roots of *Abies alba* (fir) and *Picea abies* (Norway spruce) lose Ca and Mg from the cortex, whereas Al is present. In diseased spruce trees, the cell walls of the cortex and primary xylem do not contain Ca and only small amounts of Mg. The average concentration of Al is lower than in controls from healthy trees. Chemically, this is to be expected if the solution in equilibrium with the cation exchanger approaches pH values around 4.2: dissociation of acidic groups in the cell walls reverses, thus limiting the exchange capacity; Ca and Mg are completely displaced from the exchanger surface; and polymeric Al ions are transferred to Al^{3+}, thus increasing the charge of the Al ions and decreasing the amount of Al which can be bound at the exchanger surface. Tischner et al. (1983) showed that under these conditons the cortex of spruce seedlings separate from the xylem and the formation of the endodermis is disturbed.

A further important mechanism for acid tolerant species to increase resilience is the regeneration of injured roots. This ability may become limited as more cation acids penetrate older roots also, so that finally the whole root system may be acidified.

LITERATURE CITED

Bache, B.W. 1983. The implications of rock weathering for acid neutralization. Pages 175-187 *in* Ecological effects of acid deposition. Nat. Swed. Environ. Prot. Board, Rep. SnV pm 1636.

Bauch, J., and W. Schröder. 1982. Zellulärer Nachweis einiger Elemente in den Feinwurzeln gesunder und erkrankter Tannen (*Abies alba* Mill.) und Fichten (*Picea abies* Karst.). Forstwiss. Centralbl. 101:285-294.

Hovland, J., G. Abrahamsen, and G. Ogner. 1980. Effects of artificial acid rain on decomposition of spruce needles and on mobilisation and leaching of elements. Plant Soil 56:365-378.

Johnson, N.M., Ch.T. Driscoll, J.S. Eaton, G.E. Likens, and W.H. McDowell. 1982. "Acid rain," dissolved aluminum and chemical weathering at the Hubbard Brook Experimental Forest, New Hampshire. Geochim. Cosmochim. Acta 45:1421-1437.

Killham, K., M.K. Firestone, and J.G. McColl. 1983. Acid rain and soil microbial acitivity: Effects and their mechanisms. J. Environ. Qual. 12:133-137.

Mazzarino, M.J., H. Heinrichs, and H. Fölster. 1983. Holocene versus accelerated actual proton consumption in German forest soils. Pages 113-123 *in* B. Ulrich and J. Pankrath, eds. Effects of accumulation of air pollutants in forest ecosystems. Reidel, Dordrecht, The Netherlands.

Prigogine, I.   1947.   Etude Thermodynamique des Processus Irrever-
    sibles. Desoer, Líege, Belgium.
Strayer, R.F., C.-J. Lin, and M. Alexander.   1981.   Effect of
    simulated acid rain on nitrification and nitrogen minerali-
    zation in forest soils.   J. Environ. Qual. 10:547-551.
Tamm, C.O., and G. Wiklander.   1980.   Effects of artificial acidifi-
    cation with sulphuric acid on tree growth in Scots Pine forest.
    Pages 188-189 in D. Drabløs and A. Tollan, eds.   Proc. Int.
    Conf. Ecol. Impact Acid Precip. SNSF-project, Oslo-As, Norway.
Tischner, R., U. Kaiser, and A. Hüttermann.   1983.   Untersuchungen
    zum Einfluß von Aluminium-Ionen auf das Wachstum von Fichten-
    keimlingen in Abhängigkeit vom pH-Wert.   Forstwiss. Centralbl.
    102:329-336.
Ulrich, B.   1980.   Die Bedeutung von Rodung und Feuer für die
    Bodenund Vegetationsentwicklung in Mitteleuropa.   Forstwiss.
    Centralbl. 99:376-384.
Ulrich, B.   1981a.   Ökologische Gruppierung von Böden nach ihrem
    chemischen Bodenzustand.   Z. Pflanzenernähr. Bodenk. 144:289-305.
Ulrich, B.   1981b.   Eine ökosystemare Hypothese über die Ursachen
    des Tannensterbens.   Forstwiss. Centralbl. 100:228-296.
Ulrich, B.   1981c.   Theoretische Betrachtung des Ionenkreislaufs in
    Waldökosystemen.   Z. Pflanzenernähr. Bodenk. 144:647-659.
Ulrich, B.   1983a.   A concept of forest ecosystem stability and of
    acid deposition as driving force for destabilization.   Pages
    1-29 in B. Ulrich and J. Pankrath, eds.   Effects of accumulation
    of air pollutants in forest ecosystems.   Reidel, Dordrecht,
    The Netherlands.
Ulrich, B.   1983b.   Soil acidity and its relations to acid deposi-
    tion.   Pages 127-146 in B. Ulrich and J. Pankrath, eds.
    Effects of accumulation of air pollutants in forest ecosystems.
    Reidel, Dordrecht, The Netherlands.
Ulrich, B., R. Mayer, and P.K. Khanna.   1979.   Deposition von
    Luftverunreinigungen und ihre Auswirkungen in Waldökosystemen
    im Solling.   Schriftenr. Forstl. Fak. Univ. Göttingen 58,
    Sauerländer, Frankfurt, W. Germany.

WHY MATHEMATICAL MODELS IN EVOLUTIONARY ECOLOGY?

Nils Chr. Stenseth

Department of Biology, Division of Zoology
University of Oslo
P. O. Box 1050
Blindern, N-0316 Oslo 3, Norway

INTRODUCTION

Science is the attempt to make the chaotic diversity of our
sense-experience correspond to a logically uniform system of
thought (Albert Einstein, cited from Solbrig 1981; reprinted
by permission of Blackwell, Oxford, England).

Without Ecological Theory No Ecological Science

Some ecologists develop mathematical models, other use empirical
(often descriptive) methods, still others use a combined theoretical-
empirical approach. I favour the third approach. Ecology is a
science about the real world--hence, empirical observations.
However, these observations must be placed into a broader scientific
framework--hence, theory. Without theory there is no science,
without empirical observations there is no natural science. Although
mathematics is not absolutely necessary, I argue that it helps
greatly in deducing a logically uniform system of thought--mathematics
is an important tool. In Maynard Smith's (1982a) words, "Mathematics
without natural history is sterile, but natural history without
mathematics is muddled." Or as Haila and Järvinen (1982) put it:
"Sound naturalism is to ecology what legs are to a runner; but
antitheoretical naturalists are, quite naturally, like headless
runners."

Roughgarden (1979) also described the need for mathematics:

There are two simple reasons why we cannot avoid mathematics
even if we wanted to. The first reason why mathematics is

necessary in population biology [or ecology in general] is
that the questions which define the field of population biology
are essentially quantitative to begin with. For example, we
want to know how...competition and predation influence the
population size and distribution of the interacting populations.... [This] question is a quantitative question and...
it cannot be stated precisely and answered without a mathematical
approach.

The second reason why mathematics is necessary in population
biology is that a natural population is often an entity which
exists on a larger scale than human experience.... [Hence,]
even if we know the mechanisms of an interaction between the
individuals of two populations, we cannot directly witness the
results, for the populations, of that interaction. Models
help us to infer results on the population level from information
about individuals in these populations.... [Mathematical]
models help us to infer what we cannot experience... from what
we can. (Reprinted by permission of the author.)

Unfortunately, mathematical modelling often becomes an end in
itself (e.g., Pielou 1977:107-116, 1981)--this sterile modelling
hinders much of current theoretical ecology. However, current
ecology relies too much on "handwaving" arguments and is muddled.
In this essay, I will show how mathematical models may help us ask
the appropriate questions and collect the appropriate data. To use
Maynard Smith's (1974) words, the present essay is written in "the
twin convictions that ecology will not come of age until it has a
sound theoretical basis, and that we have a long way to go before
that happy state of affairs is reached." Obviously, theory and
mathematical modelling cannot substitute data from the real world.
As Roughgarden (1983a) put it, "[theory] does not substitute for
the knowledge of actual systems any more than architectural drawings
substitute for a house. Yet we cannot build a house without architectural drawings." Thus, theoreticians or modellers must produce
results (i.e., predictions) that the empiricists can use (see Levin
1981 and Salt 1983 for relevant discussions).

What is Evolutionary Ecology?

Ecology commonly is defined as the study of the abundances and
distributions of populations (Krebs 1978a). How are populations,
communities, and ecosystems currently structured and functioning
(see Mayr 1961, Pianka 1978)? Equally important, how have modern
species evolved these features and why can certain species co-exist
and not others (e.g., Christiansen and Fenchel 1977, Roughgarden
1979, 1983b)?

Darwin (1859) was the first evolutionary ecologist. Ecology
and evolution are, he maintained, intimately linked together. How-

ever, with the boom of evolutionary genetics in the 1920s and 1930s
(e.g., Fisher 1930, Wright 1931, Haldane 1932), genetics came to
dominate evolutionary biology--ecology was considered unimportant.
This is unfortunate because without understanding how selective
pressures are determined, we cannot understand evolution:

> Since the major component of the environment of most species
> consists of other species in the ecosystem, it follows that we
> need a theory of ecosystems in which the component species are
> evolving by natural selection.... [What] we need is a theory
> which says something...about the environment (Maynard Smith
> 1982b; copyrighted 1982 by MacMillan, London, England; reprinted
> by permission of the author and *Nature*).

Hence, we cannot understand patterns in the fossil record
without understanding the corresponding ecological interactions
(see Stenseth 1984a). Because the phenotypic changes observed in
the fossil record correspond directly to the phenotypic changes
observed over the species' geographic range, ecology and paleontology
mutually supplement each other. The last decades' trend towards a
unification of the Darwinian theory of evolution and the field of
ecology is, in my mind, healthy. Even though evolutionary ecology
recently has spawned many untestable hypotheses and unwarranted
generalizations (e.g., Grant 1983), I am convinced that evolutionary
ecology at its best has resulted in a deeper understanding of how
populations, communities, and ecosystems are structured, and why
these ecological units came to be as they are. I am further convinced
that many of the unwarranted generalizations have been put forth as
a result of informal and non-mathematical analyses--it is an intrinsic property of a young and immature science.

Mathematics and Statistics in Ecology

Mathematics has never played the role in ecology that it has
in evolutionary genetics (e.g., Solomon and Walter 1977). During
the last couple of decades, there has been, however, an increase in
both volume and relevance of mathematical work in ecology. Much of
this is, I believe, due to the influence of R. MacArthur, R. May,
J. Maynard Smith, and J. Roughgarden. Their work is based on the
studies of the pioneers in theoretical and mathematical population
biology (i.e., population genetics and population ecology combined)
from the 1920s and 1930s--R.A. Fisher, J.B.S. Haldane, and S. Wright
in genetics, and V. Kostitzin, A.J. Lotka, and V. Volterra in
ecology.

I distinguish between the use of mathematical modelling and
the use of statistics. Both are important. Mathematics helps us
construct null-hypotheses to be tested against data by statistical
methods. See Strong (1980) for a discussion of the importance of
null-hypotheses in ecology and Dolby (1982) for the role of statistics

in the life sciences. The use of statistics is far more accepted
among ecologists today than is the use of mathematical modelling.
But deducing the statisticians' null-hypotheses cannot be "delegated"
to intuition or common sense. These derivations should be based on
formal reasoning. To quote Strong (1983) (albeit somewhat out of
context), "ecology needs more than common sense. 'Common sense is
that which tells you that the world is flat' (Chase 1938)."

To deduce the statisticians' null-hypotheses by formal reasoning
is, I believe, particularly important when studying the long and
short term behaviours of dynamic ecological processes. It is, for
example, tremendously difficult to deduce intuitively under which
conditions a predator-prey interaction will, in an otherwise stable
environment, lead to a stable equilibrium and under which conditions
such interaction will lead to a repeated pattern of change. With
the help of mathematical modelling, this is easier, albeit still
difficult (e.g., Lotka 1925, Volterra 1926, Maynard Smith 1974,
Roughgarden 1979).

Many empiricists feel strongly that ecological theory and
mathematical modelling have a long way to go before they can help
us understand the diversity and complexity of the natural world. A
common feeling is that theory has nothing useful to tell practicing
ecologists. Many theoretical ecologists are equally convinced that
loosely formulated concepts have received undue emphasis. These
conflicting views have resulted in a large gap between theorists
and empiricists, which is a serious obstacle to progress in ecology.
Unfortunately, "theoretical journals" such as *Ecological Modelling*,
Journal of Mathematical Biology, *Journal of Theoretical Biology*,
and *Theoretical Population Biology* exist. In my mind, theoretical
papers should be published in journals which also publish data
papers on the same topic. However, as long as theoretical papers
are rejected by the more empirical journals simply because they are
mathematical or theoretical, theoretical/mathematical journals are
needed. This is not to say that all printed theoretical papers
should have been published. Some theoretical papers do not relate
to living systems and should not be published as biological papers.
Nor should empirical papers simply reporting curiosa be published.

One of the greatest biologists in recent time, J.B.S. Haldane,
pointed out in *Possible Worlds* (1927) that to gain a proper perspective of our own world, we should try to conceive alternative realities
that may seem fantastic at first, but which nevertheless are ratio-
nally conceivable. By this view, theory and mathematical modelling
help us distinguish what is possible from what is not possible;
only what is theoretically possible, of course, exists in nature.
Data and statistical analysis help us discriminate the actualities
among the many possibilities. Failure to develop mathematical
models giving the presumed consequences of some premises (i.e., the
hypothesis) does not necessarily mean that the hypothesis is theoret-

ically impossible; but if several modellers or theoreticians try, but fail, we may distrust the plausibility of the argument.

The formulation of mathematical models confers at least two advantages. First, they prevent awkward semantic discussions, because in formulating mathematical models, our assumptions must be made explicit (but see Fig. 1). This helps clarify our thoughts. Second, as soon as one attempts to formulate a model, it becomes apparent what data we need. As pointed out above, mathematical modelling provides a basis for defining the statisticians' null-hypotheses with which the empiricists' data are to be compared (see Pielou 1977:104 for a discussion of this point).

Testing a mathematical model against data establishes whether the model provides an acceptable account of what is going on in the natural system under study (see the two stimulating papers by Quinn and Dunham 1983 and Toft and Shea 1983).

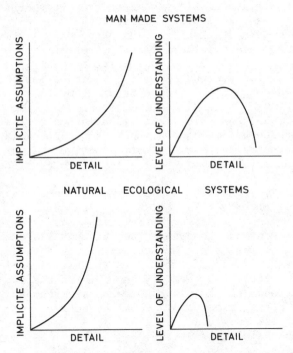

Fig. 1.   Relationships, as predicted by Bunnell (1973), between the levels of understanding for humanmade and natural systems. Compared with humanmade systems (e.g., space capsules), the peak value of understanding for ecological systems occurs at a relatively low level of detail due to our incomplete understanding of natural systems. See Agren et al. (1980) for a discussion of this issue, with conclusions similar to mine. (From Stenseth 1977a.)

WHY ARE MATHEMATICAL MODELS DEVELOPED?   A TAXONOMY OF MODELS

We have both ecological and evolutionary models. In proper
ecological models--referring to ecological time (Stenseth and
Maynard Smith 1984)--the parameters may be treated as constants
because the "adaptive peak" will be approximately fixed. However,
on the evolutionary time scale (Lawlor and Maynard Smith 1976,
Stenseth and Maynard Smith 1984), these ecological parameters
become evolutionary variables. Even though this is an important
issue of mathematical modelling in evolutionary ecology, I will not
discuss this further here (see, e.g., Reed and Stenseth 1984). In
the following, I will concentrate mainly on ecological models.

Mathematical descriptions of ecological systems may be made
for practical or theoretical purposes. I will call these "empirical"
and "theoretical" models, respectively (Stenseth 1977a, 1977b).
Other names are "inductive" and "deductive" (Watt 1961, Hempel
1965, Stenseth 1975). Maynard Smith (1974) called empirical models
"simulations." If stochastic processes are incorporated, the two
kinds of models may, following Hempel (1965) (see also Elster 1983),
be called "inductive statistical" and "deductive statistical"
models. Special cases of the former are correlation and regression
models, discussed in any standard statistics text (e.g., Draper and
Smith 1966, Snedecor and Cochran 1967). Simulation models are
therefore often nothing but "curve fitting" models.

Lotka's (1925) and Volterra's (1926) predator-prey and competition models are deductive models; their stochastic analogues (e.g.,
Pielou 1977) are typical examples of deductive statistical models.
The recruitment models for fisheries developed by Ricker (1954),
Beverthon and Holt (1957), and others (see also Clark 1976) are
inductive models. Finally, abundance models, such as the log
normal model (e.g., May 1975, Engen 1978, Ugland and Gray 1982),
are other examples of inductive statistical models even though
attempts have been made to deduce the log normal abundance model
from first principles (e.g., MacArthur 1957, Pielou 1975) and,
hence, to consider it as a deductive statistical model. Ecosystem
models (e.g., Innis 1978) commonly are nothing but curve fitting
models, even though their authors often claim them to be deductive
models. Some clear examples of proper deductive ecosystem models
do, however, exist (e.g., Kerner 1957, 1959, Patten and Auble 1980,
1981, Stenseth and Maynard Smith 1984); these I discuss below in a
separate section.

Empirical (or inductive) models make no pretence of explaining
the real operation of a system. They only pretend to faithfully
reproduce the behaviour of the system under various conditions.
Hence, they may be used for predictive purposes. If, for example,
a management agency wishes to know how many fur seals can be culled
annually from a population without threatening the population's

future survival, it would be necessary to have a detailed description of that particular population in that particular environment; hence, the model must include as much relevant detail--justified by available empirical information--as possible. At minimum, one would require age specific birth and death rates and knowledge of how these rates vary with the density of the population as well as with other features of the environment. If it included stochastic features, it would be a inductive statistical model. Elsewhere I have discussed such empirical models (Stenseth 1977b).

Theoretical (or deductive) models, on the other hand, attempt to provide insight into the organization and operation of the real world. These models, as all theory, cannot be proven correct; they can only be invalidated. They are never assumed to be valid for any particular system; they are meant to incorporate the kinds of interactions we find in real life. Earlier I have discussed this type of model in a slightly different perspective (Stenseth 1977a).

In the following, I will discuss primarily theoretical (or deductive) models within the framework of the hypothetico-deductive method (e.g., Hempel 1966)--the method of the natural sciences (see also Platt 1964, Haila 1982).

When discussing mathematical modelling, it is important to remember what Maynard Smith (1974) said in the "Introduction" to his *Models in Ecology*:

> A theory of ecology must make statements about ecosystems as a whole, as well as about particular species at particular times, and it must make statements which are true for many different species and not just for one. Any actual ecosystem contains far too many species, which interact in far too many ways, for simulation to be a practical approach. The better a simulation is for its own purposes, by the inclusion of all relevant details, the more difficult it is to generalize its conclusions to other species. For the discovery of general ideas in ecology, therefore, different kinds of mathematical description, which may be called [theoretical] models, are called for. Whereas a good simulation should include as much details as possible, a good [theoretical] model should include as little as possible. (Reprinted by permission of the Cambridge University Press, Cambridge, England.)

But remember also what the famous Danish physicist Niels Bohr is reported to have said:

> Give me one parameter and I can model a straight line. Give me two parameters and I can make a curved line. Give me three parameters and I can make an elephant. If I get an additional parameter so that I have a total of four, I can even make the elephant wag its tail.

Unfortunately, many modellers have forgotten this fundamental principle of mathematical modelling. An extreme example is provided by Swartzman and Van Dyne (1972) who developed a model for Australian arid ecosystems. Their model contained more than 1000 parameters and 100 variables, most of them estimated on the basis of some superficial field investigations!

It is important to simplify as much as possible in science and in mathematical modelling (see Fig. 1). But arriving at an acceptable degree of simplification always involves some human judgements (see, e.g., Roughgarden 1983a).

For useful introductions to the topic of mathematical modelling in ecology, see, for instance, Maynard Smith (1968, 1974), Clarke (1976), Pielou (1977), and Jeffers (1978). A variety of mathematical techniques is used in ecological modelling. However, because ecological interactions by nature are dynamic, the mathematical formulation must be dynamic. This is one of the reasons why differential and difference equations are used so commonly in ecological modelling (e.g., Lewontin 1969). We should, however, avoid the fallacy of "changing" biology to make it fit available mathematical techniques or, even worse, to fit the mathematical techniques(s) of any particular theorist. If necessary, new mathematical techniques should be developed to meet the requirements of the biological issue under study (see Goldstein 1977 and Williams 1977 for further discussion of this point).

AN IMPORTANT HISTORICAL EXAMPLE--COMPETITION MODELS AND GAUSE'S "COMPETITIVE EXCLUSION PRINCIPLE"

Lotka's (1925) and Volterra's (1926) models for competitive interactions (see also Gause 1934) are

$$dx/dt = a \cdot x - b \cdot x^2 - c \cdot x \cdot y$$
$$dy/dt = e \cdot y - f \cdot x \cdot y - g \cdot y^2 \tag{1}$$

where x and y are the densities of the two competing species, and a, b, c, e, f, and g are positive parameters defining the intra- and interspecific relationships. As is well known (e.g., Krebs 1978a, Pianka 1978), the only relative position of the two isoclines (i.e., $dx/dt = 0$ and $dy/dt = 0$) giving rise to stable co-existence is as depicted in Figure 2A. As can be seen, stable co-existence requires that $a/b < e/f$ and $e/g < a/c$.

There are two possible ways in which these inequalities can be interpreted biologically (see, e.g., Maynard Smith 1974). First, suppose that the intrinsic rates of increase of the two competing species (a and e) are equal. Then the conditions for stable co-existence become b > f and g > c (i.e., intraspecific competition must be stronger than interspecific competition). If, for example, the two competing species are limited to some degree by different resources, then these inequalities are likely to hold; however, if the two species have identical resource requirements, one of the two competing species is likely to be more efficient and is, consequently, likely to eliminate its competitor.

Another interpretation of the stability criteria derived from Figure 2A is that the joint equilibrium point should lie above the line joining the two allopatric equilibrium points. The biological significance of this can be seen more easily if the densities of each competing species are expressed as fractions of their allopatric equilibrium densities, a/b and e/g, respectively. Then a stable co-existence between the two competing species, requires--from model 1--that the total "equivalent density" at the joint equilibrium is greater than unity. This condition is easily realized if the two species use somewhat different resources.

These interpretations are of great historical interest as they formed the basis for "Gause's Competitive Exclusion Principle": two species having identical requirements and in competition for some limited resource cannot co-exist in the same habitat. Gause never claimed credit for this idea (e.g., Hardin 1960); he properly ascribed it to Lotka and Volterra who developed and analyzed model 1. Gause (1934) designed and carried out his famous experiments on various Protozoan species after reading Lotka (1925) and Volterra (1926). The mathematical arguments suggested the existence of some

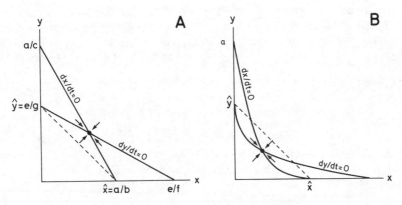

Fig. 2.   Competition between two species. (A) depicts the isoclines dx/dt = 0 and dy/dt = 0 for the linear competition model 1; the broken line joins the two allopatric equilibria $\hat{x} = a/b$ and $\hat{y} = e/g$. (B) depicts the more general, non-linear model 2.

unknown patterns which later were established by carefully designed experiments (see, e.g., Hutchison 1978). One wonders whether this famous principle ever would have been "detected" without the help of mathematical modelling. (Lack 1947 gave the credit to Gause; in the same book, Lack showed how this principle could be used to explain the distribution of Darwin's finches on the Galapogos Islands. For an account of the history of competition studies, see Hutchison 1978, Chapter 4.)

Analysis of models more general than model 1 demonstrates that the second interpretation of the stability condition (derived from Fig. 2A) is not generally valid. Before this was realized, great confusion occurred in the literature. Ayala (1969) (and some subsequent commentators) interpreted the findings in a *Drosophila* competition experiment as disproof of Gause's Competitive Exclusion Principle. However, as pointed out by Antonovics and Ford (1972) and Gilpin and Justin (1972), the disagreements between predictions derived from model 1 and Ayala's results should be viewed as an illustration of the inadequacy of the logistic model used in model 1 and as a result of the linear approximations used in that model.

A generalized version of model 1 (considered by Maynard Smith 1974) may be written as

$$dx/dt = x \cdot f_1(x,y)$$
$$dy/dt = y \cdot f_2(x,y)$$

(2)

where f_1 and f_2 are the density dependent specific growth rate functions for the two competing species. In terms of model 1, $f(x,0) = a - b \cdot x$ and $f_2(0,y) = e - g \cdot y$ (i.e., the logistic growth model). For model 2 to represent a competitive system, we must assume that $\partial f_i/\partial x < 0$ and $\partial f_i/\partial y < 0$ for $i = 1$ and 2 (e.g., Odum 1971, Christiansen and Fenchel 1977). Analysis of model 2 demonstrates that stable co-existence of the two competing species requires that

$$(\partial f_1/\partial x) \cdot (\partial f_2/\partial y) > (\partial f_1/\partial y) \cdot (\partial f_2/\partial x)$$

(3)

where the differentials are evaluated at the joint equilibrium of model 2 (see Fig. 2B). Inequality 3 (above) says that the product of the self limiting effects for each species must be greater than the product of the mutually inhibiting effects. As can be seen, the first biological interpretation based on model 1 is still valid, the second is not. However, in either case, Gause's Competitive Exclusion Principle emerges.

This sequence of theoretical and empirical studies is, in my view, a nice demonstration of how theoretical and empirical work interact to generate deepened insight into an ecological phenomenon. It exemplifies the hypothetico-deductive method of science.

SOME FURTHER EXAMPLES OF THEORETICAL MODELS

Does Competition Occur?

Currently there is a vigorous debate whether biological competition really occurs, whether it is a common phenomenon, and
whether it is an important factor in structuring ecological communities (e.g., Järvinen 1982, Schoener 1982, Roughgarden 1983a,
Simberloff 1982, 1983, Strong 1983, Connell 1983, Harvey et al.
1983). This is not the place to review this controversy. Here I
will discuss only briefly how the occurrence of competition may be
established or dismissed--a theme closely related to "why models in
ecology," even though this link is not apparent from the current
controversy.

Recently Framstad and I (1984) reviewed some models and data
pertaining to competition in small mammals. We concluded that
methods based on removal experiments, giving direct tests of competition, is preferable to indirect methods based on differences in
distribution and density. However, as presently employed, most
removal experiments are neither satisfactorily performed nor analyzed.
Several experimental plots must be compared with several control
plots to avoid spurious numerical effects; often this is not done.
In fact, several of the "musts" in experimental design (e.g.,
Clarke 1980), dictated by statistical analysis, often are violated
in ecologists' field experiments.

Furthermore, a mathematical model for the particular competitive
interaction under study must be formulated before the experimental
study; only then may the response to removing a competitor be
properly quantified and interpreted. Even a linear version of the
Lotka-Volterra competition model is better than no model! I believe
one always must have some model in mind, even if only subconciously,
to interpret the results of any experimental manipulation.

Microtine Density Cycles: Two Theories

The study of microtine density cycles has greatly influenced
ecology in general and population dynamics theory in particular--it
is difficult to imagine what ecology would be like today without
these studies. The existence of the microtine density cycle, also
is, in my mind, one of many reasons why ecologists are concerned
with questions relating to the stability-instability dicothomy.
Extensive descriptive time series data exist now on these cycles,
and an enormous number of so called hypotheses have been proposed
to explain their existence (for a review, see Stenseth 1984a). As
a theoretician I find it unfortunate, however, that those proposing
hypotheses for explaining these cycles commonly do not evaluate the
theoretical aspects of their suggestions. Hence, we often do not
know the premises for the various hypotheses, neither do we know

what must be satisfied before a particular hypothesis applies. It
is therefore difficult to test any of the available hypotheses
(Stenseth 1984a).

Population ecologists such as microtine biologists should
adopt the habit of population geneticists, who regularly present a
mathematical model whenever they present some new idea (see Provine
1977). I suggest that this lack of formal treatment is one of the
reasons why we still do not know why the microtine cycles occur in
spite of the tremendous amount of investigation into their causes.
To clarify my point, I will review briefly in this section some of
these difficulties with reference to two suggested hypotheses. In
a subsequent section, I will discuss a third hypothesis.

An entire class of hypotheses assumes the operation of intrinsic
factors only. Some of these hypotheses (or theories) assume that
cycles result because animals are behaviourally and demographically
different at various stages of the density cycle (e.g., some individuals reproduce more and are non-aggressive, whereas others reproduce
less but are more aggressive). The hypotheses differ, however, in
whether they assume the difference to be genetic (e.g., an individual
is born aggressive) or phenotypic (e.g., an individual is aggresive-
ness as a result of the environment in which it is born or lives).
That is, the behavioural and demographic differences observed at
the various phases of the density cycle may be due to the occurrence
of different kinds of animals or to the change of behaviour and
demography of any given individual. Below I will discuss examples
of both kinds--the Chitty theory and the Charnov-Finerty theory,
respectively.

Chitty's theory. All hypotheses assuming genetically fixed
behavioural and demographic differences derive from Chitty's (1960,
1967, 1970, 1977) theory. He assumed that microtine cycles occur
because of changes in the proportion of individuals with genetically
fixed traits and that density does not fluctuate if the population
is genetically monomorphic for these traits. Essentially, the
regular density cycles, he assumed, are due to (competitive) interactions between genetically fixed types of animals. Some evidence
(e.g., Krebs and Myers 1974, Krebs 1978b, Gaines 1981) suggests
that density fluctuations are accompanied by gene frequency changes
in microtine populations. However, this is not confirmation of
Chitty's idea. If a population fluctuates in numbers for purely
extrinsic reasons and if there are some loci (at which alleles are
segregating) whose fitnesses are functions of population density,
then the density fluctuations will cause gene frequency changes
(see Charlesworth and Giesel 1972); thus, the cause-and-effect
relationship may be opposite that suggested by Chitty's theory.

Unfortunately, Chitty has never provided a model that has the
suggested properties; hence, it is difficult to judge the plausa-

bility of his theory. However, some theorists have, in vain, tried to develop realistic models for this theory (for review, see Stenseth 1984a). I have been one of those.

Disregarding dispersal, I developed a general model for this hypothesis (Stenseth 1981a) which suggests that incorporating genetics into a population dynamics model contributes to stabilizing the population density rather than to destabilizing it and generating cycles. Certainly, this is what we should expect considering general ecological competition theory (e.g., Maynard Smith 1974).

Recently, Lomnicki and I (1984) also have analyzed another model for the same hypothesis; in that model, we incorporated dispersal into the original idea of Chitty--a modification which has been proposed by others (e.g., Krebs et al. 1973, Krebs 1978c). Again we found that regular density cycles do not result and that competition between different (geno-) types stabilize the population dynamics.

These unsuccessful attempts, and the results from ecological competition theory, suggest that we ought to be skeptical towards the verbal idea originally stated by Chitty. I am certain that we would have realized this much earlier if ecologists had deduced the consequences of their hypotheses by formal analyses.

However, I would do injustice to Dennis Chitty if I did not point out that he was, as far as I know, the first ecologist who emphasized that individual differences within a population might be important in generating the regular density cycles seen in microtine rodents. Population geneticists have long devoted prime interest to such differences--population ecologists have not. The work of Chitty has had an important impact on ecological thinking in this respect. Thanks to him (among others), ecologists today often incorporate individual differences--behavioural and (or) genetic--as one potentially important factor when trying to explain various population dynamics patterns. However, I believe that the tremendous amount of work devoted to testing this hypothesis that lacks a formal basis (e.g., Taitt and Krebs 1984) has slowed progress in microtine population biology.

Charnov-Finerty's theory. Charnov and Finerty (1980) suggested an alternative hypothesis to explain the occurrence of the microtine cycle. Their idea is that during periods of low density, the specific dispersal rate is low (and hence the total amount of dispersers is low), so that mainly close relatives stay together in any habitat patch (Stenseth 1983a). Based on kin selection arguments (Hamilton 1964a, 1964b), individuals surrounded by close relatives are expected to be phenotypically docile individuals. As density increases, the specific dispersal rate also increases (see Stenseth 1983a), causing non-related individuals to come into close contact

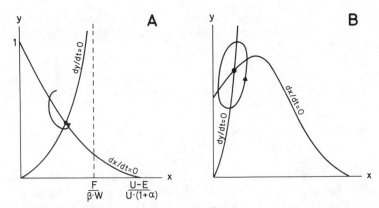

Fig. 3.   Frequency of habitat patches with docile, non-aggressive
          (x) and aggressive (y) individuals in the Charnov-Finerty
          hypothesis. E and F are the extinction rates (per patch)
          for patches with docile and aggressive individuals, respec-
          tively; U and W are the corresponding dispersal rates; and
          α and β are quantities determining how difficult it is to
          enter an occupied patch (relative to entering an empty patch)
          for docile and aggressive individuals, respectively. In (A)
          the migrating individuals do not control where they can
          settle; a stable equilibrium with damped oscillations
          results. In (B) the migrating individuals can choose the
          patches in which to settle (they prefer empty patches over
          occupied ones); a stable limit cycle may result. (Based
          upon unpublished results by Lomnicki and Stenseth 1984.)

with each other. On the basis of kin selection arguments, these
non-related individuals are expected to be phenotypically aggressive.

    This hypothesis resembles those suggested by Calhoun (1949,
1963) and Christian (1950, 1970). Unfortunately, none of these
suggestions have, until very recently, been subjected to critical
theoretical analysis. Lomnicki and I (1984) analyzed a model for
the Charnov-Finerty hypothesis. We found that regular cycles indeed may result from such intrinsic factors; however, this happens
only if dispersers prefer to enter empty patches rather than occupied
patches. Thus, only "patch choosy" species are expected to be
cyclic (Fig. 3). Furthermore, cycles are most likely to occur in
regions with large homogenous habitat patches; hence, a species is
likely to be cyclic only in parts of its range (a pattern which is
often observed; see Stenseth 1984a). We might generalize the
derived predictions by saying that cyclic populations are likely to
be found only among those species having the capacity to choose the
patch in which to settle. Several vertebrate species may have this
ability, whereas many invertebrates may have far less ability
(e.g., wind blown dispersal). Our model therefore has suggested
something specific to look for in the real world.

In Chitty's theory, there is competition between two types; in Charnov-Finerty's theory, switches occur between two behavioural "programs." The Charnov-Finerty theory is not based on competitive interaction, hence, there is no a *priori* reason to expect this type of interaction to lead to stability. Obviously, this strengthens the plausibility of the Charnov-Finerty theory.

Predator-Prey Interactions: When Do They Cause Cycles?

Comparing model predictions with data. It is old wisdom that regular density cycles may be caused by predator-prey (or, generally, trophic) interactions (Lotka 1925, Volterra 1926, Maynard Smith 1974); this is a theoretical statement about what is possible. However, whether such predator-prey interactions actually cause cycles is an empirical problem. To learn whether cycles are caused by predator-prey interaction, we must compare data with model predictions.

For example, Tanner (1975) analyzed a predator-prey model with a logistic type population dynamics for the prey in the absence of the predator, a Holling (1959) Type II functional response curve and a Leslie (1948) type predator population dynamics:

$$dx/dt = r \cdot x \cdot (1-x/K) - k \cdot x \cdot y/(x+D)$$

$$dy/dt = s \cdot y \cdot (1-\gamma \cdot y/x)$$

$$(4)$$

where x and y are the prey and predator densities, respectively, k is the maximal value of the functional response curve (Holling 1959, 1965); and γ is a factor determining the degree of density dependence in the predators' specific population growth function (see the legend to Fig. 4 for interpretation of other parameters). As can be seen from Figure 4A, a large prey carrying capacity (K) and a large ratio of maximal prey reproductive output to predator reproductive output (r/s) will cause regular density oscillations (i.e., limit cycles; see Fig. 4B). Otherwise, these interactions will give rise to stable interactions. (Other models could have been discussed; see, e.g., Caughley and Krebs 1983).

Parameter values may be estimated and compared with the predictions summarized in Figure 4A. Tanner carried out such a comparison for eight specific cases for which data were available (Table 1); his analysis suggest that only the hare-lynx cycle is driven by the predator-prey interaction.

This is another example of the interplay between theoretical and empirical work trying to improve our understanding of the real world.

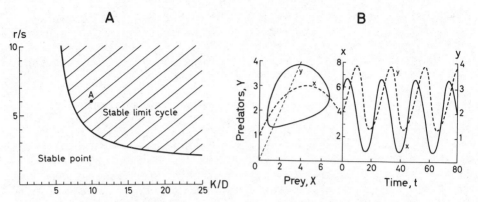

Fig. 4.   (A) depicts the stability domains of the prey-predator system
          defined by model 4 using the ratios r/s (prey/predator in-
          trinsic growth rates) and K/D (appropriately normalized value
          of the carrying capacity for the prey):   r and s are the
          maximal specific growth rates for prey and predator, respec-
          tively; K is the prey's carrying capacity; D is the half
          saturation constant (see Holling 1959, 1965).   In the
          unhatched region, the equilibrium point is stable; in the
          hatched region, stable limit cycles occur.   The figure is
          drawn for k/γ·r = 1 (see text for definition), but qualita-
          tively similar stability boundaries pertain to other values
          of this ratio.   The point labelled A in (A) shows the para-
          meter values corresponding to the example depicted in (B).
          (B, left) is the stable limit cycle for r/s = 6, K/D = 20,
          k/γ·r = 1; the dashed lines are the isoclines for prey
          (dx/dt = 0) and predator (dy/dt = 0) populations, and their
          intersection is the unstable equilibrium point.   (B, right)
          is the cyclic population numbers of prey (solid line) and
          predator (broken line) as functions of time, r·t.   If dis-
          placed, the system tends to return to these stable cyclic
          trajectories.   (Modified from May 1976.)

Some predator-prey interactions can never produce limit cycles.
It appears from the analysis of Tanner's predator-prey (or trophic)
model that such interactions may lead to limit cycles as well as to
stable equilibria; the same model exhibits both types of behaviour
depending on parameter values. This is indeed a general result and
is the main concern of Kolmogorov's theorem (e.g., May 1972, Reed
1976). However, some kinds of trophic interactions will never
produce limit cycles. Hence, it is always important to analyze
models specifically developed for any particular system of interest.
One such example is provided by the trophic type interaction between
bark beetles and a forest stand. Berryman et al. (1984) analyzed
such a model:

Table 1.   Intuitive interpretation of life history data for eight
            naturally occurring predator-prey systems (data from
            Tanner 1975; table modified from May 1976 [reprinted by
            permission of Blackwell, Oxford, England]; see those
            references for bibliographic information).

Prey-predator	Geographical location	Is K relatively large?	Approximate ratio r/s	Apparent dynamic behaviour
Sparrow-hawk	Europe	No	2	Equilibrium point
Muskrat-mink	Central North America	No	3	Equilibrium point
	Boreal North America	Yes	1	Cycles
Mule deer-mountain lion	Rocky Mountains	Yes	0.5	Equilibrium point
White-tailed deer-wolf	Ontario	Yes	0.6	Equilibirum point
Moose-wolf	Isle Royale	Yes	0.4	Equilibrium point
Caribou-wolf	Alaska	Yes	0.4	Equilibirum point
White sheep-wolf	Alaska	Yes	0.2	Equilibirum point

$$dx/dt = f(R,x) - g(x,y)$$

$$dy/dt = h(g[x,y],y)$$

(5)

where x is the biomass density of the forest stand, y is the density of bark beetles, and R is the level of basic resources available for tree growth. The function f represents the rate of forest biomass accumulation in the absence of beetles, and is assumed to be a logistic type population growth model where R determines the carrying capacity. Similarly, the beetles' population dynamics also are assumed to follow a logistic type model defined by the function h; however, in this case, g (beetle caused tree mortality) determines the carrying capacity for beetles in the forest stand being considered. The form of the lower part of the g function

(Fig. 5A) is defined by assuming that random events (lightning
strikes, root disease, etc.) acting over vast forest areas kill or
weaken a constant proportion, a, of the trees; then, susceptible
host biomass is "supplied" at a constant rate, a·x. Assuming
further that the average vigor (or resistance) of a forest stand is
inversely proportional to its biomass density and that trees of
increasing vigor are increasingly more difficult for bark beetles
to attack successfully, the upper part of the g function will be as
depicted in Figure 5A.

The resulting dynamic behaviour of model 5 is depicted in
Figure 5B. Limit cycles will never result, because of the particular
relation--forced upon the model by the biology of the system (!)--be-
tween the g and the h functions.

Counter-intuitive occurrence of cycles caused by trophic interactions. Theoretical models also are useful for deriving patterns
which might seem counter-intuitive. Models for host-pathogen
interactions provide an example. Recently Anderson and May (1980,
1981) (for review, see Anderson 1982) analyzed models for host-pathogen interactions. As expected, they found that pathogens cause
density oscillations in the host population. Such pathogen driven
density cycles arise because of the depletion (by infections and

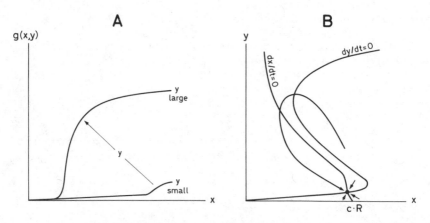

Fig. 5.   Interactions between bark beetles and a forest stand.   (A)
          depicts the tree death function g(x,y). A few severely
          weakened hosts, a·x (defining the lower linear part of the
          curve), are always available, y can exploit uninjured trees
          when x > c·R (see (B)), but large populations can exploit
          more vigorous trees when x < c·R. (B) depicts the corre-
          sponding isoclines (dx/dt = 0 and dy/dt = 0). Only one
          stable node without oscillations results for this system.
          See the text and Berryman et al. (1984) for further details.
          (Modified from Berryman et al. 1984.)

subsequent death) and renewal (by birth) of the supply of susceptible hosts. Such cycles are more likely to occur if the pathogen persists in a resistant state (e.g., spore or cyst) for long periods of time in the environment of the host. This enables the pathogen to survive periods during which the host population is below some critical infection level. Once host density rises above the critical level, an epizootic will occur followed by a density crash in the host population as a result of disease induced mortality. It is, however, somewhat counter-intuitive that viruses and parasites with very low pathogenicity may drive the host's density cycle. Empiricists would often, I believe, disregard viruses with low pathogenicity as a potential driving force of such density oscillations.

Food Webs: What Determines Their Structure?

Pimm (1979a, 1980a, 1982, 1983) has used mathematical models to understand the processes which shape food webs. His approach is based on the observation that, in the real world, community food webs seem to persist despite repeated density perturbations of the co-existing species. Communities that cannot withstand such perturbations are--according to Pimm--assumed to be eliminated. This presumption leads to "Pimm's stability hypothesis": only those food webs "producing" dynamically stable community models, as well as those being what Pimm (1979b, 1980b) calls deletion stable, are expected to exist in the real world. Hence, a class of possible food webs may be deduced given that the dynamic stability of co-existing species is the main factor "shaping" food webs. These predictions then can be tested statistically on real food webs to evaluate the theory. For instance, analysis predicts (e.g., May 1973, Pimm 1982) that connectance in the system should decrease hyperbolically with species richness--data from many ecological studies are unable to reject this hypothesis (e.g., Pimm 1982).

Pimm (1982) considers his idea to be consistent with the Darwinian theory of evolution:

If only those web structures that resist perturbations persist, a selection process is implied. Now, the process of natural selection acts upon the phenotypes of the individual themselves. Nothing I have said implies any different kind of selection in shaping food webs. Specifically, the selection of food web patterns does not create some "super-organism" with individuals consciously adopting strategies in a web's, but not their own, best interest.... [For example, species] feeding on more than one trophic level (omnivores) are relatively scarce. Such a species might be a carnivore that feeds on the plants on which its herbivorous prey also feed. The scarcity of omnivores does not mean there has been selection for carnivores that abstain from eating plants where it would be profitable for them to do so. Rather herbivores which not only suffer the attention of a predator, but also lose substantial portions of their food to the same species, are likely to be driven to extinction. (Reprinted by permission of Chapman and Hall, London, England.)

I claim, however, that Pimm implicitly assumes the existence of a selection pressure operating on communities as a whole as if it was a "super-organism" (Stenseth 1983b, 1984c). I make this claim because, under the assumption of individual selection, stability is a necessary--but not a sufficient--condition for existence (Stenseth 1983b). To exhibit long term stability, the ecosystem must be ecologically stable (as claimed by Pimm) as well as non-invadable by other species (through succession or speciation) or by other varieties of the currently co-existing species (through microevolution).

Nevertheless, I consider Pimm's food web studies a good example of empirical and theoretical/mathematical studies working together to help us disclose patterns in the living world. I believe that the earlier studies reviewed by Pimm (1982) provide one of the best examples of how our intuition may fail. Thus, in the sixties we thought that "complexity begets stability" (Elton 1958, MacArthur 1955). May's (1973) mathematical analysis demonstrated, however, that this was not necessarily correct. Rather, May demonstrated that the more complex a model ecosystem is, the less likely it is to be stable. As pointed out by Pimm (1980b), "it is not small irony that Elton's discussion features, as its first source of evidence, the fact that simple population models fluctuate. Elton did not anticipate that more complex models fluctuate even more!"

As always, we should be careful about extrapolating mathematically derived results through intuitive reasoning. If we are to use mathematics, we must do it properly and thoroughly!

APPLICATION OF THEORETICAL MODELS FOR PRACTICAL PURPOSES

In this section, I will give one example demonstrating how theoretical models may be used to help us solve practical problems. I will, as an example, discuss the application of one of the models described by MacArthur and Wilson (1967) in pest control research (Stenseth 1981b).

MacArthur and Wilson (1967:77) concluded that the chance of a single pair of individuals reaching a population density near the patch's carrying capacity (K) and taking some time (T_k) to become extinct is about $(\lambda-\mu)/\lambda$ (where λ and μ are the specific birth and death rates of the species, respectively), while that of rapid extinction is about $1 - (\lambda-\mu)/\lambda = \mu/\lambda$. A good colonizing species is, MacArthur and Wilson claimed, one maximizing the former probability while minimizing the latter. However, to design an optimal pest control strategy, I (Stenseth 1981b) argued that we should aim at minimizing $(\lambda-\mu)/\lambda$ and maximizing $1 - (\lambda-\mu)/\lambda$.

In island biogeography, we assume that natural selection molds the species to have an optimal strategy; in pest control, we aim at affecting the species and its environment to meet the criteria for defined optimality. In optimal pest control, we thus have that

$$\tau = 1 - \mu/\lambda \tag{6}$$

is to be minimized by manipulating λ and μ. Now, let μ_o be the mortality due to natural factors in a given environment and μ_c the mortality due to control treatment. Then the total realized mortality rate, μ, may be written as

$$\mu = \mu_o + \mu_c \tag{7}$$

(see Stenseth 1981b:776). Further, let $1 - q$ represent the proportional reduction of the effective reproductive output rate, λ_o, occurring under natural conditions. Then the realized birth rate, λ, under pest control is

$$\lambda = \lambda_o \cdot q \tag{8}$$

(see Stenseth 1981b:776).

The quantities μ_c and q are called the "control strategy." These quantities cannot, in general, be altered independently of each other; if we spend most of our resources (or money) on decreasing reproduction, there will be little left to increase mortality. Hence, in general, there will be some constraint function, h, defined as

$$\mu_c = h(q) \tag{9}$$

were $dh/dq > 0$ (see Stenseth 1981b). Substituting models 7, 8, and 9 into model 6 and differentiating to find the minimum of τ with respect to q, it can be shown that the optimal pest control strategy, q^* and $\mu_c^* = h(q^*)$, is found by the line through $(0, -\mu_o)$ having the steepest possible slope but touching the h function (Stenseth 1981b) (Fig. 6).

A similar analysis of a model developed by Levins (1969) yields predictions regarding how resources should be allocated (optimally) to measures affecting patch extinction (through affecting λ and μ as explained above) or to measures affecting dispersal. Combining the obtained results, I reached the following recommendation for optimal pest control (Stenseth 1981b:785):

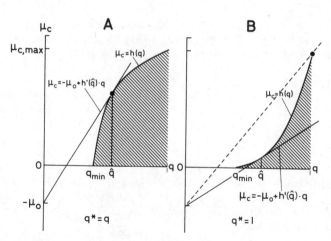

Fig. 6.   Design of optimal pest control strategies (q^*, μ_c^*) where μ_c^* = $h(q^*)$.   In (A) it is possible to reduce both reproduction and survival simultaneously; a mixed pest control strategy is optimal.   In (B) it is impossible to reduce both reproduction and survival simultaneously; a strategy reducing survival is optimal.   However, if natural survival were lower than depicted, the optimal pest control strategy might be to reduce reproduction as much as possible.   (Modified from Stenseth 1981b.)

(1)   If there are methods available to reduce immigration rates into empty patches by almost 100%, most of the limited resources should be spent on reducing immigration regardless of the demography of the pest.   According to Levins (1969), this treatment should be applied at times as variable as possible from patch to patch in the habitat complex (see also Stenseth 1977c).

(2)   If, however, no particular method to reduce immigration exists, separate predictions for r and K selected (e.g., MacArthur and Wilson 1967) species in the same habitat result:

(a)   For r selected species, most economic resources should be devoted to extinction through mortality and birth; in most cases, large resources ought to be devoted to reducing reproduction rather than increasing mortality.   Following Levins (1969), this treatment should be applied in the habitat complex with as little variation in time as possible.

(b)   For K selected pest species, large resources should, in most cases, be devoted to increasing mortality.

The purpose of this discussion has not been to discuss pest control as such, but to discuss the application of models (particu-

larly theoretical models) in applied ecological work; other examples
are provided by May (1981b). It is always worthwhile to look into
the general and theoretical ecological literature when trying to
solve practical problems. Often only slight modifications are
needed to apply general models for a practical purpose. For a
stimulating discussion relating to the use of theoretical ecology
in applied ecology, see Norton and Walker (1982) and Walker and
Norton (1982).

DO ALL MODELS NEED TO BE TESTED?

One of the purposes of mathematical modelling is to evaluate
whether the presumed consequences follow from a set of premises.
For example, do regular cycles result from the premises set forth
by Chitty (1960, 1967)? Obviously, a mathematical model developed
for this purpose does not need to be tested. Whether Chitty's
theory--should it prove theoretically plausible--is the proper
explanation for population cycles must be tested by comparing
specifically derived predictions and specifically obtained data.
Because a hypothesis only can be proven false, we will never know
whether a particular hypothesis is the correct one--the best we can
do is say that it is consistent with available data.

Maynard Smith (1982d) expressed this point well:

> I think it would be a mistake...to stick too rigidly to the
> criterion of falsifiability when judging theories in population
> biology. For example, Volterra's equations for the dynamics
> of a predator and prey species are hardly falsifiable. In a
> sense they are manifestly false, since they make no allowance
> for age structure, for spatial distribution, or for many other
> necessary features for real situations. Their merit is to
> show that even the simplest possible model of such an inter-
> action leads to sustained oscillation--a conclusion it would
> have been hard to reach by purely verbal reasoning. If,
> however, one were to apply this idea in a particular case, and
> propose, for example, that the oscillations in numbers of
> Canadian fur-bearing mammals are driven by the interactions
> between hare and lynx, that would be an empirically falsifiable
> hypothesis. (Reprinted by permission of the Cambridge University
> Press, Cambridge, England.)

Further, May's (1973) community models were never intended to
be tested against data from real ecosystems. These "caricatures of
reality" (to use May's own characterization) were developed to
investigate the earlier ecologists' intuitively deduced dogma that
"complexity begets stability". May was able to demonstrate that
increased complexity does not necessarily enhance stability. If
increased stability accompanies increased complexity, we know that

there must be something else that enhances both stability and complexity and that it is not the complexity *per se* which produces the stability.

Darwin's (1859) theory of natural selection provides another example. Assume as he did that populations consist of individuals (1) that on the average produce more offspring than is needed to replace them upon their death, (2) whose offspring resemble their parents more than they resemble randomly chosen individuals in the population, and (3) that vary in heritable traits influencing reproduction and survival. Then intuition suggests that, in an ecosystem with limited resources, those individuals producing more offspring that survive until the age of reproduction are favoured. It was this process which Darwin called "natural selection." Mathematical models have been developed (see any modern evolution text, e.g., Roughgarden 1979) to demonstrate that the above intuitive argument is logically correct. Hence, today we say that natural selection is no hypothesis; it is a logical consequence of the assumed organization of life. We need not test the idea of natural selection. What needs to be tested, by analyzing real life data, are our statements about how life is assumed to be organized as well as the predicted patterns resulting from analysis of the evolutionary models (see Maynard Smith 1978).

Theoretical models are about ideas and are developed to investigate what is possible and what is not possible. Using such analyses, we deduce patterns expected to be found in nature; these predictions need to be tested to improve our understanding of the living world. MacArthur (1972) has discussed the importance of patterns in science.

An example is the theoretical models for the evolution of competing species reviewed by Slatkin and Maynard Smith (1979). These models suggest that two potentially competing species will be more dissimilar when they co-occur in the same habitat (sympatry) than when they occur in separate habitats (allopatry). Pianka (1981) reviewed empirical examples suggesting the validity of this prediction. However, we test the general pattern, not statements about the particular numerical differences between sympatric and allopatric species found when analyzing mathematical models.

Additionally, both intuitive arguments (Lidicker 1975) and theoretical models (Stenseth 1983a) predict that the specific dispersal rate is related to the specific growth rate of the population (Fig. 7A). Data support this prediction (Fig. 7B; see Stenseth 1983a). Thus, we have no reason to reject the null-hypothesis that a positive relation exists between the specific dispersal rate and the net specific growth rate. Again we test the general pattern and not the particular numerical values; hence, there is no scale on the axis in Figure 7A (a study containing graphs without

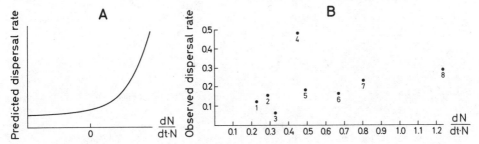

Fig. 7. Dispersal rate as a function of the specific population
growth rate, dN/(dt·N), where N is population density.
(A) depicts the expected relationship; (B) depicts the
pattern found by analyzing available data on small rodents.
In (B) 1 = *Microtus townsendii*; 2 = *M. breweri*; 3 = *M.
pennsylvanicus*; 4, 5, and 8 = *Peromyscus maniculatus*;
6 = *Synaptomys cooperi*; and 7 = *M. ochrogaster*. See Stenseth
(1983a) for theoretical deductions and bibliographic information regarding the empirical material. (Modified from
Stenseth 1983a.)

scales on the axes is not necessarily weak, as Van Valen and Pitelka
1973 have implied).

However, when patterns predicted on the basis of mathematical
models compare unfavourably with data, it is not permissible to
introduce *ad hoc* explanations for why the observations deviate from
predictions. If we do, we reject the data--not the predictions--and
do nothing but "story telling". This has been discussed, generally,
by Lewontin (1978), Gould and Lewontin (1979), and Maynard Smith
(1978).

As pointed out above, both theory and data are necessary in
ecological work. However, we must always remember that only theoretical (or deductive) models can give us insight into how the natural
system under study operates. Hence, I consider May's (1976) discussion of his time lag model for the microtine density cycle to be
rather confusing. According to May, this model suggests a possible
explanation for why microtines exhibit density cycles and this
suggestion is argued to be supported as real by data. May's model
is defined as

$$dx/dt = r \cdot (1 - x_{t-T}/K) \cdot x \qquad (10)$$

where r is the maximum net specific growth rate, T is the time
delay in response to the density regulatory factors, and K is the
carrying capacity. This model is, as a theoretical model, fine;
however, the biological interpretation May draws from his analysis
seems unwarranted. May found that if a time lag equals approxima-

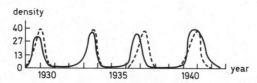

Fig. 8. Shelford's (1943) data on the collared lemming (*Dicrostonyx groenlandicus*) population in the Churchill area in Canada (solid line), compared with a theoretical curve deduced from model 10 (dashed line); the time lag, T, in model 10 is set equal to 0.72 yr. (Modified from May 1976.)

tely 9 months, the model provides a good fit to Shelford's (1943) data on the collared lemming (*Dicrostonyx groenlandicus*) (Fig. 8). May interpreted the 9 month time lag, found by curve fitting, as the length of the winter period during which, he supposed, there is no breeding. However, because winter breeding is a common phenomenon and reproduction only ceases for fairly short periods during spring and fall, May's interpretation is inappropriate (Stenseth 1977d; see also Finerty 1980 for additional critical remarks about this model).

A THEORY FOR ECOSYSTEMS

It is curious that while the political ecological movement is very concerned about "the balance of nature" and the preservation of existing communities, professional ecologists have only hazy ideas of how this stability has been maintained in the past (if it has!) and how it is maintained today. Essentially, this is because we have no theory for the long and short term dynamic behaviours of communities. We need such a theory in ecology; we need something similar to the Darwinian theory of evolution. My hunch is that such an ecological theory will have to be based on the Darwinian theory of evolution--it will be a Darwinian theory for evolution in ecosystems. Such a theory also must be formulated to be consistent with the theory of thermodynamics (e.g., Margalef 1978, 1984). I cannot imagine how such a theory ever could be formulated without the help of mathematics. However, I am uncertain what kind of mathematical models would be most helpful in solving the various ecological problems we want to solve; should they be simple analytical models or complex detail rich models? But this I know: "ecology will not come of age until it has a sound theoretical basis," and "that happy state of affairs" will be reached by using a multitude of mathematical approaches. Hence, I advocate a pluralistic view on the kind of mathematics to be applied in ecology.

A theory for ecology should pay special attention to species-species interactions, biotic-abiotic interactions, time lags in feedback reactions, the greatly different time scales for various ecological processes, and spatial and temporal heterogeneity of habitats (e.g., Southwood 1977, Margalef 1978, this volume; see also Price, this volume, and Remmert, this volume, for some further comments).

It has been fairly popular in the ecological literature (e.g., Odum 1971) to compare each component species in an ecosystem with organs of an organism, and to regard the ecosystem as a "super-organism" (e.g., Odum 1971; see Egerton 1973 for a discussion of this holistic view). The most extreme of these views is the Gaia hypothesis (Lovelock 1979; see also Lovelock and Margulis 1974a, 1974b). The Gaia (Greek for "earth") hypothesis states that

> ...the physical and chemical condition of the surface of the Earth, of the atmosphere, and of the oceans has been and is actively made fit and comfortable by the presence of life itself. This is in contrast to the conventional wisdom which held that life adapted to the plantary conditions as it and they evolved their separate ways (Lovelock 1979; reprinted by permission of the Oxford University Press, Oxford, England).

I can do no better than quote Dawkins (1982):

> The fatal flaw in Lovelock's hypothesis would have instantly occurred to him if he had wondered about the level of natural selection process which would be required in order to produce the Earth's supposed adaptations. Homeostatic adaptations in individual bodies evolve because individuals with improved homeostatic apparatuses pass on their genes more effectively than individuals with inferior homeostatic apparatuses. For the analogy to apply strictly, there would have to have been a set of rival Gaias.... In addition we would [further] have to postulate some kind of reproduction, whereby successful planets spawned copies of their life forms on new planets.

> I am not, of course, suggesting that Lovelock believes it happened like that. He would surely consider the idea of interplanetary selection as ludicrous as I do. Obviously he simply did not see his hypothesis as entailing the hidden assumptions that I think it entails.... [It has] all the notorious difficulties of "group selection" [Maynard Smith 1976].

> The Gaia hypothesis is an extreme form of what...I...call the "BBC Theorem." The British Broadcasting Corporation is rightly praised for the excellence of its nature photography, and it usually strings these admirable visual images together with a serious commentary. Things are changing now, but for years the dominant message of these commentaries was one that had

been elevated almost to the status of a religion by pop "ecology." There was something called the "balance of nature," an
exquisitely fashioned machine in which plants, herbivores,
carnivores, parasites, and scavengers each played their appointed
role for the good of all. The only thing that threatened this
delicate ecological china shop was the insensitive bull of
human progress, the bulldozer of..., etc. The world needs
patient, toiling dung beetles and other scavengers, but for
whose selfless efforts as the sanitary workers of the world...,
etc. Herbivores need their predators, but for whom their
populations would soar out of control and threaten them with
extinction, just as man's population will unless..., etc. The
BBC Theorem is often expressed in terms of the poetry of webs
and networks. The whole world is a finemeshed network of
interrelationships, a web of connections which it has taken
thousands of years to build up, and woe betide mankind if we
tear it down..., etc. (Reprinted by permission of Freeman,
New York, NY USA.)

(See also Van Valen 1983).

   I believe that the Odum school of thought in ecology (Odum
1969, 1971, Patten and Odum 1981) is based essentially on a version
of the Gaia hypothesis. Unfortunately, the Darwinian theory of
evolution (and what we know today about the organization of life)
does not warrant drawing such an analogy between a community, an
ecosystem, or some other more comprehensive unit of organization,
and a single individual organism. It is the genetic identity
between true organs which causes their cooperation. No such genetic
identity exists, however, between the various species in a community
or an ecosystem; hence, such "cooperation" is unlikely to evolve by
natural selection. Based on natural selection--or rather kin
selection (e.g., Hamilton 1964a, 1964b)--we should expect cooperation
between the various organs in a body to take place to make the
whole individual function as well as possible. Individual organisms
should evolve improved homeostatic properties to contribute to
increasing individual fitness which is maximized over evolutionary
time (e.g., Roughgarden 1979). No similar justification is possible
for assuming that co-existing species should cooperate for the
benefit of the entire system, to result in some homeostatic properties
for (for example) the ecosystem!

   These arguments reject the ecosystem approach of Odum and
others. Unfortunately, I cannot refer to another fully developed
and more appropriate approach toward the development of a theory
for ecology (but see Kerner 1957, Patten 1982, Patten and Auble
1981, Stenseth and Maynard Smith 1984). However, I am convinced
that such a theory would, in addition to the Darwinian theory of
evolution, have to be composed of Pimm's (1982) stability hypothesis,
the island biogeography theory of MacArthur and Wilson 1967), the

concept of limiting similarity proposed by MacArthur and Levins (1967), and Van Valen's (1973) Red Queen view of evolution (see Stenseth 1983b). Maynard Smith and I have suggested parts of such a Darwinian theory for the dynamic behaviour of ecosystems (Stenseth and Maynard Smith 1984). Reed and I (1984) have taken another route towards the same goal (see also Allen 1976, Schaffer 1977). Roughgarden and his coworkers (e.g., Roughgarden 1977, 1979, 1983b, Roughgarden et al. 1983, Rummel and Roughgarden 1983) have taken a slightly different but comparable approach. Post and Pimm (1983) have taken still another similar approach.

Such a theory of ecosystems--or of ecology in general--would be invaluable for several reasons. It would help ecologists as the Darwinian theory of evolution helps evolutionists to organize their data on populations, communities, and ecosystems (e.g., energy flow, food webs, etc.). Such a theory could permit us to say what is true, given certain assumptions about how life is organized. Indeed, it might help us diminish the discrepancy between what laymen believe professional ecologists say and what professional ecologists actually can say. A theory for the dynamic behaviour of ecosystems would also help us sort out the controversy between the gradualists and the punctuationists (Gould 1980, Williamson 1981a, 1981b, Jones 1981, Maynard Smith 1982b, 1983)--currently one of the most intense controversies in evolutionary biology (see Stenseth 1983b, 1984b for further discussion of this point).

Finally, having such a theory, we might be better prepared for solving several practical problems, such as understanding what makes a species rare in some ecosystems and abundant (and thus often a pest species) in others (e.g., Cherrett and Sagar 1977). We also might be able to answer why some species seem unable to establish themselves in some ecosystems, whereas they have no difficulties in other systems. Such a theory would help us understand why some ecosystems (e.g., temperate) are more stable than others (e.g., arctic). Specifically, it would help us understand why some ecosystems are more prone to biological invasion than others. Thus, we might be better able to design appropriate conservation strategies for ecosystems believed to be threatened.

THE RELATION BETWEEN THEORY AND PRACTICE IN ECOLOGY

Theory without relation to empirical observations becomes only a game which can teach us nothing about the living world (see INTRODUCTION). As Mao Dzedong (1937) once said:

If we have a theory that is correct, but only speak about it, leave it in a box and never put it into practice, then this theory regardless of how good it is, is of no importance whatsoever. *Knowledge begins with practice, theoretical*

> *knowledge is acquired through practice, and it must also
> return to practice* (my italics).

Theoretical work in ecology must always start with natural
history observation. Such theoretical work would then, if successful,
suggest further, broader natural history observations. This is the
essence of the hypothetico-deductive method of science.

Haldane (1937) expressed similar views in the same year Mao
Dzedong wrote the above cited words:

> No scientific theory is worth anything unless it enables us to
> predict something which is actually going on. Until that is
> done, theories are a mere game of words, and not such a good
> game as poetry.

Hence, there must be a dialectic interaction between theory
and practice. Christiansen and Fenchel (1977)--the former predominantly a theorist and the latter predominantly an empiricist--stated
this very nicely in their book entitled *Theories of Populations in
Biological Communities*:

> Many biologists, especially among those who have been trained
> to distinguish details and to study the diversity of life (and
> this is after all a necessary component of the training of a
> good biologist since one of our most important goals is to
> explain the diversity of life), discard an approach like the
> one adopted in this book as an oversimplification. We acknowl-
> edge (and enjoy) the immense complexity of nature; still we
> feel that the theory presented here constitutes an, albeit
> tiny, element of a true understanding of the real world beyond
> a purely descriptive approach. (Copyrighted 1977 by Springer,
> Heidelberg, W. Germany; reprinted by permission.)

An Optimal Research Strategy

Should an ecologist be a pure theoretician or a pure empiricist,
or should he or she adopt a mixed research strategy? (Compare the
INTRODUCTION.) Fretwell (1972) discussed this issue thoroughly in
the preface to his book entitled *Populations in a Seasonal Environment*.

Let x_1 be a student's scientific success as judged by a theoretician and let x_2 be his or her scientific success as judged by an
empiricist. That is, x_1 measures competence in theoretical ecology,
whereas x_2 measures competence in descriptive ecology and data
gathering ("field" competence). Fretwell assumed that theoreticians do not judge data gathering ability, and vice versa. Hence,
to be favourably judged a student must commit to a period of study
and independent research. However, because time is limited, this

striving for competence in theory often forces neglect of field development, and vice versa. It follows that theoretical and empirical competence (as judged either by only theorists or by only empiricists) must be inversely related. The relationship between x_1 and x_2 is analogous to the evolutionary biologists' concept of "fitness set" (e.g., Levins 1968).

Two kinds of fitness sets are possible. If it is easy to acquire empirical as well as theoretical techniques, a curve like that depicted in Figure 9A results. However, if this is not easy, a curve as depicted in Figure 9B results. Fretwell argues--correctly, I believe--that it is the latter kind which is the more appropriate in ecology. Theoretical competence expands rapidly as new mathematical techniques are added to one's repertoire; doubling the number of techniques at one's disposal more than doubles competence. A student who devotes half time to theory and the other half to data gathering probably knows much less than half the theoretical or empirical techniques that the specialist knows, who concentrates on just one or the other.

Scientific success may now be measured either by prestige or progress. Prestige derives from the judgements of our superiors encountered at meetings or at other social or professional events and from reactions to publications. A proportion, p, of these encounters are with theorists and the rest, 1 - p, are with empiricists. The scientific prestige, Pr, received by a scientist having theoretical competence, x_1, and empirical competence, x_2, may be defined as

$$Pr = p \cdot x_1 + (1 - p) \cdot x_2. \tag{11}$$

The best strategy for maximizing one's prestige may be found by rewriting model 11 as

$$x_2 = Pr/(1 - p) - p \cdot x_1/(1 - p). \tag{12}$$

This corresponds to what Levins (1968) called the "adaptive function." The slope of the adaptive function depends upon the proportion of theorists and empiricists we meet (see model 12 and Fig. 10). Hence, one should specialize in whichever speciality one encounters most frequently.

Fretwell argued that the adaptive function for optimizing one's research strategy with respect to maximal scientific progress (SP) is a curved line (Fig. 11). The same level of progress achieved by two moderately skilled specialists (either a pure theorist or a pure empiricist) is equal to the progress achieved by a person moderately skilled in both theory and fieldwork. Such a "generalist" researcher would be able to explain much of the data he or she

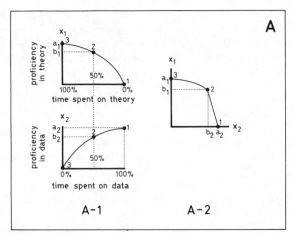

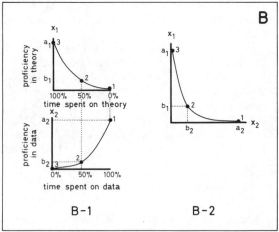

Fig. 9.   How to be a good scientist.   In (A-1), proficiency is plotted
          against percentage of time invested.   The time axes are in
          opposite directions because more time spent on data means
          less time spent on theory.   The upper graph yields values for
          theory (x_1), the lower for data (x_2).   For example, at 100%
          time spent on data (point 1), data proficiency is a_2; at 0%
          time spent on theory, proficiency is zero.   For each percent-
          age of time, there is a particular proficiency in x_1 and in
          x_2.   For example, at 50% time, proficiency is b_1 or b_2 (dot-
          ted lines).   These x_1 and x_2 values can be plotted against
          one another (A-2).   At $t_2 = 100\%$ ($t_1 = 0\%$), $x_1 = 0$ and $x_2 =$
          a_2.   These values yield point 1 of (A-2).   At $t_2 = 50\%$ ($t_1 =$
          50%), $x_1 = b_1$ and $x_2 = b_2$; these values yield point 2.   When
          all time is spent in theory ($t_1 = 100\%$) and none in data
          ($t_2 = 0\%$), $x_1 = a_1$ and $x_2 = 0$ (point 3).   Similar interpreta-
          tions apply to (B-1) and (B-2).   See the text for details.
          (Modified from Fretwell 1972.)

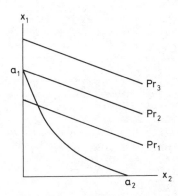

Fig. 10. Prestige lines (Pr) or adaptive functions (straight lines) and fitness set (curved line). The highest prestige attainable is Pr_2, which is achieved if one studies only theory and attains $x_1 = a_1$, leaving $x_2 = 0$. (Modified from Fretwell 1972.)

collected, and would be able to test most aspects of whatever theory he or she could develop (or find in the literature). Because the specialist field person always has some unuseable data and the pure theorist always has some untestable models, their extra competence in these areas is not efficiently used; hence, some intermediate specialization seems profitable (see Fig. 11).

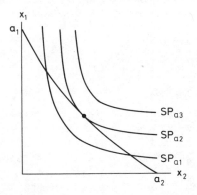

Fig. 11. Hypothetico-deductive (H-D) progress at various levels of x_1 and x_2. Progress adaptive function (SP; curved lines) is plotted with the fitness set (straight line). PS_{a2} is the highest rate of H-D progress that intersects the fitness set (that is, which has at least one attainable combination of x_1 and x_2). (Modified from Fretwell 1972.)

This may be difficult to achieve. In discussing evolutionary
work, Maynard Smith (1977) claimed that

> ...the kind of scientist who is good at developing clear
> theories often finds it difficult to remember facts, whereas
> those who know the facts tend to jib at the algebra. It seems
> to me that there is no single idea in biology which is hard to
> understand in the way that ideas in physics can be hard. If
> biology is difficult, it is because of the bewildering number
> and variety of things one must hold in one's head. (Copyrighted
> 1977 by Pergamon, Oxford, England; reprinted by permission.)

However, I believe with improved training of ecology students
in both theory (including proper training in mathematics and statistics) and empirical work, the state of affairs described by Maynard
Smith may easily change.

A Plea

As concluded by Fretwell (1972 [p. xviii]),

> ...there are two strategies available: specialized training
> and mixed training. The first optimizes prestige, the second
> progress. One can satisfy his ego or his curiosity, but not
> both. The mixed-strategy scientist may be criticized (justly!)
> for incompetence by both pure theorists and pure data biologists,
> as he proceeds to make satisfying advances in the science.
> The specialist will be frustrated by drawers full of unpublished
> data or untested theories. (Copyrighted 1972 by Princeton
> University Press, Princeton, NJ USA; reprinted by permission.)

I am, as Fretwell, convinced that our ignorance about how the living world is organized and operates could be reduced greatly if
most of us adopted such a mixed theorist-empiricist research strategy.

Here is my plea: when a new idea or speculation--both essential
to scientific progress--is put forth, one (or more) precisely
formulated mathematical model should be presented and analyzed.
Those suggesting a new idea, hypothesis, or theory ought to demonstrate its plausibility themselves. Furthermore, empiricists as
well as theorists should adopt the method of the experimental
scientists, which is to vary one factor at a time in a system that
is as simple as possible. Unfortunately, ecologists only rarely
proceed this way. This is probably one reason why molecular biologists have been more successful in answering their questions than
ecologists have been in answering theirs. I do not believe that
this difference in scientific progress is because molecular biology
is simpler than ecology; they are equally complex but refer to
different levels of organization. For a stimulating discussion
relating to this subject, see MacFadgen (1975).

THE SCIENCE OF ECOLOGY: WHERE DO WE COME FROM AND WHERE DO WE GO?

Obviously it is easier to report "where we come from" than "where we are going." As Humphrey Lyttleton is reported to have said (after Elster 1983), "If I knew where jazz was going, I'd be there already." However, from the history of ecology, we may get some hints of where we should go in order to maximize our collective scientific progress.

Darwin is the greatest biologist who ever lived. He suggested a theory by which many observations from the living world may be organized--by this theory, facts do not remain simply disconnected curiosa. He not only formulated the modern theory of evolution, he also founded modern evolutionary ecology (see Chapter 3 in *The Origin*) (Maynard Smith 1982c).

The weak point in Darwin's theory of natural selection was the lack of understanding of heredity; this was changed by Weismann's (1883) conceptual framework separating the germe- and soma-line and by Mendel's (1866) laws of heredity. After Weismann, we can say that natural selection is the only process which brings about adaptations (Maynard Smith 1982d). An ecological theory overlooking this fact would, I believe, be inferior to one which incorporates it.

The 1920s and 1930s were a great period in theoretical ecology in general and in population biology in particular. The years 1923-1940 have been called "The Golden Age of Theoretical Ecology" (Scudo and Ziegler 1978). After this period, there was stagnation in (theoretical) ecology.

However, since the sixties, ecological research has progressed quickly. The ecological works of Odum (1969, 1971), MacArthur (1972), and May (1973, 1981a), and the evolutionary studies of Williams (1966), Maynard Smith (1972, 1975, 1982d), and Roughgarden (1977, 1979) have contributed greatly to this recent progress. Ecological research has ceased to be purely descriptive; instead it has become more problem oriented. Originally, ecological information was used mainly to help taxonomists (systematicists) classify specimens; a species' ecology was used mainly as a tool for the taxonomists. Now ecology stands by itself as a scientific field equal to other biological disciplines.

Odum and MacArthur may be regarded as the founders of two important ecological schools of thought. In both schools, mathematical models play a central role. The latter school uses primarily simple mathematical models; the Odum school, on the other hand, applies very complex mathematical models. As opposed, I believe, to the Odum school, the MacArthur school uses natural history observations in the field and in the laboratory as the basis of its models; natural history observations are the stimulus for theory development.

This history of modern ecology from Darwin until today is depicted in Figure 12 (see also Hutchison 1975, 1978, Diamond 1978, Jackson 1981).

In evolutionary biology, there were two somewhat independent developments. The one led by Fisher (1930), Wright (1931), and Haldane (1932) became the field of evolutionary genetics and was rather weak in ecological understanding. The other led by Dobzhansky (1937), Huxley (1942), Mayr (1942), Simpson (1944), and Stebbins (1950) came to be called "the modern synthesis" and was rather weak in formal population genetics theory. The two approaches were, however, unified in the late sixties by the work of Williams and MacArthur.

The mathematical developments by Pearl (1925), Lotka (1925), Volterra (1926), and Kostitzin (1937) were first in the more "pure" ecology branch of the history of evolutionary ecology. Their work stimulated experimental laboratory work (e.g., Gause 1934, Crombie 1946, Park 1948, 1954) to test the ideas generated by earlier theoretical work. Subsequently, this experimental work stimulated field studies (see, e.g., Lack 1947, Hutchinson 1978, MacArthur 1972).

Of the many scientific achievements that have brought us where we are today, I will single out five recent contributions:

(1) The re-interpretation of evolutionary theory (Williams 1966) with the emphasis on individual selection (as Darwin originally suggested, but was later often forgotten; e.g., Wynne-Edwards 1962) as opposed to group selection (population, species, community, or ecosystem).
(2) The theory of island biogeography (MacArthur and Wilson 1967, MacArthur 1972).
(3) The development of the kin selection theory; Hamilton's two articles (1964a, 1964b) entitled "The Genetical Evolution of Social Behaviour" unified genetics, ethology, and Darwinism.
(4) The development of the concept of Evolutionarily Stable Strategy (ESS) (Maynard Smith and Price 1973, Maynard Smith 1982d). This concept made it easier to predict which properties organisms are expected to have under various conditions.
(5) May's (1973) theoretical studies demonstrating that complexity does not necessarily beget increased stability of ecosystems (see Maynard Smith 1974 and Roughgarden 1979).

Because of this recent progress, I am convinced that mathematical modelling can help us understand the living world better. Furthermore, I believe that we will achieve this goal if we study individual

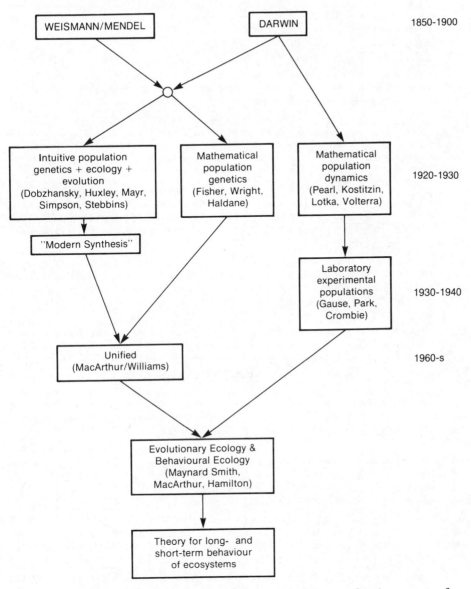

Fig. 12. The conceptual development of modern evolutionary ecology. See text for discussion.

species in particular ecosystems--and if we study how these species interact with each other. That is, I favour an approach where we develop models for simple single species interactions before we develop models for many species interactions. We have to learn to walk before we can run!

SUMMARY AND CONCLUSIONS

Developing mathematical models in ecology has several advantages:
(1) verbal and vague ideas about how the living world is organized
and functions must be specified if a mathematical model is to be
developed; (2) many semantic misconceptions are thus avoided;
(3) formulating an idea mathematically or analyzing a mathematical
model reveals what empirical information is necessary for further
progress and suggests how data can be analyzed to improve understanding; and (4) analysis of mathematical models provides a basis
for deriving the statisticians' null-hypotheses.

Above I have argued that ecology cannot develop into a mature
science until a better theoretical foundation is developed. Ecology ought to be similar to population genetics where it is difficult
to do empirical work without understanding the subject's mathematical
and theoretical basis.

In particular, we need a theory for ecosystems. Such a theory
for the long and short term dynamics of ecological systems would
help us (1) organize our data on ecosystems (e.g., energy flows,
food webs, etc.), (2) sort out the controversy among paleoecologists
between the "punctuationistic" and "gradualistic" views of evolution,
(3) understand better what makes certain ecosystems easier to
invade than others, and (4) explain why some invading species
become pests in certain ecosystems but not in others.

We might have to look for new mathematical methods or approaches
to develop such theories. However, we must always aim at making
the mathematics tractable (and, hence, making analysis possible),
while not letting the mathematics force a certain set of biologically
unreasonable assumptions upon us.

Usually, mathematical models are most effectively developed
and interpreted by researchers with personal experience from field
or laboratory work; mathematicians are, however, often essential to
help simplify the original mathematical models and to assist with
their analyses. It is, in my mind, greatly damaging to progress in
ecology that a steadily increasing gap is developing between theoretical and mathematical ecology on the one hand, and empirical ecology
on the other; such a gap also fosters much semantic confusion.

I hope we soon will see evolutionary ecology--using both
mathematical and empirical techniques--become a unified and mature
science. I believe this is many ecologists' dream after the Decade
of the Environment. I believe we still have a long way to go--but
I do think we can see the goal.

ACKNOWLEDGEMENTS

I thank the organizers (Bourliere, Golley, and Cooley) of the Belgium meeting in April 1983 for inviting me--and thereby forcing me--to summarize my thoughts on "mathematical models in evolutionary ecology." The final version of this chapter was written after a series of lectures on the included topics at the Zoological Institute of Academia Sinica, Beijing, China. I believe the chapter was improved because of that experience. Hence, I express my thanks to the Academies in China, Sweden, and Norway and to Professors Ma (Beijing, China), Brinck (Lund, Sweden) and Semb-Johansson (Oslo, Norway) for making this visit possible. My work upon which this paper is based has--at different stages--been supported by grants from NFR, SJFR, FRN, and The Ecological Committee of NFR (all in Sweden), NAVF and NLVF (both in Norway), the Nordic Council for Ecology, and the Zoological Institutes of the Universities of Oslo (Norway) and Lund (Sweden); I thank them all. Liss Fusdahl and Tove Valmot provided superb secretarial help even under tight time schedules.

REFERENCES CITED

Agren, G.I., F. Andersson, and T. Fagerström. 1980. Experiences of ecosystem research in the Swedish coniferous forest project. Pages 591-596 *in* T. Persson, ed. Structure and function of northern coniferous forests--an ecosystem study. Ecol. Bull. (Stockholm) 32.

Allan, P.M. 1976. Evolution, population dynamics, and stability. U.S. Proc. Nat. Acad. Sci. 73:665-668.

Anderson, R.M. 1982. Theoretical basis for the use of pathogens as biological control agents of pest species. Parasitology 84:3-33.

Anderson, R.M., and R.M. May. 1980. Infections diseases and population cycles of forest insects. Science 210:658-661.

Anderson, R.M., and R.M. May. 1981. The population dynamics of microparasites and their invertebrate hosts. Proc. R. Soc. London 291B:452-524.

Antonovics, J., and H. Ford. 1972. Criteria for the validation or invalidation of the competitive exclusion principle. Nature 237: 406-408.

Ayala, F.J. 1969. Experimental invalidation of the principle of competitive exclusion. Nature 224:1076-1079.

Berryman, A.A., N.C. Stenseth, and D.J. Wollkin. 1984, *in press*. Metastability of forest ecosystems infected by bark beetles. Res. Pop. Ecol.

Beverton, R.J.H., and S.J. Holt. 1957. On the dynamics of fish populations. Minist. Agric. Fish. Food (London). Fish. Invest. Ser. 2(19).

Bunnell, F. 1973. Decomposition: Models and the real world.
    Bull. Ecol. Res. Comm. (Stockholm) 17:407-415.
Calhoun, J.B. 1949. A method for self-control of population
    growth among mammals living in the wild. Science 109:333-335.
Calhoun, J.B. 1963. The social use of space. Physiol. Mammal.
    1:1-187.
Caughley, G., and C.J. Krebs. 1983. Are big mammals simply little
    mammals writ large? Oecologia 59:7-17.
Charlesworth, B., and J.T. Giesel. 1972. Selection in populations
    with overlapping generations. II. Relationship between gene
    frequency and demographic variables. Am. Nat. 106:388-401.
Charnov, R., and Finerty. 1980. Vole population cycles: A case
    for kin-selection? Oecologia 45:1-2.
Chase, S. 1938. The tyranny of words. Harcourt, Brace, New York,
    NY USA.
Cherrett, J.M., and G.R. Sagar, eds. 1977. Origins of pest,
    parasite, disease and weed problems. Blackwell, Oxford,
    England.
Chitty, D. 1960. Population processes in the vole and their rele-
    vance to general theory. Can. J. Zool. 38:99-113.
Chitty, D. 1967. The natural selection of self-regulatory behav-
    iour in animal populations. Proc. Ecol. Soc. Aust. 2:51-78.
Chitty, D. 1970. Variation and population density. Symp. Zool.
    Soc. Lond. 26:327-333.
Chitty, D. 1977. Natural selection and the regulation of density
    in cyclic and non-cyclic populations. Pages 27-32 in B.
    Stonehouse and C. Perrins, eds. Evolutionary ecology. MacMil-
    lan, London, England.
Christian, J.J. 1950. The adreno-pituitary system and population
    cycles in mammals. J. Mammal. 31:247-259.
Christian, J.J. 1970. Social subordination, population density
    and mammalian evolution. Science 168:84-90.
Christiansen, F.G., and T.M. Fenchel. 1977. Theories of popula-
    tions in biological communities. Springer, Berlin, W. Germany.
Clark, C.W. 1976. Mathematical bioeconomics. The optimal manage-
    ment of renewable resources. Wiley, New York, NY USA.
Clarke, G.M. 1980. Statistics and experimental design. Arnold,
    London, England.
Connell, J.H. 1983. On the prevalence and relative importance of
    interspecific competition: Evidence from field experiments.
    Am. Nat. 122:661-696.
Crombie, A.C. 1946. Further experiments on insect competition.
    Proc. R. Soc. London 133B:76-109.
Darwin, C. 1859. The origin of species. Murray, London, England.
Dawkins, R. 1982. The extended phenotype. Freeman, San Francisco,
    CA USA.
Diamond, J.M. 1978. Niche shifts and the rediscovery of inter-
    specific competition. Am. Sci. 66:322-331.
Dobzhansky, T. 1937. Genetics and the origin of species. Columbia
    Univ. Press, New York, NY USA.

Dolby, G.R. 1982. The role of statistics in the methodology of the life sciences. Biometrics 38:1069-1083.

Draper, N.R., and H. Smith. 1966. Applied regression analysis. Wiley, New York, NY USA.

Egerton, F.N. 1973. Changing concepts of the balance of nature. Quart. Rev. Biol. 48:322-350.

Elster, J. 1983. Explaining technical change. Cambridge Univ. Press, Cambridge, England.

Elton, C.S. 1958. The ecology of invasions by animals and plants. Methuen, London, England.

Engen, S. 1978. Stochastic abundance models. Chapman and Hall, London, England.

Finerty, J.P. 1980. The population ecology of cycles in small mammals. Yale Univ. Press, New Haven, CT USA.

Fisher, R.A. 1930. The genetical theory of natural selection. Clarendon, Oxford, England.

Framstad, E., and N.C. Stenseth. 1984, in press. Habitat selection and competitive interactions: Review of some ecological and evolutionary models with data pertaining to small rodents. Acta Zool. Fenn.

Fretwell, S.D. 1972. Populations in seasonal environments. Princeton Univ. Press, Princeton, NJ USA.

Gaines, M.S. 1981. Importance of genetics to population dynamics. Pages 1-27 in M.H. Smith and J. Joule, eds. Mammalian population genetics. Univ. of Georgia Press, Athens, GA USA.

Gause, G.F. 1934. The struggle for existence. Williams and Wilkins, Baltimore, MD USA.

Gilpin, M.E., and K.E. Justin. 1972. Reinterpretation of the invalidation of the principle of competitive exclusion. Nature 236:273-301.

Goldstein, R.A. 1977. Reality and models: Difficulties associated with applying general models to specific situations. Pages 206-215 in D.L. Solomon and C. Walter, eds. Mathematical models in biological discovery. Lect. Notes Biomath. 13.

Gould, S.J. 1980. Is a new and general theory emerging? Paleobiology 6:119-130.

Gould, S.J., and R.C. Lewontin. 1979. The spandrels of San Marco and the Panglossion paradigm: A critique of the adaptationist program. Proc. R. Soc. London 205B:581-598.

Grant, P.R. 1983. Conclusion: Lizard ecology, viewed at a short distance. Pages 411-417 in R.B. Huey, E.R. Pianka, and T.W. Schoener, eds. Lizard ecology. Studies of a model organism. Harvard Univ. Press, Cambridge, MA USA.

Haila, Y. 1982. Hypotetico-deductivism and the competition controversy in ecology. Ann. Zool. Fenn. 19:255-264.

Haila, Y., and O. Järvinen. 1982. The role of theoretical concepts in understanding the ecological theater: A case study on island biogeography. Pages 261-278 in E. Saarinen, ed. Conceptual issues in ecology. D. Reidel, Dordrecht, The Netherlands.

Haldane, J.B.S. 1927. Possible worlds. Chatto and Windus, London, England.

Haldane, J.B.S. 1932. The causes of evolution. Longmans, Green, New York, NY USA.

Haldane, J.B.S. 1937. Adventures of a biologists. Harper, New York, NY USA.

Hamilton, W.D. 1964a. The genetical evolution of social behaviour I. J. Theor. Biol. 7:1-16.

Hamilton, W.D. 1964b. The genetical evolution of social behaviour II. J. Theor. Biol. 7:17-52.

Hardin, G. 1960. The competition exclusion principle. Science 131:1292-1297.

Harvey, P.H., R.K. Colwell, J.W. Silvertown, and R.M. May. 1983. Null models in ecology. Annu. Rev. Ecol. Syst. 14:189-211.

Hempel, C. 1965. Aspects of scientific explanations. Free Press, New York, NY USA.

Hempel, C. 1966. Philosophy of natural science. Prentice-Hall, Englewood Cliffs, NJ USA.

Holling, C.S. 1959. Some characteristics of simple types of predation and parasitism. Can. Entomol. 91:385-398.

Holling, C.S. 1965. The functional response of predators to prey density and its role in mimicry and population regulation. Mem. Entomol. Soc. Can. 45:1-60.

Hutchinson, G.E. 1975. Variations on a theme by Robert MacArthur. Pages 493-521 in M.L. Cody and J.M. Diamond, eds. Ecology and evolution of communities. Harvard Univ. Press, Cambridge, MA USA.

Hutchinson, G.E. 1978. An introduction to population ecology. Yale Univ. Press, New Haven, CT USA.

Huxley, J.S. 1942. Evolution. The modern synthesis. Allen and Unwin, London, England.

Innis, G.S., ed. 1978. Grassland simulation model. Springer, Berlin, W. Germany.

Jackson, J.B.C. 1981. Interspecific competition and species' distributions: The ghosts of theories and data past. Am. Zool. 21:889-902.

Järvinen, O., ed. 1982. Deducing interspecific competition from community data. Ann. Zool. Fenn. 19:239-356.

Jeffers, J.N.R. 1978. An introduction to systems analysis: With ecological applications. Arnold, London, England.

Jones, J.S. 1981. An uncensored page of fossil history. Nature 293: 427-428.

Kerner, E.H. 1957. A statistical mechanistics of interacting biological species. Bull. Math. Biophys. 19:121-146.

Kerner, E.H. 1959. Further considerations on the statistical mechanistics of biological associations. Bull. Math. Biophys. 21:217-255.

Kostitzin, V.A. 1937. Biologie mathematique. Collin, Paris, France.

Krebs, C.J. 1978a. Ecology: The experimental analysis of distri-
    bution and abundance. Harper and Row, New York, NY USA.
Krebs, C.J. 1978b. Aggression, dispersal, and cyclic changes in
    populations of small mammals. Pages 49-60 in L. Kramer, P.
    Pliner, and T. Allowy, eds. Aggression, dominance and individual
    spacing. Plenum, New York, NY USA.
Krebs, C.J. 1978c. A review of the Chitty hypothesis of popula-
    tion regulation. Can. J. Zool. 56:2463-2480.
Krebs, C.J., M.S. Gaines, B.L. Keller, J.H. Myers, and R.H. Tamarin.
    1973. Population cycles in small rodents. Science 179:35-41.
Krebs, C.J., and J.H. Myers. 1974. Population cycles in small
    mammals. Adv. Ecol. Res. 8:268-299.
Lack, D. 1947. Darwin's finches. Cambridge Univ. Press, Cambridge,
    England.
Lawlor, L.R., and J. Maynard Smith. 1976. The coevolution and
    stability of competing species. Am. Nat. 110:79-99.
Leslie, P.H. 1948. Some further notes on the use of matrices in
    population mathematics. Biometrika 35:213-245.
Levin, S.A. 1981. The role of theoretical ecology in the descrip-
    tion and understanding of populations in heterogeneous envir-
    onments. Am. Zool. 21:865-876.
Levins, R. 1968. Evolution in changing environments. Princeton
    Univ. Press, Princeton, NJ USA.
Levins, R. 1969. The effect of random variations of different
    types on population growth. U.S. Proc. Nat. Acad. Sci. 62:1061-
    1065.
Lewontin, R.C. 1969. The meaning of stability. Pages 13-24 in
    Diversity and stability in ecological systems. Brookhaven
    Symp. Biol. 22, U.S. Dep. of Commerce, Springfield, VA USA.
Lewontin, R.C. 1978. Fitness, survival and optimality. Pages
    3-21 in D.H. Horn, R. Mitchell, and C.R. Stairs, eds. Analysis
    of ecological systems. Ohio State Univ. Press, Columbus, OH
    USA.
Lidicker, W.Z., Jr. 1975. The role of dispersal in the demography
    of small mammals. Pages 103-128 in K. Petrusewicz, F.B.
    Golley and L. Ryszkowski, eds. Small mammals: Productivity
    and dynamics of populations. Cambridge Univ. Press, Cambridge,
    England.
Lomnicki, A., and N.C. Stenseth. 1984. Can dispersal drive the
    microtine density cycle? Manuscript. Available from the
    authors.
Lotka, A.J. 1925. Elements of physical biology. Williams and
    Wilkins, Baltimore, MD USA.
Lovelock, J.E. 1979. Gaia. Oxford Univ. Press, Oxford, England.
Lovelock, J.E., and L. Margulis. 1974a. Atmospheric homeostasis
    by and for the biosphere: The Gaia hypothesis. Tellus 26:1-10.
Lovelock, J.E., and L. Margulis. 1974b. Homeostatic tendencies of
    the earth's atmosphere. Origin of Life 1:12-22.
MacArthur, R.H. 1955. Fluctuations of animal populations, and a
    measure of community stability. Ecology 36:533-536.

MacArthur, R.H. 1957. On the relative abundance of bird species.
    U.S. Proc. Nat. Acad. Sci. 43:293-295.
MacArthur, R.H. 1972. Geographical ecology: Patterns in the
    distribution of species. Harper and Row, New York, NY USA.
MacArthur, R.H., and R. Levins. 1967. The limiting similarity,
    convergence, and divergence of coexisting species. Am. Nat.
    101:377-385.
MacArthur, R.H., and E.O. Wilson. 1967. The theory of island bio-
    geography. Princeton Univ. Press, Princeton, NJ USA.
MacFadyen, A. 1975. Some thoughts on the behaviour of ecologists.
    J. Anim. Ecol. 44:351-363.
Mao Dzedong. 1937. On practice. Beijing, China.
Margalef, R. 1978. Life-farms of phytoplankton as survival alter-
    natives in an unstable environment. Oceanol. Acta 1:493-509.
May, R.M. 1972. Limit cycles in predator-prey communities.
    Science 177:900-902.
May, R.M. 1973. Stability and complexity in model ecosystems.
    Princeton Univ. Press, Princeton, NJ USA.
May, R.M. 1975. Patterns of species abundance and diversity.
    Pages 81-120 in M.L. Cody and J.M. Diamond, eds. Ecology and
    evolution of communities. Harvard Univ. Press, Cambridge, MA
    USA.
May, R.M. 1976. Models for single populations. Pages 4-25 in
    R.M. May, ed. Theoretical ecology: Principles and application.
    Blackwell, Oxford, England.
May, R.M., ed. 1981a. Theoretical ecology: Principles and applica-
    tions. Blackwell, Oxford, England.
May, R.M. 1981b. The role of theory in ecology. Am. Zool. 21:
    903-910.
Maynard Smith, J. 1968. Mathematical ideas in biology. Cambridge
    Univ. Press, Cambridge, England.
Maynard Smith, J. 1972. On evolution. Edinburgh Univ. Press,
    Edinburgh, Scotland.
Maynard Smith, J. 1974. Models in ecology. Cambridge Univ.
    Press, Cambridge, England.
Maynard Smith, J. 1975. The theory of evolution. Penguin, Balti-
    more, MD USA.
Maynard Smith, J. 1977. Limitation of evolutionary theory. Pages
    235-242 in R. Duncan and M. Weston-Smith, eds. The encyclopedia
    of ignorance. Pergamon, London, England.
Maynard Smith, J. 1978. Optimization theory in evolution. Annu.
    Rev. Ecol. Syst. 9:31-56.
Maynard Smith, J. 1982a. Storming the fortress. N. Y. Rev., 13
    May:41-42.
Maynard Smith, J. 1982b. Evolution--sudden or gradual? Pages
    125-128 in J. Maynard Smith, ed. Evolution now. MacMillan,
    London, England.
Maynard Smith, J. 1982c. Introduction. Pages 1-6 in J. Maynard
    Smith, ed. Evolution now. MacMillan, London, England.

Maynard Smith, J.   1982d.   Evolution and the theory of games.
    Cambridge Univ. Press, Cambridge, England.
Maynard Smith, J.   1983.   Current controversies in evolutionary
    biology.   Pages 273-286 *in* M. Grene, ed.   Dimensions in
    Darwinism: Themes and counter theory.   Cambridge Univ. Press,
    Cambridge, England.
Maynard Smith, J., and G.R. Price.   1973.   The logic of animal con-
    flict.   Nature 246:15-18.
Mayr, E.   1942.   Systematics and the origin of species.   Columbia
    Univ. Press, New York, NY USA.
Mayr, E.   1961.   Cause and effect in biology.   Science 134:1501-1506.
Mendel, G.   1866.   Versuche über Pflanzenhybriden.   Verh. Natur-
    forsch. Ver. Brunn 4:3-17.
Norton, G.A., and B.H. Walker.   1982.   Applied ecology: Towards a
    positive approach.   I. The context of applied ecology.   J.
    Environ. Manage. 14:309-324.
Odum, E.P.   1969.   The strategy of ecosystem development.   Science
    164:262-270.
Odum, E.P.   1971.   Fundamentals of ecology, 3rd ed.   Saunders,
    Philadelphia, PA USA.
Park, T.   1948.   Experimental studies of interspecific competition.
    I. Competitions between populations of flour beetles *Tribolium
    comflusum* Duval and *I. castaneum* Herbst.   Physiol. Zool.
    18:265-308.
Park, T.   1954.   Experimental studies of interspecific competition.
    II. Temperature, humidity, and competition in two species of
    Tribolium.   Physiol. Zool. 27:177-238.
Patten, B.C.   1982.   Environs: Relativistic elementary particles
    for ecology.   Am. Nat. 119:179-219.
Patten, B.C., and G.T. Auble.   1980.   System approach to the concept
    of niche.   Synthese 43:155-181.
Patten, B.C., and G.T. Auble.   1981.   System theory of the ecological
    niche.   Am. Nat. 118:345-369.
Patten, B.C., and E.P. Odum.   1981.   The cybernetic nature of
    ecosystems.   Am. Nat. 118:886-895.
Pearl, R.   1925.   The biology of population growth.   Knopf, New
    York, NY USA.
Pianka, E.R.   1978.   Evolutionary ecology.   Harper and Row, New
    York, NY USA.
Pianka, E.R.   1981.   Competition and niche theory.   Pages 167-196
    in R.M. May, ed.   Theoretical ecology: Principles and applica-
    tion.   Blackwell, Oxford, England.
Pielou, E.C.   1975.   Ecological diversity.   Wiley, New York, NY
    USA.
Pielou, E.C.   1977.   Mathematical ecology.   Wiley, New York, NY
    USA.
Pielou, E.C.   1981.   The usefulness of ecological models: A stock-
    taking.   Quart. Rev. Biol. 56:17-31.
Pimm, S.L.   1979a.   Complexity and stability: Another look at
    MacArthur's original hypothesis.   Oikos 33:351-357.

Pimm, S.L.   1979b.   The structure of food webs.   Theor. Pop. Biol.
     16:144-158.
Pimm, S.L.   1980a.   Properties of food webs.   Ecology 61:219-225.
Pimm, S.L.   1980b.   Food web design and the effects of species
     deletion.   Oikos 35:139-149.
Pimm, S.L.   1982.   Food webs.   Chapman and Hall, London, England.
Pimm, S.L.   1983.   Food webs, food chains and return times.   Pages
     89-99 in D.R. Strong and D.S. Simberloff, eds.   Ecological
     communities: Conceptual issues and the evidence.   Princeton
     Univ. Press, Princeton, NJ USA.
Platt, J.R.   1964.   Strong inference.   Science 146:347-353.
Post, W.M., and S.L. Pimm.   1983.   Community assembly and food web
     stability.   Math. Biosci. 64:169-192.
Provine, W.B.   1977.   Role of mathematical population genetics in
     the evolutionary synthesis of the 1930's and 40's.   Pages 1-30
     in D.L. Solomon and C. Walter, eds.   Mathematical models in
     biological discovery.   Lect. Notes Biomath. 13.
Quinn, J.F., and A.E. Dunham.   1983.   On hypothesis testing in eco-
     logy and evolution.   Am. Nat. 122:602-617.
Reed, J.   1976.   Noen matematiske modeller i økologi.   Nord. Mat.
     Tidsskr. 24:77-100 (In Norwegian).
Reed, J., and N.C. Stenseth.   1984, in press.   On evolutionarity
     stable strategies (ESS).   J. Theor. Biol.
Ricker, W.E.   1954.   Stock and recruitment.   J. Fish. Res. Board of
     Can. 11:559-623.
Roughgarden, J.   1977.   Coevolutions in ecological systems: Results
     from "loop analysis" for purely density-dependent coevolution.
     Pages 449-517 in F.B. Christiansen and T.M. Fenchel, eds.
     Measuring selection in natural populations.   Lect. Notes
     Biomath. 16.
Roughgarden, J.   1979.   Theory of population genetics and evolut-
     ionary ecology: an introduction.   MacMillan, London, England.
Roughgarden, J.   1983a.   Competition and theory in community eco-
     logy.   Am. Nat. 122:583-601.
Roughgarden, J.   1983b.   The theory of coevolution.   Pages 33-64 in
     D.J. Futuyma and M. Slatkin, eds.   Coevolution.   Sinauer,
     Sunderland, MA USA.
Roughgarden, J., D. Heckel, and E.R. Fuentes.   1983.   Coevolu-
     tionary theory and the biogeography and community structure of
     Anolis.   Pages 371-410 in R. Huey, E.R. Pianka, and T.W.
     Schoener, eds.   Lizard ecology.   Harvard Univ. Press, Cambridge,
     MA USA.
Rummel, J., and J. Roughgarden.   1983.   Some differences between
     invasion-structured and coevolutionary-structured competitive
     communities: A preliminary theoretical analysis.   Oikos 41:
     477-486.
Salt, G.W.   1983.   Roles: Their limits and responsibilities in
     ecological and evolutionary research.   Am. Nat. 122:697-703.
Schaffer, W.M.   1977.   Evolution, population dynamics and stabi-
     lity: A comment.   Theor. Pop. Biol. 11:326-329.

Schoener, T.W. 1982. The controversy over interspecific competition. Am. Sci. 70:586-595.

Scudo, F.M., and J.R. Ziegler. 1978. The golden age of theoretical ecology: 1923-1940. Lect. Notes Biomath. 22:1-490.

Shelford, V.E. 1943. The relation of snowy owl migration to the abundance of the collared lemming. Auk 62:592-594.

Simberloff, D.S. 1982. The status of competition theory in ecology. Ann. Zool. Fenn. 19:241-253.

Simberloff, D.S. 1983. Competition theory, hypothesis-testing and other community ecological buzzwords. Am. Nat. 122:626-635.

Simpson, G.G. 1944. Tempo and mode in evolution. Columbia Univ. Press, New York, NY USA.

Slatkin, M., and J. Maynard Smith. 1979. Models of coevolution. Quart. Rev. Biol. 54:233-263.

Snedecor, G.W., and W.G. Cochran. 1967. Statistical methods, 6th ed. Iowa State Univ. Press, Ames, IA USA.

Solbrig, O.T. 1981. Energy, information and plant evolution. Pages 274-299 in C.R. Townsend and P. Calow, eds. Physiological ecology. Blackwell, Oxford, England.

Solomon, D.L., and C. Walter, eds. 1977. Mathematical models in biological discovery. Lect. Notes Biomath. 13.

Southwood, T.R.E. 1977. Habitat, the temple for ecological strategies? J. Anim. Ecol. 46:337-366.

Stebbins, G.L. 1950. Variation and evolution in plants. Columbia Univ. Press, New York, NY USA.

Stenseth, N.C. 1975. Energy models for individual small rodents and its significance to general population theory. Pages 283-299 in F.E. Wielgolaski, ed. Fennoscandian tundra ecosystems, part 2: Animals and systems analysis. Springer, Berlin, W. Germany.

Stenseth, N.C. 1977a. Modelling the population dynamics of voles: Models as research tools. Oikos 29:449-456.

Stenseth, N.C. 1977b. Forecasting of rodent outbreaks: Models and the real world. Eur. Plant Prot. Organ. Bull. 7:303-315.

Stenseth, N.C. 1977c. On the importance of spatio-temporal heterogeneity for the population dynamics of rodents: Towards a theoretical foundation of rodent control. Oikos 29:545-552.

Stenseth, N.C. 1977d. Evolutionary aspects of demographic cycles: The relevance of some models of cycles for microtine fluctuations. Oikos 29:525-538.

Stenseth, N.C. 1981a. On Chitty's theory for fluctuating populations: The importance of polymorphisms in the generation of regular cycles. J. Theor. Biol. 90:9-36.

Stenseth, N.C. 1981b. How to control pest species: Application of models from the theory of island biogeography in formulating pest control strategies. J. Appl. Ecol. 18:773-794.

Stenseth, N.C. 1983a. Causes and consequences of dispersal in small mammals. Pages 63-101 in I.R. Swingland and P.J. Greenwood, eds. The ecology of animal movements. Oxford Univ. Press, Oxford, England.

Stenseth, N.C. 1983b. A coevolutionary theory for communities and
    food web configurations. Oikos 41:487-495.
Stenseth, N.C. 1984a, in press. Mathematical models of microtine
    cycles: Models and the real world. Ann. Zool. Fenn.
Stenseth, N.C. 1984b, in press. Are geographically widely distrib-
    uted species in general phenotypically uniform? Some comments
    on the concept of evolutionary stasis. Paleobiology.
Stenseth, N.C. 1984c, in press. The structure of food webs predicted
    from optimal food selection models: An alternative to Pimm's
    stability hypothesis. Oikos.
Stenseth, N.C., and J. Maynard Smith. 1984, in press. Coevolution
    in ecosystems: Red Queen evolution or stasis? Evolution.
Strong, D.R. 1980. Null hypotheses in ecology. Synthese 43:
    271-286.
Strong, D.R. 1983. Natural variability and the mainfold mechan-
    isms of ecological communities. Am. Nat. 122:636-660.
Swartzman, G.L., and G.M. Van Dyne. 1972. An ecologically based
    simulation-optimization approach to natural resource planning.
    Annu. Rev. Ecol. Syst. 3:347-398.
Taitt, M., and C.J. Krebs. 1984, in press. Population dynamics
    and cycles. In R.H. Tamarin, ed. Biology of New World Microtus.
    Am. Soc. Mammal. Spec. Publ. 8.
Tanner, J.T. 1975. The stability and the intrinsic growth rates
    of prey and predator populations. Ecology 56:855-867.
Toft, C.A., and P.J. Shea. 1983. Detecting community-wide pat-
    terns: Estimating power strengthens statistical inference.
    Am. Nat. 122:618-625.
Ugland, K.I., and J. Gray. 1982. Lognormal distributions and the
    concept of community equilibirum. Oikos 39:171-178.
Van Valen, L.M. 1973. A new evolutionary law. Evol. Theor.
    1:1-30.
Van Valen, L.M. 1983. How pervasive is coevolution? Pages 1-20
    in M.H. Nitecki, ed. Coevolution. Univ. of Chicago Press,
    Chicago, IL USA.
Van Valen, L.M., and F.A. Pitelka. 1973. Commentary--intellectual
    censorship in ecology. Ecology 55:925-926.
Volterra, V. 1926. Fluctuations in the abundance of species, con-
    sidered mathematically. Nature 118:558-560.
Walker, B.H., and G.A. Norton. 1982. Applied ecology: Towards a
    positive approach. II. Applied ecological analysis. J.
    Environ. Manage. 14:325-342.
Watt, K.E.F. 1961. Mathematical models for use in insect pest
    control. Can. Entomol. Suppl. 93.
Weismann, A. 1883. Über die Vererbung. Gustav Fischer, Jena,
    Germany.
Williams, G.C. 1966. Adaptation and natural selection. Princeton
    Univ. Press, Princeton, NJ USA.
Williams, M.B. 1977. Needs for the future: Radically different
    types of mathematical models. Pages 225-240 in D.L. Solomon
    and C. Walter, eds. Mathematical models in biological discov-
    eries. Lect. Notes Biomath. 13.

Williamson, P.G. 1981a. Paleontological documentation of speci-
    ation in Cenozoic molluscs from Turkana basin. Nature 293:
    437-444.
Williamson, P.G. 1981b. Morphological stasis and developmental
    constraints: Real problems for neo-Darwinism. Nature 294:
    214-215.
Wright, S. 1931. Evolution in Mendelian populations. Genetics
    16:97-159.
Wynne-Edwards, V.C. 1962. Animal dispersion in relation to social
    behaviour. Oliver and Boyd, Edinburgh, Scotland.

A GUIDELINE FOR ECOLOGICAL RESEARCH

M. Godron

Botanical Institute
163 Rue Auguste-Broussonnet
34000 Montpellier, France

INTRODUCTION

The major ecological questions are as numerous as ecologists, and each ecologist could draw up a long list of urgent problems. Simple compilation of such a list would lead to a grab bag, where acid rain, the wells of the Sahel, or polymetallic nodules would be neighbors of Melville's white whale and Prevert's raccoon. To walk with security in such a labyrinth, one must seize Ariadne's thread and follow it to its end, at the risk of naively underlining the obvious and of posing once again Olbers's paradoxes. This approach requires a chain of arguments, often academic, which will be re-stated here only to clarify the vocabulary used. The reader, therefore, is advised to read the text rapidly to seize the essential structure, before returning to the major points.

All ecological problems are facets of the same reality which, basically, is thermodynamic in nature. Ecologists are becoming convinced of this unity of their discipline, thanks to the work accomplished in the last 20 years or so. However, a program of research oriented toward this fundamental aspect would have little chance of interesting a government, or even an international organization, today. But it is possible to draw from Carnot's principle a coherence applicable to all biological systems: these systems evolve according to a probabilist rule which directs them to levels of metastability (Section 1) and allows generalizations about most of the problems of applied ecology that must be resolved during the next decades.

The first type of system which emphasizes this rule is that of relatively mature, barely productive biogeocenoses, which risk

289

being de-stabilized, or even destroyed, if they are submitted to
excessive action (Section 2). The second example is that of intermediary communities, intensively used, which are the source of most
of the globe's food (Section 3). The third example is that of
"degraded" communities (Section 4).

1. FUNDAMENTAL EQUILIBRIUMS

     Biological systems normally evolve toward an accumulation of
information (i.e., neguentropy), but they do not escape the second
principle of thermodynamics. In fact, the system constituted by
the Earth and the Sun is essentially an isolated system (i.e.,
closed to outside exchanges of matter and energy) inside which
entropy increases inexorably, as in every isolated system. The
biosphere is a fraction of this system, and entropy can decrease
here because:

(1) the biosphere sets aside information coming from the "high
     quality" energy of short wavelength radiation from the sun
     (cf. the paradox of Schrodinger);
(2) the biosphere possesses feedback mechanisms, allowing metastable
     equilibriums (Points C, D, and E of Fig. 1) more and more
     distant from the most stable equilibrium (Point B of Fig. 1),
     which is the minimal potential energy state.

     These equilibriums are established at the molecular level, but
they produce macroscopic structures (individuals, biogeocenoses,
landscapes), and the laws of thermodynamics link these levels of
integration. Boltzmann has shown how the innumerable interactions
of the myriad molecules of a gas are governed by simple macroscopic
laws, such as Mariotte's law which involves the temperature and
pressure of gases.

     Similarly, in the thermodynamic system that includes the
biosphere, each biogeocenoses has a well known logistical tendency
(schematized in Fig. 2) whereby the accumulation of phytomass,
which corresponds to an accumulation of information, tends towards
a ceiling that depends mainly on climatic conditions (temperature
and precipitation). In fact, this ceiling is sometimes a provisional
level (corresponding, e.g., to Point D of Fig. 1), which later can
have access to a higher level (Point E of Fig. 1). However, the
important point is that it constitutes a dynamic equilibrium, where
the increase in biomass (or increment, or even "true" production)
is zero.

     This schema is banal and accepted by everyone, but its utility
appears when it is applied to real examples. Then it becomes
obvious that serious gaps still exist in our ecological knowledge--
gaps that must be filled rapidly.

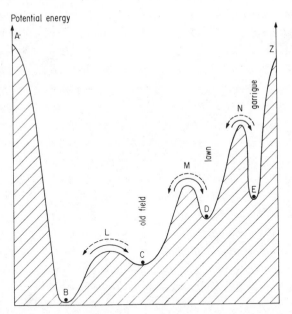

Fig. 1. Potential energy and types of stability: A, L, M, N, Z =
unstable; C, D, E = metastable; B = most stable. Transitions from B to C, D, or E are not easy; they increase
the degree of metastability. Transitions from E to D, C,
or B are easy; they correspond to a loss of potential
energy. (Reprinted by permission of Centre d' Etudes
Phytosociologiques et Ecologiques Louis Emberger, Montpellier, France.)

In particular, as soon as man tries to set aside useful production from a mature ecosystem, he de-stabilizes the system at least
a little, but the system tends to return to equilibrium along a
slope which is not very different from the "natural" slope (Fig. 2).
This slope is minimal in pioneering biogeocenoses and also in those
that are close to the ceiling. It is maximal in intermediary
systems, where man can profit from an important acquisition (humus,
microflora, root systems, genetic diversity, nitrogen equilibrium,
etc.). Urban ecosystems are somewhat special, but they can be
studied similarly by emphasizing their openness.

Practically, then, a good set of ecological problems can be
reduced to the following question: How can man displace the equilibriums of natural systems towards other metastable systems, which
would then permit him to harvest the phytomass and zoomass which he
needs each year?

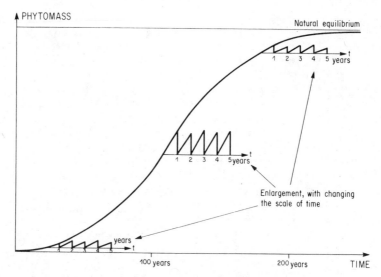

Fig. 2.   After de-stabilization, the accumulation of phytomass in
a biogeocenoses tends to return to equilibrium. (Reprinted
by permission of Centre d' Etudes Phytosociologiques et
Ecologiques Louis Emberger, Montpellier, France.)

The principal task of theoretical ecologists is to specify the
nature of these equilibriums (and the spatial structures that they
imply); the more applied problems that urgently need to be solved
all derive from this general question. If they appear very different
from each other, it is only because they do not concern the same
fraction of the curve in Figure 2. Thus, problems involving the
preservation of nature concern, above all, the high part of this
curve, which characterizes mature and barely productive biogeocenoses.

2.   MATURE AND BARELY PRODUCTIVE BIOGEOCENOSES

The increment (biomass at time 2 minus biomass at time 1) of
mature vegetation that has achieved its equilibrium (e.g., a primary
forest) is zero. This affirmation does not contradict those of the
authors of world models (Lieth and Box 1977) if vocabulary is
precisely defined.

The useful production of a biocenose is the quantity of biomass
harvested in a known area during a determined period of time (e.g.,
a hectare of grain can have a useful production of 60 quintals per
year). On the contrary, in an integral reserve, useful production
is zero; gross production (mass of the photosynthetized material)
is not zero, but it is rarely known with precision. Ecologists are
more interested in net production, which is usually defined as the

sum of the annual litter fall and the increment (Box 1978). For an integral reserve, the net production is positive, but the increment certainly is zero.

When man removes some biomass from a mature vegetation, the internal forces of the system tend immediately to re-establish the equilibrium profile of the vertically distributed biomass. (This profile often seems to be a logarithmic function in the form of a "piédestal de colonne," analogous to the piezometric equilibrium of the atmosphere [Godron 1975].) The vegetation then gives a positive true production, which is exactly equal to the phytomass taken away as long as the system has not gone beyond its elastic deformation limit (e.g., so long as it has not exceeded its metastatic capabilities). Two problems thus merit attention: the vertical profiles of biomass and elasticity.

The Vertical Profiles of Biomass

The vertical profiles of biomass are not being actively studied at present. Ecologists do not seem to have a clear opinion on the subject, because the mass of ligneous portions and, especially, the bases of trunks are frequently underestimated. Besides the previously discussed results, the only point that seems accepted in this area today is that old Mediterranean "maquis" of oaks are far from a piezometric equilibrium.

To specify this general characteristic of biogeocenotic equilibriums, the first step would be to combine the cubage tariffs of foresters and the allometric relations calculated by ecologists (Newbould 1967, etc.) and then to compute vertical profiles to derive more general rules. One of the immediately useful results of this work would be to specify the typology of forests, which is being improved in several countries (Canada, Mexico, France, etc.) by linking ligneous production to simple structural characteristics. In mature steppe vegetations, this work could be carried out rapidly, but it would have to be complemented by an analysis of the horizontal structure, because these types of vegetation generally are very heterogeneous.

Next, it would be tempting to find out if a regular relationship links the form of the profile and the dynamics of the vegetation (using the exact sense of the word "dynamic," which includes knowledge of the forces driving the system). The answer which appears most realistic today is that an ecological system passes several times from a piédestal profile (a phase of dynamic stabilization) to a chapiteau profile (during a phase of production), then again to a piédestal profile, etc. Thanks to research oriented in this direction one can hope to obtain two practical results:

(1) rapid estimation of the state of a vegetation and of its place
     in the cinematic succession of vegetation types;
(2) comparison of the harvested production with the potential
     production (harvestable biomass compatible with the "normal
     state").

The particular cases just discussed involve primarily the
equilibriums of phytomasses. An analogous reflection should be
undertaken to link the latter with the equilibriums of zoomasses.
Here it is already known that setting aside aged animals can lead
to more productive equilibriums than spontaneous equilibriums, but
more general laws have to be found.

Elasticity (Rather Than Resilience)

The beginning of Section 2 emphasized that biomass production
compensates for material removed only when the system stays within
the limits of elastic deformations (Fig. 3). In ecology, the idea
of resilience has been used inappropriately so often that the
vocabulary in this area has become babble (Westman 1978) and precau-
tions must be taken to do better. The type of stability that the
observed system possesses leads to two crucial problems:

(1) Stable ecological systems are obviously not static, because
     they include thermodynamic cycles. Is the intensity of these
     cycles linked to their resistance to disturbances, to total
     biomass, to structural diversity, to stratification, etc.?
(2) Mechanical systems deformed beyond their limits of elasticity
     (but within their "rupture loads") keep the same coefficient
     of elasticity (lines OM and PN are parallel in Fig. 3). Their
     domain of elasticity becomes greater, while their ductility
     diminishes. Is it the same in ecological systems?

These problems can be summed up in one sentence: Following
what point of deformation does an ecological system cease to be
itself? As long as this question is not answered, advice given to
decision makers by ecologists cannot be convincing.

3. PRODUCTIVE COMMUNITIES

Since the neolithic revolution, man has known that he can
rejuvenate ecological systems in order to take his nourishment from
them and free himself from the hazards of hunting and gathering.
Later, he learned how to build durable agrarian civilizations.
Today, the alimentary equilibrium of humanity does not rest totally
on autarkic familial agriculture, and the part played by the great,
very artificialized agricultural systems is becoming more and more
important because it involves international political equilibriums
(Lesourne et al. 1979). Two rather contradictory tendencies appear,
therefore, with increasing clarity.

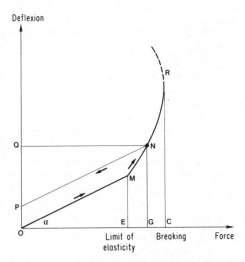

Fig. 3.   Limits of elasticity of a system.   (Reprinted by permission
of Centre d' Etudes Phytosociologiques et Ecologiques
Louis Emberger, Montpellier, France.)

Increase in Artificialization

Intensification of agriculture appears feasible in several
regions, and agricultural yields continue to increase in numerous
cases (Manitoba, the Parisian Basin) even though farmers are confused
about risking decreased fertility (this idea should be clarified,
however, because recent research shows that it withstands analysis
by agronomists).

The problem is thus to calculate the price (energetic, financial,
and social) paid to obtain this increase in yield. The experiments
of Pimentel and others have drawn attention to this point. However,
they obviously cannot provide clear conclusions because they reduce
everything (including the farmer's fatigue or his competence) to
kilocalories, as if a calorie of petroleum had the same value as a
calorie of alfalfa or mutton. Here, ecology can, less than ever,
dissociate itself from its etymological Siamese twin, economics.
Also, biogeochemical equilibriums must be integrated in a more
global model, a task which ecologists seem capable of doing.

Under-Used Lands

At the opposite extreme, territories which had been cultivated
for several centuries are often abandoned because they are barely
profitable. This abandonment entails a rural exodus which is
costly for the society, even when it corresponds to a gain in
financial productivity for the worker. This imbalance is particularly
evident in Europe, but it is now reaching countries which are
barely industrialized, where it leads to the growth of shanty towns.

The vocation generally assigned to these territories is forestry, because foresters usually are discreet and are accustomed to content-ing themselves with little. It would be better to understand that the wood needs of human populations must be satisfied through several routes and not to consider the forest as a reject of agricul-ture. For this, models of regional management must be constructed to optimize the partitioning of earth's resources.

Several types of models can be used to attain this objective. Experience has shown that too complex models are hardly useful. One must not fear, then, to choose simple conversational models that a manager can understand without difficulty, and for which he or she can supply data applicable to a specific situation. For example, a model of tensors (i.e., array of matrices) has been useful for the barren moors and forests of Lozere (Pillet 1981).

4. "DEGRADED" ENVIRONMENTS

The lowest part of the curve shown in Figure 2 represents the functioning of "pioneer" communities, which are extremely rare if one gives this term its exact meaning. In fact, communities in intensely degraded environments are often considered pioneers, even though they have an important reservoir of seeds available in the soil (and eventually a root system); by contrast, they generally suffer ionic imbalances. Whatever the case may be, these are poorly understood communities, difficult and discouraging to classify. However, their areas increase year by year, particularly in tropical countries and in arid zones. The study of their dynamics is much more difficult in that the force of elasticity which tends to lead them toward maturity is often relaxed (the slope of the curve is close to zero). This case shows perfectly that knowledge about the forces which make biogeocenoses evolve is one of the most important for future studies.

SUMMARY AND CONCLUSIONS

Research carried out during the years from 1960-1980 has permitted a better understanding of the functioning of numerous biogeocenoses (particularly those that are the least disturbed). But we lack information about the essentials, that is, knowledge of the dynamics (and not only the cinematics) of these ecological systems. In fact, describing the functions of a system permits one to know the cinematic view of it, but this is not sufficient for determining the forces that make it evolve. Thus, provision for the future becomes a lottery game in which the only sure thing is that there will be losers.

So that ecology might become the knowledge of the "economy of nature" (and thus become a synonym of "bionomy" again), it is now necessary to study, in priority:

(1)   the feedbacks which control the natural stability of mature biogeocenoses (old forests, unburnt wooded savannas, marshes, etc.);
(2)   the modalities of the artificial stability of intensively exploited biogenocenoses (economic externalities);
(3)   the limits which slow down the evolution of very degraded biogeocenoses;
(4)   more generally, the relationships between the spatial structures of biogeocenoses and their organization.

The most serious gap is the lack of convergence of ecological works towards the whole, where all the problems evoked above find cohesion. This gap can be filled only by very coordinated research, oriented towards knowledge of the metastability of ecological systems.

LITERATURE CITED

Box, E. 1978. Geographical dimensions of terrestrial net and gross primary productivity. Radiat. Environ. Biophys. 15:305-322.

Godron, M. 1975. Préservation, classification et évolution des phytocénoses et des milieux. Biol. Contemp. II, 1:6-14.

Lesourne, J. 1979. Face aux futurs. Org. Coop. Dev. Econ., Paris, France.

Lieth, H., and E. Box. 1977. The gross primary productivity pattern of land vegetation. Trop. Ecol. 18:109-115.

Newbould, P. 1967. Methods for estimating the primary production of forests. Blackwell, Oxford, England.

Pillet, Ph. 1981. Organisation et évolution des unités écologiques du parc national des Cévennes. Application à l'aménagement et à la gestion du territoire du Mt Lozère (Bougès nord). Bull. Ecol. 12(2-3):187-224.

Westman, W.E. 1978. Measuring the inertia and resilience of ecosystems. Bioscience 28:705-710.

SIMPLE FACTS ABOUT LIFE AND THE ENVIRONMENT NOT TO FORGET IN PREPARING SCHOOLBOOKS FOR OUR GRANDCHILDREN

Ramon Margalef

Department of Ecology
University of Barcelona
Barcelona, Spain

Ecology is becoming a very sophisticated science, at least in print. Complicated models of ecosystems are proposed; they may be helpful for interpolation, but their extrapolation rarely leads to accurate predictions. Usually the system is driven by inputs from outside that cannot be predicted from inside. We need to develop a whole spectrum of models of different scales and then to explore the regularities in the behavior of the whole set. Theoretical ecology, in its attempt to isolate relations and phenomena and to formulate testable hypotheses, offers an easy target for criticism; certain mathematical developments are highly unrealistic, perhaps justifying a factitious rule of thumb that any expression in ecological theory more than four inches long is false.

It is necessary to go back to fundamentals--look at nature and state real and interesting problems in a simple and general form, if possible. Ecological systems can be considered as physical systems, and ecology has to derive much inspiration from physics. Further, mathematical models of ecological systems are acceptable only when they include all the constraints imposed by the physical world (e.g., ecosystems extend in space, space imposes constraints on interactions, individuals are not differentiable). Human affairs also are relevant in the reconstruction of ecology or in the construction of a new ecology, not only for the usual motivations, expressed as concern about pollution and environmental protection, but principally because humans have become a paradigm for transportation and use of external energy, as well as for modes of cultural evolution inside the ecosystem. The comparison between the branching of a tree and the development of a highway system is a case in point. I would dispense with so many assessments of environmental impact, give more attention to the consequences of environmental change,

and analyze the results of unintended experiments done with or
without the blessings of environmental agencies. But here I especially want to belabor the role of external energy and the principles
of thermodynamics and information.

EXTERNAL ENERGY

     The main focus of scientific concern in ecology has been
shifting around the concepts of matter, energy, and organization or
information. Recognition and analysis of structure have always
been emphasized, but this approach is difficult and often gets only
lip service. Interest in biomass is obvious and relatively old.
More recently, internal or endosomatic energy has become the focus
of interest--witness the many quantitative models of ecosystems,
with fluxes and storages. Its counterpart, external or exosomatic
energy, often is seen only as one more property of the physical
environment. Humanized systems help in understanding the role of
external energy in ecosystems, although usually that role is more
than "production" and "consumption" of energy by humans. Nevertheless, we are more conscious of its role after recognizing the
importance of exometabolic energy in our own modern lives (Odum
1971).

     Biogeochemical cycles are governed by external energy. Rainfall,
erosion, and transportation on land, the very essence of rivers and
marshes, and mixing and upwelling in the sea provide living images
of the workings of external energy. In more quantitative terms, a
relationship exists in terrestrial vegetation of roughly 50 parts
of energy used in evapotranspiration to one part of energy used in
photosynthesis. Further examples supplement the same fundamental
concept: yield of cropland is related to support energy (work,
fertilizer, water supply); primary production of phytoplankton, in
oceans and lakes, depends on the work done locally by wind or by
kinetic energy introduced elsewhere. In mapping heat exchange in
any engine or over the oceans (Fig. 1), the places where work is
done or where primary production is supported are clearly apparent.
Such relations open new perspectives for space surveys that match
images of wavelengths reflected by chlorophyll with corresponding
pictures of heat exchanged and work done. The best predictor of
primary production of seaweeds along coasts is the amplitude of
local tides. The inescapable conclusion is that endosomatic energy,
that is, energy in primary production, is supported by a larger
amount of external energy. If this is so, should we, in any general
ecological approach, include some estimate of the ratio between
significant exosomatic versus endosomatic energy? Or should we
perhaps care less about the limit between the two categories of
energy, and consider the exact limits of endosomatic energy (i.e.,
primary production) not particularly relevant?

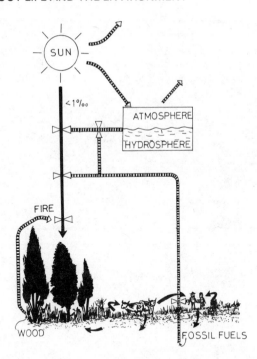

Fig. 1. Yearly net oceanic heat gain and yearly fluxes of latent
         heat and sensible heat in watts/m^2. The most productive
         areas are well defined, overlying appropriate values of
         heat exchange. (From Hastenrath and Lamb 1978; reprinted
         by permission of the University of Wisconsin Press, Madison,
         WI USA.)

     In schoolbooks, the ancient picture of energy flow in the
ecosystem has to be completed, introducing a whole circuitry for
external energy (Fig. 2), following, more or less, H.T. Odum.
Energy that is internal, derived from primary production and stored
as biomass in one system or part of one system, can operate as
exosomatic energy for another system or for another part of the
same system, as in muscular work, forest fires, and use of fossil
fuels. In all these situations, it is easy to see how internal
energy interacts with potential flows of external energy through a
number of switches and multipliers. Consequently, humans are not
only the exploiters--the plunderers--as pictured in schoolbooks.
Humans have moved close to the hub of a complex net of energy
distributors. A man moves a lever in a bulldozer, which uses
fossil fuel, moves a lot of earth, and builds a dam, or changes the
course of a river, and, in any case, introduces a new succession
under a different energetic pattern. The concept of multiplier and
accelerator is indispensable nowadays. The ultimate example is the
"red button" that starts a nuclear war.

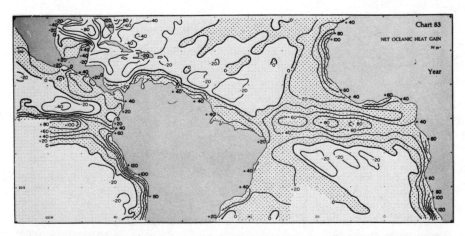

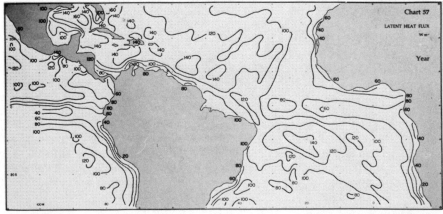

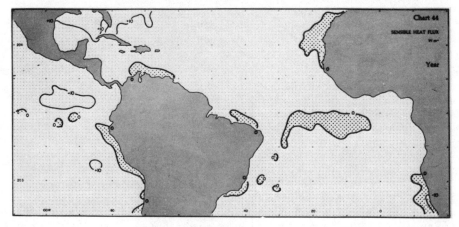

Fig. 2.   The usual flow of energy from the sun and through the
           ecosystems has to be supplemented by the effects of
           external energy that regulates the flow of endosomatic
           energy.

ALTERNATIVE APPROACHES TO ECOSYSTEM MODELLING

External energy is very much associated with transportation and space. In my opinion, most models, even those that accept compartments, deal poorly with real space. To introduce external energy properly, we have to start with models that accept space and movement. The work of Riley et al. (1949) presented a simple and elegant model, still unsurpassed, for plankton. I have attempted to build on it (Margalef 1978) and the essentials of a possible development are sketched in Table 1. Consideration of space is essential, requiring that the parameters concerning the distribution of kinetic energy in the environment and the relative distributions of organisms and factors of production be expressed in the form of spectra. This leads also to the consideration of the ecosystem as sets of Chinese boxes. With a fixed volume of water, developments can look somewhat deterministic, or internally determined. More important are the boundary conditions, as these include unpredictable events that drive our small system of reference to new situations. This is obvious in phytoplankton communities. Small volumes of water can be compared to, and modelled as, culture flasks in the laboratory, but the same volumes of water are moved up and down and mixed in the whole water mass. Both the expression of external energy, as turbulent diffusion, and the expression of covariance among the reactants or direct factors of production require a definite, measured reference volume.

If we trust this model, production depends mainly on mechanical energy and on the coincidence in the distribution of the immediate factors of production, that is, on the covariance of the distribution of organisms, nutrients, and light. Dependence on scale is obvious; on a large scale, the marine areas richer in assimilable nitrogen and phosphorus compounds sustain, on the average, more phytoplankton, but, on a smaller scale, patches with exhausted nutrients contain more plankton and may be close to other patches in which nutrients have not been totally used and which, consequently, contain smaller populations. Because the availability of nutrients in water depends mostly on mixing and turbulence, a double dependence—direct and indirect—on turbulence could be assumed; perhaps production could be related to the square of the turbulence. This relationship is not the case, however, because the distribution of light is independent of turbulence, photons cannot be pushed mechanically, and the vertical distribution of light cannot be uniform. Consequently, covariance in the distribution of the factors of production cannot be limited by turbulence. Actually, covariance of the reactants means simply the presence of nutrients and cells in the photic zone.

The spectral approach leads to some insights that may prove fruitful. In water, the relation between available energy and size of the reference cell may follow the Kolmogorov regularity of -5/3

Table 1.   Starting with the model proposed for phytoplankton (Margalef
           1978), a more general expression can be developed.   Taking
           derivatives in relation to time and space, succession and
           stratification can be described.[a]

---

1. Model proposed for phytoplankton

    $dB/dt = rB - V'(dB/dx) + A'(d^2B/dx^2)$

    $rB = -V(dS/dx) + A(d^2S/dx^2)$

            $A = A' + a$ and $V = V' + V$

    $dB/dt = -V(ds/dx) + A(d^2S/dx^2) - (V-v)(dB/dx) + (A-a)(d^2B/dx^2)$

    $(dB/dt) + V'(dB/dx) = -V(dS/dx) + A((d^2S/dx^2) + (d^2B/dx^2)) - a(d^2B/dx^2)$

    net   loss of   use of   effect of lack of   effect of
    increase   cells   nutrients   conformity in distri- anchoring
                                         bution of nutrients   cells in
                                         and plants   large eddies
    (production)   (flow)   (covariance in   (lagging
                                          distributions)   behind)

2. Generalized model for any ecosystem[b]

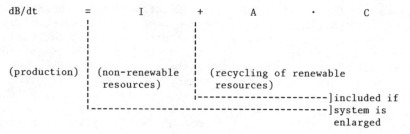

    In a system with internal cycling only[c]

. $dB/dt = A \cdot C$

3. Model with reference to space and time

    Succession

        d^2B/dt^2 = $(dA/dt)C$ + $(dC/dt)A$

        deceleration   decay of   ecological
                              turbulence   segregation

Table 1 (continued)

Vertical organization (aquatic ecosystems)

$$\partial^2 B/\partial t \partial z \quad = \quad (dA/dz)C \quad + \quad (dC/dz)A$$

| decrease of productivity | dampening of turbulence going downward | decrease of covariance dependent on extinction of light |

---

[a]Definitions are as follows:
    a = difference between diffusion of water and particles.
    A'= turbulence, referring to organism.
    A = turbulence, referring to water.
    B = biomass.
    C = covariance in distribution of reactants.
    I = inflow of new material.
    r = rate of net growth.
    S = substrate or nutrient concentration.
    v = difference between speed of water and particles.
    V'= speed of organism along x.
    V = speed of nutrients along x.
[b]In this model, A and C are spectral expressions.
[c]In the model $dB/dt = A \cdot C$, A^2 cannot be substituted for $A \cdot C$, because C is not highly correlated with A, since photons cannot be pushed mechanically; A can be compared to temperature, C to entropy of distributions.

power. Covariance in the distribution of reactants could follow some similar law, and perhaps the spectra of diversity provide some hint about relative distributions of organisms. The living part of the ecosystem effectively manipulates the whole setting through growth and mobility--active and passive. In pelagic ecosystems, transport is left to the water, with a small contribution from the animals. In terrestrial ecosystems (Fig. 3), the productive units are linked through a transport system. This transport system is made up of relatively non-biodegradable material. Seasonal vegetation is ready for transport and support from the beginning of a new vegetative period. This characteristic imparts vegetation an advantage in natural selection similar to that enjoyed by animals which can move around and accumulate reserves for the hard times.

It is easy to relate the approach of Table 1 to the usual way of modelling ecosystems as a set of binary interactions among components. One only has to write

$$dN_i/dt = \sum_{j=1} f(R_j N_i) + \sum_{j=1} f(N_i N_j)$$

where R_i represents random variables, not determinable from inside (i.e., either non-renewable resources or unpredictable impacts of

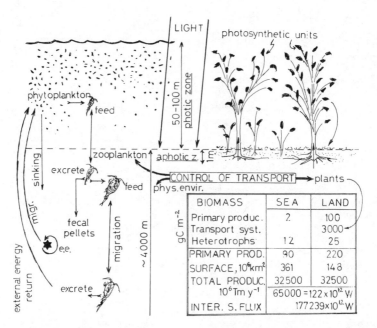

Fig. 3.   Comparison between oceanic and terrestrial ecosystems.
The main differences are the connection of the productive
units in a common system of support and transport in the
terrestrial ecosystems and the vertical dimension of the
space populated by heterotrophs.

energy), and the summation f(N.N.) represents the internal workings
in a unit volume, where the reactions are dependent on the distribution and the probabilities of encounter of the reactants and on the
local climate of external energy. Internal recycling is modelled
using this approach.

MODELLING AND THE SECOND PRINCIPLE

Most of the models proposed in theoretical ecology perform
like clockwork, even when they pretend to be stochastic. If random
events follow the laws of probability, at some level the models
become deterministic. Should stochasticity be related to the
discontinuity and individual sizes of organisms? In responding to
this question, models show a serious disregard for thermodynamics.
Constancy in the properties of the components and constancy in the
parameters are too easily assumed. No machine turns exactly the
same way twice, much less an organism or an ecosystem. According
to Konrad Lorenz, life is a process of acquired knowledge. I would
prefer to write that any exchange of energy results in an increase
of entropy, and appears as an increase in information. Perhaps

there is a similar failure in the consideration of economic systems, evident in recent history. If I engage in some transaction today and try to re-enact it tomorrow, it will not be the same, because both partners are no longer the same--our experiences are a bit more divergent. Our first transaction generated some information that spread and probably was distributed in an unequal way.

We should not look at the entropy laws as "forbidden" signs, but as rules of construction. In an ecosystem, the decay of energy is associated with ecological segregation (bottom of Table 1). We must pay more attention to the conversion of energy into structure, or to the equivalence between entropy and information. Any input may act as a disorganizer, or may be used to build further organization, according to the degree of organization attained by the system. An example is the role of the support and transport systems of terrestrial plants, able to use and change the flow of energy in the environment.

The view supported by these considerations is helpful also because it enlarges the concept of segregation. The world appears to be made of misplaced things. In the oceans, where there is light, there are no nutrients, and where nutrients accumulate, no light is available. The shift towards isolation of reactants proceeds at any level. The concept of segregation takes a cosmic dimension if it is true that matter and anti-matter persist only if kept apart, like fire and the powder keg.

SPECTRA AND INTENSITY OF EXCHANGE ACROSS BOUNDARIES

Most expressions used as ecological models start with dN/dt. Ecologists, in general, have been more reluctant to the introduction of space. Even using the model of Riley et al. (1949) as a starting point, space enters through the back door. Consideration of space is necessary, but brings with it many new problems. Measures of transport, coefficients of turbulent diffusion, covariance in the distribution of reactants (species, nutrients, light), as well as diversity and connectivity of the assemblages of species are spectral quantities, or nothing. Models of ecosystems have to be constructed recognizing the spectral nature of all or most of the variables. When related to man, the distinctions between non-renewable and renewable resources and between what is internal and what is driven from outside express the same reality at a relatively large scale. By enlarging the volume of reference, non-renewable resources become assimilated in what is then an internal cycle.

A spectrum specifies the value of a given variable as a function of the space over which it has been measured or averaged. The shape of the spectra is related more to the functional and historical properties of ecosystems than to the actual composition of species.

In Table 1, P = I + A · C, where I is "new production" depending on
non-renewable resources. Because both A and C are approximately
functions of size, in the form ~ $L^{5/3}$, clearly the fraction of new
production is related to the size of system under consideration,
that is

$$I/P = 1 - (1/P)L^{5/3}.$$

The problem has been discussed by Platt et al. (1981), extrapolating
from microcosms to natural, large ecosystems and appreciating
geometrical scales in ecology. We are led, of course, to considerations in the realm of fractals.

One class of ecosystems recycle most materials locally and can
be modelled as a set of prisms, extended in the dimension defined
by light and gravity and relatively independent of its neighbors
because lateral or horizontal exchange is not important or is
symmetrical across the vertical boundaries. If function and structure
are relatively uniform in different areas, the number of local
variants can reach infinity, and "small is beautiful." This is the
model that fits stable oceanic waters with dinoflagellate populations
or some forms of rain forests. Other kinds of ecosystems are based
on a more important horizontal transport--along a slope on land, in
a river, in the coupling between an oceanic upwelling area and the
offshore "blue water." Such systems would lose their characteristics
after the introduction of vertical barriers and use much energy in
transportation, and the slogan "big is powerful" applies to them.
Both patterns are recognized as well in human societies; some
humans are linked directly to the complex of local resources; thus,
they no not need transportation and communication systems, or they
can survive their decay. More affluent countries invest external
energy to keep horizontal communications active and are dependent
on the preservation of a considerable exchange among sections
distant in space. In horizontal transport and stability, a country
works like any ecosystem. Everything that exists is stable in some
measure, but may be stable in different ways. Persistence of
specified properties can survive dissection of the system into vertical prisms of smaller sections--at least as a thought experiment--or,
if the system is divided in this way, it collapses. Of course,
there is a continuum in the range of the possible qualities of
divisibility and closure, but, unfortunately, different persons
with different philosophies have attached the same label of stable
to both and alternative ends of the whole range. A rain forest can
be declared stable, until humans come with ax, fire, or bulldozer;
or a wheat field can be declared stable, until humans abandon the
field, go elsewhere, and the local vegetation enters a process of
reconstruction and succession. The concept of stability is dependent
on the whole setting and on the spectrum of the organization. To
borrow models made of balls and cups from physics is no easy solution: a ball can rest on the bottom of the cup and return to the

same spot if displaced. But the stability position is different if
the system is subjected to centrifugal force, or if there is a
slave-genius with the job of returning the ball to a marked space
each time the ball moves away.

Ecologists may be more interested in the contents of small
volumes, or more interested in the boundaries between the spots.
To draw boundaries to define areas or systems has many qualities of
an arbitrary enterprise (Canny 1981). Boundaries must be chosen
with a certain reference in mind (e.g., as continuous and normal in
each elementary unit of some selected property to the steepest
possible gradient). Ecologists are aware that boundaries can be
sharp sometimes, suggesting a high "interface tension," or wandering,
passive, indefinite, of low energy--in short, fractal. These two
classes of boundaries correspond approximately to the "limes convergens" and "limes divergens," respectively, of Dutch authors. The
recognition and constructivity of boundaries (boundaries inside
boundaries, extremely convoluted boundaries) are problems related
to the spectral measurement of ecosystem variables. Low energy
boundaries, approaching their vanishing points, provide a good
opportunity for introducing dynamics to the consideration of fractals
(Mandelbrot 1977).

EXPLOITATION AND STRESS

Boundaries that tend to be short, high tension boundaries, or
limes convergens, are active because exchange across them is asymmetric. Such are the boundaries between a pond and the surrounding
land and between plankton and benthos. Most often such asymmetrical boundaries are sites of exploitation--one side of the boundary
feeds on part of the organic matter produced on the other side.
This is one pattern of coupling between parts of ecosystems. The
benthic community, for instance, exploits the plankton community,
and the exploitation is mostly, but not completely, passive.

An eutrophic lake is force fed with sewage and enriched water
from cultivated land. It seems more common to say that the lake is
stressed than to say that the lake exploits its surroundings. The
distinction is, in part, a question of who is in control, a subtlety
perhaps more important than the actual direction of the net flow of
matter and energy. In addition to the eutrophic lake, other pertinent
examples come to mind. Deciduous trees produce and shed litter and
regulate large populations of bacteria, fungi, and animals. Although
hosts feed their parasites, hosts usually are in control of the
local systems, in part because they are individually larger and
live longer. Stress in this sense is not a profound concept, but
this distinction helps in the comparison of eutrophic lakes, upwelling
areas, litter communities, cropland, etc.

The factors of stress are unforeseen or unpredictable from inside the stressed system, in an evolutionary or adaptive sense. The reaction of the stressed system includes an acceleration of internal cycling and an increased output to neighboring compartments or systems through some external loop or through the deposition and immobilization of part of the material from the cycle in the boundaries. Typically, stressed ecosystems export oxygen and nitrogen to the atmosphere and phosphate compounds and organic carbon to the sediment or to the soil. Consequently, an ecosystem under stress is a paradigm of the workings of the primitive biosphere, of the accretion of the atmosphere, and of a biological membrane as well (Margalef and Estrada 1981). Eutrophic lakes are most interesting ecological experiments. Separation of materials at the boundaries is another form of ecological segregation. Clearly this effect prevents the full advantage of any process of fertilization on production. In effect, returning to Table 1 and remembering that energy may not be important,

$$P/I = 1 - A/I \text{ (segregation).}$$

Ecosystems are exploited as well as stressed by man (overfertilizing, pollution). Systems working under stress, as upwelling systems do, may be very productive and exploitable. But the condition of exploitability probably is related more to the unpredictability of inputs, when it does not allow an historical process of gradual building of complex systems. As a result, the system can be exploited--with fluctuating intensity--without altering its working structure very much. Successful cultures, such as temperate zone agriculture, rice, fish ponds, and shrimp ponds in Ecuador, take profit from this principle. All these systems behave as stressed systems; it is well known that crop soils lose a large fraction of the added fertilizers through denitrification and accumulation of insoluble phosphate. On the other hand, enhanced productivity in eutrophic lakes does not keep pace with the phosphorus load, because part of this load goes to the sediment through the workings of complex mechanisms, summarized as "segregation" in the last expression. If the process of segregation accelerates with time, the ratio P/I, or use of the inputs, decreases. Contrary to what is believed sometimes, phosphate in sediment and development of macrophytes help dampen the consequences of eutrophication.

INTERACTIONS AND ORGANIZATION

I feel that in this area it may be creative to return to the naturalist's approach. This means to be receptive to all sorts of stimuli and insight, to store them and let them mature in the mind without immediately turning the observations into a paper. In other words, this is a plea for more fieldwork and relatively less computer time. Look at the organisms and try to pose relevant questions. Organisms should not be considered exclusively as bearers of specified atoms or molecules, but as subjects of natural selection as well.

Usually, the different species present in an ecosystem are joined by relations that are symbolized by arrows, thus expressing the asymmetry of the interactions. A big question is how to improve on the common Volterra-Lotka tradition. It is not a matter of small changes in algebra; the criticism must cut deeper. Volterra was a genius but his contributions, including his contributions to ecology, have been followed more to the letter than in spirit. Interactions are studied in binary systems that behave as negative feedback loops with oscillatory properties, depending on inertial mass and information, and thus are specifically asymmetrical. By a slight of hand, ternary systems involved in competition also have been reduced to binary systems with a positive feedback loop, resulting from the composition of two negative loops. The elementary binary relations are combined into a larger net of relations, always based on matter and energy exchange, in the form of food or trophic nets. Usually, the relations between pairs of species have no other qualifications, and it seems unjustified to compute "indirect dependencies" among species that are not directly related.

Theoretical ecologists have been curiously, and perhaps studiously, ignorant of space and thermodynamics. Each binary system, each pair of interacting species, is an energy gate, and energy goes one way. The oscillator composed by cat and mice is supported by energy flowing through the population of mice. This fundamental asymmetry is not properly reflected in the usual expressions that include only a "natural" mortality for the cat and a cat induced mortality for the mice. I agree that the purpose or the result of the usual mathematical treatment is restricted and well defined, but I complain that this specified purpose is not more interesting to the biologist.

Decay of energy is associated with an increase of information. This term perhaps is too general and imprecise in the present context. Usable information is materialized at the genetic level and at the cultural (non-hereditary) level. In addition, there are changes that require a new context to qualify as information. "Cultural" is meant in a very broad sense: a developed fungus or a tree begins with a spore or a seed, respectively, and ends in a different and infinitely variable structure that reflects its long history of interaction with the environment; it is never inherited in its detail. Much information at the level of the ecosystem has similar rich and unique configurations of low energy. The accumulation of this new information does not fit a map of energy decay exactly. Consider the interaction between a predator and its prey; interactions may be unique--and fatal--events for the prey, but may be routine for the predator. The mouse runs for its life, the cat

just for a meal. Each shot kills an Indian, the movie hero survives
a thousand shots. The student of behavior conditions the rat, the
reverse is more doubtful.

This asymmetry in responses is obvious in the relations between
insects and birds. The evolution of colours and behavior of insects
follows a rather strict path of genetic determination; in birds,
the evolution of learning capacity has been obvious. As a consequence
of the nature of such binary relations, any food web is anisotropic
from the points of view of thermodynamics and information theory.
Energy enters one end of the net, where the primary producers are,
but information, in the form of special cultural information and
its support, increases more rapidly at the other end. An external
criterion for assessing the capacity of the different food links
for accumulating information might be the size of the individuals
and the lengths of their lives. The total energy exchanged in an
ecosystem sets the upper limit of the possible acquisition of
information. Perhaps its increase is distributed among the species
in proportion to the product of individual mass times length of
life of each of them. The aptitude to accumulate total information,
and its partitioning into different stores, may be correlated to
the position of the respective species in an ordination according
to the axis of r-K strategies. If these are general properties of
systems like food nets, one could conclude that any system with
discrete individuals, related the way organisms are, has to serve
as a matrix for the evolution of organisms able to learn, that is,
blessed or cursed with consciousness. Further, it is impossible to
unify the concept of niche. The ecological niche of a grass and of
a large cat comprise different descriptors and in practice are
incommensurable. Obviously, models that include only the numbers
of individuals of the different species and use their products as
the unique measure of interaction may miss important phenomena.

Animals are only part of the heterotrophs; trophic nets based
on animals are surrounded by a complex cloud of bacteria and other
small heterotrophs. Large, macrophagous animals are only a small
fraction of all the animals. But such large animals are interesting,
may exert a control disproportionate to their biomass (e.g., vertebrates retain large amounts of phosphorus in their skeletons), and
provide the main support for the evolution of a cultural information
channel and its genetically determined basis.

Trophic nets are traditionally represented by graphs, with
arrows signaling the flows. The arrows point in one direction,
setting constraints on the building of the net and on its final
properties. Food nets are never totally connected, otherwise they
would be like short-circuited electrical circuits. It is helpful
to consider the development of a trophic net from an undifferentiated
stage in which many connections are possible. But in real ecosystems,
the number of links of interaction among species is much lower than

the number of total possible interactions. Connectivity is the fraction of the possible connections that is actually realized (Di Casti 1979, Margalef and Gutiérrez 1983). Cutting potential links generates a hierarchy not only in ecological systems, but in human social systems as well.

Diversity, if expressed in appropriate units (Margalef and Gutiérrez 1983), represents the upper limit of connectivity. If the number of elements is N, possible interactions is a function of N^2, but real interactions are a power less than 2, and, consequently, a fractal concept. Space constraints are reason enough for such limitations in connectivity. The same reasoning can be applied to individuals and to species. The Lotka-Volterra approach shows that highly diverse communities, if totally connected, are subject to decay (May 1972). But actual communities are much less connected (Pimm 1982). The discontinuous nature of interactions between individuals is a supplementary source of stability. Similarly, separate sexes decrease the connectivity inside one species, thus, gonochorism is a stabilizing factor.

The lack of superposition between the places of exchange of matter and energy and the places where information increases, as well as the need to deal properly with competition, suggests the advisability of using ternary (not binary) systems as the elementary unit to build models of ecosystems. The different combinations of the possible relations would become more explicit, as well as the distinction between positive and negative feedback loops. If organisms belong to the same reproductive team, they have a common information pool and converge (maintenance of one species); however, if they have a differential accretion of information, they segregate and diverge as a result of competition. The latter usually is expressed as the splitting of information along different "qualities" (e.g., r-K). Organisms in a net of relations in an ecosystem not only are junctions for transference of matter and energy, but also centers for decisions concerning information. This distinction would be better served out of the constraints of elementary binary relations, but three body problems and three persons games are most intractable. We cannot ignore or shrug off the problem. Binary models are too small to offer an appropriate grasp of the construction and working of actual ecosystems.

The usual presentations of the concept of fitness are too narrow for the needs of ecology. In fact, bacteria have all the essential cellular machinery and are so happy. Why then do they evolve? Short food chains can provide the maximum biomass. Why lengthen food chains? The point that I want to make is that the increase of individual size and the lengthening of individual life provide new paradigms of fitness that need to be understood in the framework of thermodynamics of non-equilibrium, open systems, that large entities are usually out of equilibrium with the close environ-

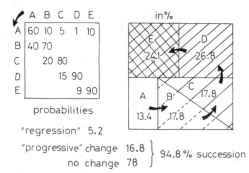

Fig. 4. Probabilities of transition among a set of five communities or stages in ecosystem development (left); and relative extensions occupied by the different communities at equilibrium (right). The figures below the illustrations represent percentages of the different kinds of observable changes.

ment, and that information, however difficult to grasp, has to be introduced into models, although at present no way exists to measure it appropriately. But even if we continue dealing with numbers as we have until now, my wish is that more emphasis will be placed on expressions of turnover, on the product of individual mass times average life span for the species, or on the ratio (material lost to the prey minus material gain to the predator) / (square of the fluctuation time in the predator-prey system) in binary systems.

SUCCESSION AND HISTORY

I do not feel like a diplodocus in the science of ecology when endorsing a quasi Clementsian model of succession (Clements 1916), however unpopular it may appear now (Drury and Nisbet 1973, Margalef 1974, 1980, Odum 1969). And I identify the increase in information preserved and forwarded in relation to the energy exchanged--or the entropy produced--as the guiding principle in succession (Matsuno 1978). This also would agree with the trends expected in non-equilibrium and open systems. But I recognize that it is easy to falsify any theory of succession. We need only to look at the changes going on everywhere, at different speeds and in almost all possible directions. As Figure 4 shows, slow and relatively ordered change (succession) follows a certain path, but any other change is possible, and realistic transition probabilities can be assigned to each event. The operation of the accepted transition probabilities can lead to a steady state, in which all the possibilities are realized over extended surfaces and for variable periods of time. Such is the appearance of the biosphere, but in such a patchy situation, an overwhelming proportion of transition events support the classic concept of succession.

A careful look at the examples of elementary change might lead
to a classification of the events into two groups: fast changes
with an apparently inefficient use of resources and slow changes in
which the rules expected from the behavior of open systems almost
hold. The ecologist can easily produce many examples--for instance,
the droppings of a cow in a mountain pasture. Seen from the continuum of the pasture, the event happens at random and in an unforseen
place which receives an input of chemical energy in the form of a
relatively homogeneous material with uniformly distributed reactants.
In the first stages, energy is exchanged rapidly, entropy increases
accordingly, and competition leads to the dominance of a small
number of species. As time advances, the system becomes more
heterogeneous, competition favours species relatively parsimonious
in the use of energy, and the system goes the way expected of a
dissipative system out of equilibrium (Glansdorff and Prigogine
1971, Margalef 1982) until it blends with the large and enclosing
system of the meadow. Another instructive example concerns planktonic change: every mixing event, under the work of wind, sharply
introduces a segment of succession, and usually includes the rapid
development of populations of diatoms in the mixed water, followed
by stratification and the predominance of swimming organisms.

The most general model accepts that the history of the ecosystem
is punctuated by unpredictable events--usually as inputs of energy--
with a distribution that can approach a spectrum of white noise.
The changes due to the impact, and the reconstruction a short time
later, are not symmetrical. It is hard to find regularity in the
destruction following the impact, but reconstruction is slower and
follows the regularities of secondary succession. In fact, all
successions proceed through a steeplechase of disturbances ("random
density-independent disasters"). The ecosystem is a black box that
adds such events and gives an output. It can be assumed that each
perturbation has an effect that decreases exponentially with time
and that each class of the catastrophic impacts may be destructive
for some species but not for others. Motility, homeostatis, accumulation of reserves may help in surviving disturbance. Evolution
probably attempts to cope with higher and higher levels of change,
as required, but extinction is always sure after the most destructive
events, which are rare.

The contrast between sharp, energetic change and slow change
that brakes itself is also present in inorganic systems. For
instance, in cycles of sedimentation (Bouma cycles), every cycle
starts with coarse materials, gradually passes to fine grained
materials, and comes back to coarse materials in a sharp transition
associated with an input of external energy.

Thus, there is a difference between relatively short periods
of history making, not predictable from inside the system of reference, and long periods of history telling, in which the system

behaves more regularly in that its behavior can be predicted from
the boundary conditions and thermodynamics of open systems. Without
apologies for my propensity of associating classical or Clementsian
succession with the period of history telling, I cannot help but
wonder why the label progressive is often attached to it--but
perhaps this is the reflection of an anarchist at heart.

Plant ecologists have always maintained that succession slows
down and that its rate of slowing down decreases. This can be
written in a more dignified form introducing Liapunov's expressions,

$$\partial/\partial t = \sum_{i=1} \frac{N_i}{N} \frac{d(N_{it} - N_{i(t-1)})^2}{dt}$$

N_{it} = number of individuals of species i at time t

N = total number of individuals

or in any other suitable expression. Expressions of this kind are
spectral, related to the sampling space in which N and N_i have been
counted. On a large scale, convergence is likely, but on a small
scale this is not so sure.

SUCCESSION AND EVOLUTION

If ecosystems repeat sets of transitions with great probability,
we can expect that many opportunities for selection along the same
tracks exist and, consequently, what I call an adherence of evolution
on succession. Then we should expect that the selected genotypes
often will embody kinds of adaptations that refer to subsequent
stages of succession. Such would be the reason of the common
evolutionary trends towards a larger size, longer life, smaller
number of better protected offspring. Not only do we need a complementary "theory of ecosystems, in which the component species are
evolving by natural selection" (Maynard Smith 1982), but also a
theory of selection in which co-evolution within a non-static
ecosystem plays a due and important role. The classical concept of
fitness does not explain the drop of the quotient, production/bio-
mass (P/B), with succession; thermodynamics does. We could propose
a rule for the admission or rejection of new genotypes or new
species, depending on whether P/B in the system drops or not, but
humans can dominate nature with fire and ax. Surely it is impossible
to maximize fitness for everybody, if fitness has to be defined in
relation to each species. In the long run, neutral and dormant
genes may be decisive. If a species is to continue its existence,
it must evolve as rapidly as possible forever. But if a species
persists, its rate of reproduction will decline as low as possible.
These are the two versions (pessimistic/optimistic?) of the Red
Queen hypothesis that, combined, make it not much more than a

verbal construction. But the organization of hierarchical systems
requires that some elements be able to evolve faster than others--RNA
faster than DNA, parasites faster than hosts, mimics faster than
models.

In broader terms, many of the problems common to succession
and evolution can be stated in terms of the relations between
different hierarchical levels in any system. If the model is
reduced to the essential, the highest or encompassing level is a
large structure that evolves in a process of self organization, and
is made of self replicating units that constitute the inferior
level. The behavior of the larger system defines the way in which
the selection of the replicable elements occurs. In ecology, the
larger systems are ecosystems, and the replicable components are
the individuals of the different species. But equivalent and
corresponding structures can be discovered in many other systems,
for instance, in language or discourse, comprised of words, or in
electronic circuits, made of electronic components. I have found
the consideration of electronic systems extraordinarily enlightening
for exploring the relations between the evolution of parts and the
evolution of circuits, as well as in the study of constraints that
limit the connectivity among the parts in the actual circuits. It
is not farfetched to see a model of the evolution of eukaryota in
the evolution of integrated circuits. No less interesting is the
analysis of mechanical systems made of standard parts that can be
taken apart and used to build other models, as in the "Meccano"
construction toys. They are also exceedingly useful for exploring
the statistical aspects of diversity and connectivity.

The different levels of organization are separately typified
and classified, and are studied by different sciences: ecosystems
by ecology, organisms by other biological sciences. In this discussion, I consider only two levels. The link between both requires
that more attention be paid to processes than to structures. The
top level--ecosystems, languages, electronic circuits--changes
slowly in a way typified by ecological succession. The elements at
the lower level--individuals, words, electronic components--are
replaceable and replicable, and can evolve rapidly through Darwinian
evolution. It appears, and this is extremely important, that the
workings of the organization at the higher level brake rather than
enhance the speed of evolutionary change in the components. The
components, as individuals or words, being discontinuous and mass
produced, are always out of equilibrium. Never is an individual
totally or passively adapted to its environment, and selection and
evolution work on that, just as a creative writer fights with the
available words.

Ecosystems are the workshops of evolution; any ecosystem is a
selection machine working continuously on a set of populations. No
constant environment exists for use as a reference to qualify

fitness, and selection in a system is more comparable to a game of
musical chairs than to numerous pigeonholes addressed as ecological
niches. An evolutionary strategy rarely can be stable; at best it
will adhere to succession, at worst it will be just a figment of
the imagination. Evolution trapped in succession can be envisioned
as a set of escalators (Fig. 5), each representing one particular
site under succession. The possibilities are either to step out to
a matching spot on the next escalator, or to drift along with the
successional change, slowing down evolution and (or) giving it the
appearance of orthogenetic gradualism. Darwin was aware that
evolution could go faster than it usually goes in nature, and paid
particular attention to organisms evolving out of the old context--
plants and animals under domestication. Kimura (1968) and others
have shown that change in genetic material can go faster than
change in phenotypes. Consequently, macroevolution poses no problem,
because organisms have the capacity to evolve rapidly; the real
question is why does microevolution generally run so slowly. My
feeling is that the answer to this venerable question--now in terms
of gradualism versus punctuated equilibria--has to be found in the
field of ecology and in the appropriate understanding of how the
processes of change and succession are interconnected in the different
system levels.

Fig. 5.   System of escalators as a model for adherence of evolution
          on succession. Each population is pushed up the escalation
          by the local trends of change. To keep the same environment,
          organisms shift to a lower step on the neighboring escalator.
          With catastrophic change, i.e., falling off the escalator,
          organisms adjust to an empty space with less or different
          constraints and start again ("punctuation").

CONCLUSIONS

In future presentations of ecology, I would emphasize the following concepts:

(1) External energy is a key factor. The development of switches and multipliers controlled by humans explains our success and provides a more complete framework for the consideration of present and future menaces to our survival.
(2) Machines and ecosystems cannot turn exactly the same way twice. Ecological equilibria are as impossible as stable economies. Ecological modelling has to respect the fundamental properties of the physical world.
(3) The problem of space and boundaries must be faced realistically. The most significant variables in ecology should be expressed in spectral form.
(4) Ecological systems are often dissected and analyzed in binary components; this has a long tradition in ecological theory. Relevant information appears in places different from where most energy decays. Extensive food nets are characteristically anisotropic from the points of view of history, information processing, and thermodynamics. The reduction to binary systems, as mechanical oscillators, was a first step, but it is necessary to improve on that.
(5) Changes in ecosystems can proceed in different directions, but it is important to identify two major kinds of change: unpredictable and rather brutal changes due to impacts (usually energetic) from outside (history making), and slow changes, controlled mostly from inside, which follow the expected changes of open systems trying to close themselves (history telling). After the revolution, the growth of a bureaucracy.
(6) These modes of change might be associated with modes of evolution--macroevolution or punctuated evolution and microevolution. It must be understood that succession usually acts as a brake rather than as a stimulus to species evolution.

LITERATURE CITED

Canny, M.J. 1981. A universe comes into being when a space is severed: Some properties of boundaries in open systems. Proc. Ecol. Soc. Aust. 11:1-11.
Clements, F.E. 1916. Plant succession. Carnegie Inst., Washington, DC USA.
Di Casti, J. 1979. Connectivity, complexity, and catastrophe in large-scale systems. Wiley, Chichester, England.
Drury, W.H., and J.C.T. Nisbet. 1973. Succession. J. Arnold Arbor. 54:331-368.
Glansdorff, P., and I. Prigogine. 1971. Thermodynamics theory of structure, stability and fluctuations. Wiley, New York, NY USA.

Hastenrath, S., and P.J. Lamb. 1978. Heat budget atlas of the
    tropical Atlantic and eastern Pacific oceans. Univ. Wisconsin
    Press, Madison, WI USA.
Kimura, M. 1968. Evolutionary rate at the molecular level.
    Nature 217:624.
Mandelbrot, B. 1977. Fractals, form, chance and dimension.
    Freeman, San Francisco, CA USA.
Margalef, R. 1974. Ecologia. Omega, Barcelona, Spain.
Margalef, R. 1978. Life-forms of photoplankton as survival alterna-
    tives in an unstable environment. Oceanol. Acta 1:493-510.
Margalef, R. 1980. La biosfera entra la termodinámica y el juego.
    Omega, Barcelona, Spain.
Margalef, R. 1981. Asimetrias introducidas por la operación de la
    energía externa en secuencias de sedimentos y de poblaciones.
    Acta Geol. Hisp. 16:35-38.
Margalef, R. 1982. Instabilities in ecology. Pages 295-306 in J.
    Casas-Vázquez and G. Lebon, eds. Stability of thermodynamic
    systems. Springer, Berlin, W. Germany.
Margalef, R., and M. Estrada. 1981. On upwelling, eutrophic
    lakes, the primitive biosphere, and biological membranes.
    Coastal upwelling. Pages 522-529 in Coastal and estuarine
    sciences, I. Am. Geophys. Union, Washington, DC USA.
Margalef, R., and E. Gutiérrez. 1983. How to introduce connectance
    in the frame of an expression for diversity. Am. Nat. 121:601-
    607.
Matsuno, K. 1978. Evolution of dissipative systems: A theoretical
    basis of Margalef's principle on ecosystems. J. Theor. Biol.
    70:23-31.
May, R.M. 1972. Will a large complex system be stable? Nature
    238:413-414.
Maynard Smith, J. 1982. Evolution and the theory of games.
    Cambridge Univ. Press, Cambridge, England.
Odum, E.P. 1969. The strategy of ecosystem development. Science
    164:262-270.
Odum, H.T. 1971. Environment, power, and society. Wiley Intersci.,
    New York, NY USA.
Pimm, S.L. 1982. Food webs. Chapman and Hall, London, England.
Platt, T., K.H. Mann, and R.E. Ulanowicz. 1981. Mathematical
    models in biological oceanography. UNESCO Press, Paris,
    France.
Riley, G.A., H. Stommel, and D.F. Bumpus. 1949. Quantitative
    ecology of the plankton of the Western North Atlantic. Bull.
    Bingham Oceanogr. Collect. 12:1-169.

SUMMARY

Prepared by the Workshop Participants

Written by J. Melillo, T. Rosswall, and W. Wiebe

Three major topics were emphasized at the International Association for Ecology (INTECOL) workshop in Louvain-la-Neuve, Belgium,
7-9 April 1983: (1) important areas for ecological research in the
1980s; (2) establishing a structure to foster long term ecological
research; and (3) general concerns about how ecology is practiced
and suggestions of how these practices could be improved.

IMPORTANT RESEARCH AREAS

Information Transfer

Traditionally, we describe ecological systems by quantifying
the transfer of energy and materials in biological systems. This
basis for ecosystem description is incomplete because it does not
consider the transfer of information among the components of a
system. The nature of information transfer takes many forms, for
example, pollination, animal behavior, intraspecific and interspecific chemical signals, and genetic transfer such as occurs through
plasmids.

Element Interactions

Studies of energy flow and nutrient dynamics at various levels
have been common. There is a growing awareness that element interactions need to be considered at a mechanistic level and that
energy flow and nutrient cycles cannot be understood independent of
each other. The study of such interactions is important at different
levels of organization--from the global system, where increased CO_2
concentrations may lead to a change in plant nutrient uptake, to

321

the ecosystem, where, for example, the relative concentrations of various elements in litterfall may be important in regulating certain characteristics of the ecosystem, to the species-physiological level, where carbon and nutrient partitioning needs to be understood.

Comparative Ecology

Understanding the organizing factors in communities is essential for understanding the distribution and abundance of all organisms. Types of interactions and their relative importance must be studied in greater detail, including (1) the roles of competition, mutualism, predation, parasitism, amensalism, etc.; (2) the synergistic and inhibitory effects that other species have on species-species interactions; and (3) the changes in magnitude and types of inter-actions through time. The approach should be strongly comparative, with comparisons made among taxa and kingdoms, functional groups, and ecosystem types. The method should include experiments in the field to determine the mechanisms involved. Close attention should be paid to phenotypic and genotypic variations within and between populations, which modify interactions, and to genetic variations, which, through natural selection, produce various types of inter-actions and thus various consequences. In addition, emphasis should be placed on understanding mechanisms in community ecology to provide an effective link between organismal and ecosystem ecology.

Human Impact on the Biosphere

Topics deserving special consideration include (1) the alteration of major biogeochemical cycles (carbon, nitrogen, phosphorus, sulfur) at the regional and global levels; (2) the circulation, distribution, and effect of toxic substances in the biosphere (magnitudes, pathways, rates); (3) the reduction of productivity of ecosystems as a consequence of poor management (e.g., from tropical forests to *Imperata* grasslands); and (4) the reduction of biotic diversity. Emphasis should be on limits to production and other processes.

Models of Ecosystems

Ecosystem models should be made consistent with Darwin's theory of evolution and with thermodynamics; at least ecosystem models should not be in conflict with these theories. The development of ecological theory should be deduced from concepts related to species-species interactions, biotic-abiotic interactions, non-biotically induced changes of the physical environment, effects of time lags, and occurrence of various processes in ecosystems at greatly different rates.

Examination of Long Lived Phenomena

Ecological phenomena should be examined on time scales appropriate to their behavior. Most often this means that longer time periods of study are needed. Specific areas of concern are studies of chemical balances in soil, population sizes of plants, animals, and microorganisms, turnover patterns, recognition of major types of natural architecture, and studies of life cycles. Although many advances have been made in identifying animal, plant, and microbial interactions, the lack of resources for long term studies has hindered the field examination of the mechanisms that control the interactions. New kinds of sites should be developed and funding agencies should be encouraged to offer more support for long term studies (see below).

ESTABLISHMENT OF LONG TERM ECOLOGICAL STUDY SITES

Many ecological phenomena, such as succession or the occurrence of stochastic events, can be understood only on a long term basis. These studies are important in many areas of ecological research and encompass both the analyses of historical records and long term field observations over decades or even centuries, as has been done, for example, in agricultural systems at the Rothamsted Station (UK).

Relatively few sites have been established for long term ecological studies. Additional long term study areas should be designated; each should be "permanently" controlled and sufficiently large to include a whole system (e.g., a whole watershed). Research practices on these sites should include (1) routine collections of basic data (i.e., climatological, biogeochemical, demographic, taxonomic); (2) use of standardized methods; (3) specification of paleoecological and land use histories; and (4) vertical integration of studies (i.e., microbiologists, plant physiologists, population biologists, behavioral ecologists, ecosystem scientists, theoretical ecologists should be encouraged to work together).

GENERAL CONCERNS ABOUT RESEARCH PRACTICES

More attention should be paid to spatial and temporal scale heterogeneity in the design of research and the interpretation of results. Experimental manipulation in ecology should be encouraged, not to replace careful observations of patterns in nature, but to allow an understanding of mechanisms responsible for these patterns.

Careful definition of terms (such as stability) must be made. Taxonomy and general ecology must be supported because they are essential to all specialized areas of ecological research. All

subdisciplines of ecology must strive to interact to maintain the
holistic approach and to advance the science.

ACKNOWLEDGEMENTS

We wish to express our gratitude to all of the following for
their contributions to that workshop and to this resulting publication: to F. Bourliere, President of INTECOL, who worked with us to
organize and facilitate the workshop; to the 15 discussants, who
presented stimulating papers and exchanged information across a
wide range of topics; to the NATO Ecosciences Panel, who sponsored
the workshop; to T. Truelsen and C. Sinclair of the Ecosciences
Panel, who assisted in arranging the workshop and who welcomed the
participants on behalf of NATO; to Ph. Lebrun, President of the
Belgium Ecological Society, who welcomed the participants on behalf
of the host country; to the staff of the Hotel ETAP who provided
efficient and courteous service and a friendly atmosphere; to the
the Institute of Ecology, University of Georgia, who provides
INTECOL administrative offices and support services; to Kula Campbell,
who typed the manuscripts both skillfully and good naturedly; to
Hedy Carpenter, who proofread them in like manner; and to Pat Vann,
Associate Editor at Plenum, who smoothed the way for this publication.

June H. Cooley and Frank B. Golley

WORKSHOP PARTICIPANTS

Joachim Adis
Max Planck Institute for Limnology
Tropical Ecology Working Group
D-2320 Plön, West Germany,
in cooperation with
Instituto Nacional de Pesquisas da
    Amazônia (INPA)
Caixa Postal 478, 69000 Manaus/AM, Brazil
(Convênio INPA/Max Planck)

Francois Bourliere
15 Avenue de Tourville
F-75007 Paris, France

June H. Cooley
Institute of Ecology
University of Georgia
Athens, Georgia 30602 USA

Henri Décamps
Centre d'Écologie des Ressources
    Renouvelable
29 Rue Jeanne-Marving
31055 Toulouse Cedex, France

M. Godron
Botanical Institute
163 Rue Auguste-Broussonnet
3400 Montpellier, France

Frank B. Golley
Institute of Ecology
University of Georgia
Athens, Georgia 30602 USA

Peter J. Grubb
Botany School
Downing Street
Cambridge CB23EA, England

Ph. Lebrun
Universite Catholique de Louvain
Ecologie Animale
Place Croix de Sud 5
1348 Louvain-la-Neuve, Belgium

327

Pierre Legris

Centre d'Etudes de Geographie Tropicale
Domaine Universitaire de Bordeaux
F-33405 Talence Cedex, France

Ramon Margalef

Department of Ecology
University of Barcelona
Barcelona, Spain

Jerry M. Melillo

The Ecosystem Center
Marine Biological Laboratory
Woods Hole, Massachusetts 02543 USA

H.A. Mooney

Department of Biological Sciences
Stanford University
Stanford, California 94305 USA

Tor Arve Pedersen

Agricultural College of Norway
Department of Microbiology
N-1432 Aas-NLH, Norway

Peter W. Price

Department of Biological Sciences
Box 5640
Northern Arizona University
Flagstaff, Arizona 86011 USA

Oscar Ravera

Department of Physical and Natural
    Sciences
Commission of the European Communities
Joint Research Center
21010 Ispra (Varese), Italy

Hermann Remmert

Fachbereich Biologie
Philipps-Universität
D-3550 Marburg/Lahn, FRG

Thomas Rosswall

Department of Microbiology
Swedish University of Agricultural
    Sciences
S-750 07 Uppsala, Sweden

Nils Chr. Stenseth

Department of Biology
Division of Zoology
University of Oslo
P.O. Box 1050, Blindern
N-0316 Oslo 3, Norway

Truels Truelsen

NATO Scientific Affairs Division
B-1110 Brussels, Belgium

Bernhard Ulrich

Institut für Bodenkunde und Waldernährung
Universität Göttingen
Bügenweg 2
D-3400 Göttingen, FRG

Eddy van der Maarel

Institute of Ecological Botany
Box 559, S-751 22
Uppsala, Sweden

W.J. Wiebe

Department of Microbiology
University of Georgia
Athens, Georgia 30602 USA